Almost Periodic Oscillations and Waves

Constantin Corduneanu

Almost Periodic Oscillations and Waves

Constantin Corduneanu
The University of Texas at Arlington
Department of Mathematics
411 S. Nedderman Dr.
Arlington, TX 76019
USA
and
Romanian Academy
Bucharest, Romania

e-ISBN: 978-1-4419-1890-1   e-ISBN: 978-0-387-09819-7
DOI: 10.1007/978-0-387-09819-7

Mathematics Subject Classification (2000): 11K70, 34C27, 35B15

Color illustration: watercolor by the late German mathematician, Lothar Collatz (1910–1990).

Printed on acid-free paper

springer.com

Preface

The concept of almost periodicity was introduced into mathematics by the Danish mathematician Harald Bohr, the brother of the physicist Niels Bohr. He wrote a series of papers during the period 1923–1925 in which the fundamentals of the theory of almost periodic functions can be found. Harald Bohr published a book (see the references) that synthesizes the basic knowledge on the subject, at least for the initial period of its development. Bohr's theory quickly attracted the attention of very famous mathematicians of that time, among them V. V. Stepanov, H. Weyl, N. Wiener, S. Bochner, A. S. Besicovitch, and J. von Neumann.

The applications of the new theory to various fields appeared shortly after its beginnings, and results concerning the almost periodicity of solutions of various classes of differential equations became known during the late 1920s. One of the first results was due to Bohr and Neugebauer and stated that the bounded solutions of the differential system $x' = Ax + f(t)$ are necessarily almost periodic when f is. The case of solutions to partial differential equations did not come much later in the discussion, and C. F. Muckenhoupt approached the case of hyperbolic equations (the vibrating string equation). Therefore, both oscillations and waves of almost periodic nature were investigated in the early stages of this new theory. A historical remark seems to be in order in this context related to a short mention H. Poincaré made in his famous treatise on celestial mechanics (1893). He obtained a series of sine functions with rather arbitrary frequencies and proposed the right method of developing a function in such a series by showing the way to calculate the coefficients. This method is simply based on the concept of mean value for an almost periodic function—just a small step from the core of the theory! An inspection of the literature in the applied fields shows that the simple use of the term almost periodicity is often avoided. This occurs despite the fact that the use of almost periodic functions or processes appears in a natural way in the solutions of various problems. A recent example is the publication *Encyclopedia of Vibration*, edited by S. G. Braun et al. (2002), which does not dedicate a single entry to the concept of almost periodic vibrations.

On the other hand, journal papers authored by engineers sporadically use this concept, which is so often absent in monographs, treatises, or textbooks.

Our main goal in writing this book is to make the concept of almost periodicity more familiar to its natural users: applied mathematicians, various categories of scientists who study oscillations or waves, engineers who are either researchers or interested in these forms of motion so often encountered in their fields of specialization, and – why not – graduate students in science and engineering, who will be the future users of these topics.

In the introduction, we illustrate the fact that almost periodic oscillations or waves are more often encountered in the study of various phenomena than the rather special case of periodic ones. Classical problems such as oscillations of a pendulum or the vibrations of an elastic string are sufficient to emphasize the fact that almost periodicity has a much higher occurrence than simple periodicity.

The body of the book consists of six chapters, Chapters 2–4 dealing with the concept of almost periodicity in various senses and stresses the idea of organizing the almost periodic functions in Banach function spaces, while Chapters 5–7 contain applications of the general theory to oscillations and waves. More precisely, the theory of oscillations, related to systems that can be described by a finite set of parameters, is first illustrated in the linear case (Chapter 5), and then the nonlinear case (Chapter 6). The final chapter is dedicated to wave theory, with the main concern being periodic or almost periodic waves. The quasi-periodic waves are summarily surveyed in the last section of the book, pinpointing the fact that they are under close scrutiny in contemporary research. Their particular interest stems from the fact that they are approximating the almost periodic ones in various ways and therefore can be regarded as approximate solutions of the almost periodic ones. Most of the topics included in the book have been taught to graduate students at the University of Texas at Arlington as part of the course Generalized Fourier Analysis and Applications. I thank my former students Amol Joshi and Rohan D'Silva for providing their class notes.

I take this opportunity to express my thanks to my colleagues A. Korzeniowski (University of Texas at Arlington) and Liviu Florescu and Ştefan Frunză (University Al.I. Cuza of Iaşi) for fruitful discussions concerning various topics and for bibliographical references. Also, my thanks to Mrs. Elena Mocanu, the secretary of the Mathematical Institute O. Mayer, with the branch of Iaşi of the Romanian Academy. She has deployed hard work and used her skills diligently in order to provide adequate form to the manuscript. Finally, it has been a real pleasure to cooperate with Springer in the persons of Mrs. Ann Kostant and Mrs. Elizabeth Loew in presenting the manuscript in the required format. I thank them for their amiability, patience, and understanding.

November 2007   *C. Corduneanu*

Contents

1

Introduction

Oscillatory motion, encountered in dynamical systems with a finite number of degrees of freedom, and *wave* propagation phenomena, in continuum mechanics, electromagnetic theory, and other fields, have been investigated under various circumstances by mathematicians, physicists, and engineers. In most papers and books dealing with oscillations or waves, the periodic case has been illustrated more frequently. The case of almost periodic oscillations or waves has also been considered by many authors, but it has been primarily considered in journal papers or monographs. It is surprising that the almost periodic case has not attracted the attention of authors of textbooks dedicated to these topics. As an example, we mention the well-written and comprehensive textbook by K. U. Ingard [52] addressed to undergraduate students. But even graduate textbooks avoid or expeditiously mention the almost periodic case.

The occurrence of almost periodic oscillations or waves is actually much more common than for periodic ones. It is our aim to present in this book, which we would like to serve as a graduate textbook on oscillations and waves, the basic facts related to the almost periodic case. We have taught this topic for several years and have received positive feedback from students.

In order to achieve our goal of presenting the basic facts related to almost periodic oscillations and waves, we will need some preliminaries, which will be treated in Chapter 2. These include the concept of metric space and some related topics. These introductory concepts will enable us to provide a solid mathematical foundation for the theory of several classes of almost periodic functions and their applications in physics and engineering.

Oscillations

The word *oscillation* suggests the motion of a pendulum and in its simplest realization is described by a mathematical model known as the *harmonic oscillator*.

C. Corduneanu, *Almost Periodic Oscillations and Waves*,
DOI 10.1007/978-0-387-09819-7_1, © Springer Science+Business Media, LLC 2009

If we consider the rectilinear motion of a material point of mass m without friction that is attracted by a fixed point on the line of motion with a force proportional to the distance to this fixed point, then Newton's law of dynamics $(ma = f)$ leads to the following second-order ordinary differential equation:

$$m\ddot{x}(t) = -kx(t). \tag{1.1}$$

In equation (1.1), $x(t)$ represents the abscissa of the point in motion, while $k > 0$ is the proportionality constant. We must assume $k > 0$ because the fixed point of attraction is taken as the origin on the line of motion, and only in this case is the origin attracting the point of mass m. Of course, $\ddot{x}(t)$ stands for the acceleration $a = \frac{dv}{dt}$, where the velocity $v = \frac{dx}{dt} = \dot{x}(t)$.

Let us now consider equation (1.1) and denote $\omega^2 = \frac{k}{m}$. Then equation (1.1) becomes

$$\ddot{x}(t) + \omega^2 x(t) = 0, \tag{1.2}$$

which is a homogeneous second-order differential equation with constant coefficients. It is well known that $\cos \omega t$ and $\sin \omega t$ are linearly independent solutions of equation (1.2). Therefore, the general solution of equation (1.2) is $x(t) = c_1 \cos \omega t + c_2 \sin \omega t$, with arbitrary constants c_1, c_2. If one denotes $c_1 = A \cos \alpha$, $c_2 = -A \sin \alpha$, which means $A = (c_1^2 + c_2^2)^{1/2}$, $\alpha = -\arctan\left(\frac{c_2}{c_1}\right)$, $c_1 \neq 0$, and $\alpha = -\left(\frac{\pi}{2}\right) \operatorname{sign} c_2$, $c_1 = 0$, these notations allow us to write the general solution of equation (1.2) in the form

$$x(t) = A \cos(\omega t + \alpha). \tag{1.3}$$

Formula (1.3) shows that any motion of the harmonic oscillator is a *simple harmonic* of frequency $2\frac{\pi}{\omega}$. The positive number A is called the *amplitude* of the oscillatory motion described by equation (1.3). The number α is called the *phase* of those oscillations. Both A and α depend on the initial conditions $x(0) = x_0$ and $v(0) = v_0$. In terms of x_0 and v_0, the amplitude and the phase are expressed by $A = (x_0^2 + v_0^2 \omega^{-2})^{1/2}$, $\alpha = -\arctan(\frac{v_0}{\omega} x_0)$.

Equation (1.2) describes the motion of the material point without friction and *free* of any external force. When other forces are involved, we have to modify equation (1.2) accordingly. We are led in this manner to deal with the so-called *forced oscillations*, as opposed to the *free oscillations* described by equation (1.2). The new equation will have the form $m\ddot{x}(t) = -kx(t) + F(t)$, with $F(t)$ designating the external force acting on the point at the moment t. Denoting $f(t) = m^{-1}F(t)$, the equation of forced oscillations becomes

$$\ddot{x}(t) + \omega^2 x(t) = f(t), \tag{1.4}$$

in which $f(t)$ is the density of the external force, or the force on a unit of mass of the moving point.

In mathematical terms, equation (1.4) for forced oscillations of the material point is an inhomogeneous equation of the second order. Using the variation of parameters method (Lagrange), the general solution of equation (1.4) can be expressed by the formula

$$x(t) = A\cos(\omega t + \alpha) + \omega^{-1}\int_0^t f(s)\sin\omega(t - s)ds. \qquad (1.5)$$

Formula (1.5) can describe both oscillatory and nonoscillatory motions. For instance, for $f(t) \equiv 1$, one obtains

$$x(t) = A\cos(\omega t - \alpha) + \omega^{-2}(1 - \cos\omega t),$$

which obviously describes periodic oscillatory motions of the material point for any A and α. On the other hand, if we choose $f(t) \equiv K\sin\omega t$, $K \neq 0$, one easily finds from equation (1.5) that

$$x(t) = A\cos(\omega t + \alpha) - \frac{K}{2\omega}t\cos\omega t. \qquad (1.6)$$

Formula (1.6) shows that none of the solutions of equation (1.4) can be periodic in t. They have an oscillatory character, but the amplitude can grow beyond any limit. Such oscillations, occurring in the so-called cases of *resonance*, are not of interest in applications.

Equation (1.4) can possess solutions that describe bounded oscillations that are not necessarily periodic. Investigating such examples will lead us to classes of oscillations that are not periodic but *almost periodic*.

Let us choose $f(t) = K\sin\gamma t$ with γ a number incommensurable with ω (i.e., $\frac{\gamma}{\omega}$ = irrational). Then, the general solution of equation (1.4) is

$$x(t) = A\cos(\omega t + \alpha) + K(\omega^2 - \gamma^2)^{-1}\sin\gamma t. \qquad (1.7)$$

Both terms on the right-hand side of equation (1.7) represent simple harmonic oscillations, and their sum also represents an oscillatory motion. Under the hypothesis of incommensurability of the numbers ω and γ, which is the same as incommensurability of the frequencies of the harmonic oscillations in equation (1.7), we can show that $x(t)$ given by equation (1.7) is not periodic in t. Shifting to the complex-valued functions, we will show that $c_1 e^{i\omega t} + c_2 e^{i\gamma t}$, $c_1, c_2 \neq 0$, cannot be periodic. In other words, if

$$c_1 e^{i\omega t} + c_2 e^{i\gamma t} \equiv c_1 e^{i\omega(t+T)} + c_2 e^{i\gamma(t+T)},$$

then

$$c_1(1 - e^{i\omega T})e^{i\omega t} + c_2(1 - e^{i\gamma T})e^{i\gamma t} \equiv 0,$$

and taking into account the fact that $e^{i\omega t}$ and $e^{i\gamma t}$ are linearly independent, we must have $1 - e^{i\omega T} = 0$, $1 - e^{i\gamma T} = 0$. This implies $i\omega T = 2\pi ki$, $i\gamma T = 2\pi mi$, with k and m integers, and $\frac{\omega}{\gamma} = \frac{k}{m}$. This contradicts the hypothesis that ω and γ are incommensurable.

The discussion above illustrates the fact that we can generate oscillatory solutions to equation (1.4), that are not periodic. As we shall see later, such oscillatory solutions belong to the class of *quasi-periodic* motions. This type of oscillation is a special case of almost periodic oscillation.

A more general situation can be illustrated by choosing in equation (1.4) $f(t) = K_1 \sin\gamma_1 t + \cdots + K_m \sin\gamma_m t$. The solution to be added to $A\cos(\omega t + \alpha)$ is $\sum_{k=1}^{m} K_k(\omega^2 - \gamma_k^2)^{-1}\sin\gamma_k t$, which is also generally quasi-periodic.

Another choice for $f(t)$ in equation (1.4), which takes us even closer to the concept of almost periodicity, is provided by

$$f(t) = \sum_{j=1}^{\infty} K_j \sin\gamma_j t, \qquad (1.8)$$

where

$$\sum_{j=1}^{\infty} |K_j| < \infty. \qquad (1.9)$$

Of course, we assume γ_j, $j \geq 1$, to be arbitrary real numbers. If the distribution of γ_j on the real line is such that

$$\sum_{j=1}^{\infty} |K_j| |\omega^2 - \gamma_j^2|^{-1} < \infty, \qquad (1.10)$$

then the series

$$\sum_{j=1}^{\infty} K_j(\omega^2 - \gamma_j^2)^{-1}\sin\gamma_j t \qquad (1.11)$$

is a candidate for a solution to equation (1.4). Indeed, for formula (1.10), we can infer the absolute convergence of the series (1.11). In general, the functions represented by the series (1.11) may not be twice differentiable, as required by equation (1.4). If the series

$$\sum_{j=1}^{\infty} |K_j| |\gamma_j^2| |\omega^2 - \gamma_j^2|^{-1}$$

converges, then the function represented by the series (1.11) will be a solution of equation (1.4). As we shall make precise later, the series (1.11) provides an *almost periodic* solution for which the Fourier series is absolutely convergent. The space of these functions will be denoted by $\mathrm{AP}_1(R, \mathcal{C})$ (see Chapter 2).

Remark 1.1. It is easy to provide conditions under which both formula (1.10) and the series (1.11) hold true. For instance, if $|\gamma_j| \longrightarrow \infty$ as $j \longrightarrow \infty$ and formula (1.9) is verified, both conditions required for convergence will be true. It suffices to note that $|\omega^2 - \gamma_j^2|^{-1} \longrightarrow 0$ as $j \longrightarrow \infty$, while $\gamma_j^2 |\omega^2 - \gamma_j^2|^{-1} \longrightarrow 1$ as $j \longrightarrow \infty$.

In concluding these considerations on the possible motions of the harmonic oscillator, particularly in the case of forced oscillations, we see that various types of oscillatory motions can be displayed, both periodic and — more generally — *quasi-periodic* or *almost periodic*. Such oscillations occur in the simplest model of the harmonic oscillator (which is the same as the mathematical pendulum), but they are also present in many other mechanical (and not only mechanical!) systems. We shall deal with such systems in the forthcoming chapters.

Waves

The usual mathematical tool in dealing with wave propagation in continuum media is the so-called *wave equation*. In one space dimension, it is written in the form

$$u_{tt} = a^2 u_{xx} + f(t,x), \qquad (1.12)$$

with u the unknown function, $a^2 = $ const., and $f(t,x)$ a given function representing forces occurring in the phenomenon of wave propagation. There are auxiliary conditions to be imposed on $u = u(t,x)$ in order to determine a single solution of equation (1.12). There are usually two *initial conditions* of the form

$$u(0,x) = u_0(x), \qquad u_t(0,x) = u_1(x), \qquad (1.13)$$

to be satisfied for those x in the domain of definition for the solution, and one also imposes *boundary value conditions*. They can have various forms, depending on the interpretation of equation (1.12). For instance, when equation (1.12) describes the vibrations of an elastic string occupying at rest the segment $0 \leq x \leq \ell$ of the x-axis, with its ends fixed at $x = 0$ and $x = \ell$, the boundary value conditions look like

$$u(t,0) = 0, \qquad u(t,\ell) = 0, \qquad t \geq 0. \qquad (1.14)$$

More complex conditions than those in (1.14) can be dealt with in looking for solutions of equation (1.12), and we shall later consider such conditions.

Before moving further, we note that the homogeneous equation associated to equation (1.12) is

$$u_{tt} = a^2 u_{xx}, \qquad (1.15)$$

and its general solution can be represented by d'Alembert's formula

$$u(t,x) = f(x + at) + g(x - at), \qquad (1.16)$$

where f and g are arbitrary functions of one variable twice continuously differentiable on the real axis.

Formula (1.16) emphasizes the so-called *solitary waves*, represented by $v(t,x) = f(x \pm at)$. We are not going to deal with this kind of wave here.

If in equation (1.16) we fix $x \in R$, then assuming f and g are periodic functions, or even almost periodic, then equation (1.16) defines a periodic or an almost periodic function. The almost periodic case will be clarified later, while the periodic case is obvious from equation (1.16). In order to obtain a periodic or almost periodic wave, it suffices to allow x to belong to an interval, say $x_1 \leq x \leq x_2$. Then, the family of periodic or almost periodic functions given by equation (1.16), depending on the parameter $x \in [x_1, x_2]$, provides the parametric representation of a periodic or almost periodic wave.

We will now return to equation (1.15) and consider the initial conditions (1.13) and boundary value conditions (1.14). The method we will use to find the solution is that of separation of variables, also known as Fourier's method.

Let us first look for a special solution in which the variables x and t are separated, namely

$$u(x, t) = X(x)T(t), \tag{1.17}$$

where $X(x)$ and $T(t)$ are to be determined by substitution into equation (1.15). We find

$$X(x)T''(t) = a^2 X''(x)T(t), \tag{1.18}$$

which by excluding the zero solution leads to the equality

$$\frac{T''(t)}{a^2 T(t)} = \frac{X''(x)}{X(x)}. \tag{1.19}$$

Because $x \in [0, \ell]$ and $t \in R_+$ are independent variables, keeping one of the variables fixed leads us to

$$\frac{X''(x)}{X(x)} = -\lambda, \qquad \frac{T''(t)}{a^2 T(t)} = -\lambda, \tag{1.20}$$

with $\lambda = \text{const.}$ Hence, $X(x)$ and $T(t)$ must be solutions of the equations

$$X''(x) + \lambda X(x) = 0, \tag{1.21}$$
$$T''(t) + \lambda a^2 T(t) = 0. \tag{1.22}$$

The conditions (1.14) lead to the following conditions for $X(x)$:

$$X(0) = 0, \qquad X(\ell) = 0. \tag{1.23}$$

The initial conditions (1.13) cannot be used at this moment, and we shall try to satisfy them later.

Therefore, the function $X(x)$ in equation (1.17) must be subject to equations (1.21) and (1.23). In other words, we have a second-order linear differential equation with the boundary value condition (1.23). This type of problem constitutes one of the simplest cases of the so-called Sturm–Liouville problems, amply treated in any textbook on ordinary differential equations. It is shown that the only possible values of λ for which the problem has nonzero solutions are the *eigenvalues* $\lambda = \lambda_n = \frac{n^2 \pi^2}{\ell^2}$, $n = 1, 2, \ldots$, with corresponding solutions, called *eigenfunctions*,

$$X(x) = X_n(x) = \sin \frac{n\pi}{\ell} x, \quad n = 1, 2, \ldots. \tag{1.24}$$

Returning to equation (1.22) and substituting into λ the eigenvalues indicated above, we obtain the second-order equations

$$T''(t) + \frac{n^2\pi^2a^2}{\ell^2}\, T(t) = 0, \tag{1.25}$$

whose general solutions are given by

$$T(t) = T_n(t) = A_n \cos\frac{n\pi a}{\ell}t + B_n \sin\frac{n\pi a}{\ell}t, \tag{1.26}$$

with A_n, B_n, $n = 1, 2, \ldots$, arbitrary constants. Therefore, from equations (1.17), (1.21), (1.22), (1.24), and (1.26) we obtain for equation (1.15) the solution

$$u(t, x) = u_n(t, x) = \left(A_n \cos\frac{n\pi a}{\ell}t + B_n \sin\frac{n\pi a}{\ell}t\right)\sin\frac{n\pi}{\ell}x, \tag{1.27}$$

which satisfies the boundary value conditions (1.14). We cannot expect to satisfy the initial conditions (1.10) using particular solutions of the form (1.27). But relying on the linearity of equation (1.14), we can obtain new solutions of this equation by using the particular solutions (1.27). Indeed, we notice first that any finite sum of solutions in equation (1.17) is also a solution of conditions (1.13). In other words, any function of the form

$$u(t, x) = \sum_{k=1}^{n} u_k(t, x) \tag{1.28}$$

is a solution of conditions (1.14) because each term of the sum is a solution. Still, we won't be able to satisfy the initial conditions of the general form (1.13) unless $u_0(x)$ and $u_1(x)$ are trigonometric polynomials with adequate exponentials. It is natural to take one more step and consider, if legitimate, solutions of the form

$$u(t, x) = \sum_{k=1}^{\infty} u_k(t, x). \tag{1.29}$$

This is obviously possible if the series in equation (1.29) is uniformly convergent, together with the series obtained by differentiation term by term, twice (in either x or t). We shall discuss this problem in some detail in what follows. Proceeding formally, if we impose the initial conditions (1.13) on the function (1.29), we obtain the following relations:

$$u(0, x) = u_0(x) = \sum_{k=1}^{\infty} A_k \sin\frac{\pi}{\ell}x, \qquad 0 \leq x \leq \ell, \tag{1.30}$$

$$u_t(0, x) = u_1(x) = \sum_{k=1}^{\infty} \frac{\pi a}{\ell} B_n \cos\frac{n\pi}{\ell}x, \qquad 0 \leq x \leq \ell. \tag{1.31}$$

The equations (1.30) and (1.31) display on the right-hand side Fourier series (of sine or cosine only!). Since $u_0(x)$ and $u_1(x)$ are specified functions, we can determine A_k and B_k, $k \geq 1$, such that equations (1.30) and (1.31) are valid. That is, Euler's formulas for Fourier coefficients lead to

$$A_k = \frac{2}{\ell} \int_0^\ell u_0(x) \sin \frac{k\pi}{\ell} x \, dx, \tag{1.32}$$

$$\frac{k\pi a}{\ell} B_k = \frac{2}{\ell} \int_0^\ell u_1(x) \cos \frac{k\pi}{\ell} x \, dx. \tag{1.33}$$

Based on Bessel's inequalities for the Fourier coefficients, we can write

$$\sum_{k=1}^\infty A_k^2 < +\infty, \qquad \sum_{k=1}^\infty k^2 B_k^2 < +\infty. \tag{1.34}$$

Actually, the first inequality (1.34) can be substantially improved. We will show that the inequality

$$\sum_{k=1}^\infty k^2 A_k^2 < +\infty \tag{1.35}$$

holds true. Indeed, starting from the obvious inequality

$$\int_0^\ell y'^2(x) dx \geq 0, \tag{1.36}$$

valid for any continuously differentiable function $y(x)$ on $[0, \ell]$, we substitute

$$y(x) = u_0(x) - \sum_{k=1}^n A_k \sin \frac{k\pi}{\ell} x, \tag{1.37}$$

with A_k given by equation (1.30). We must assume that $u_0(x)$ is continuously differentiable on $[0, \ell]$. The result of substituting equation (1.37) is

$$\sum_{k=1}^n k^2 A_k^2 \leq \frac{2\ell}{\pi^2} \int_0^\ell u_0'^2(x) dx \tag{1.38}$$

after elementary calculations and taking into account equation (1.32). Since n is arbitrarily large in formula (1.38), one obtains formula (1.35).

If we now consider $u(t, x)$ given by the series (1.29) and take into account solutions (1.27), one easily derives that the series (1.29) is uniformly and absolutely convergent, together with the first two derivatives (term by term), in either x or t. This remark justifies the fact that the series (1.29) represents the solution of the problem (1.13), (1.14), (1.15). It is obvious that this solution, regarded as a function of t with fixed x, is *periodic* (because the sines and cosines are periodic in t with period $\frac{2\ell}{a}$).

It will suffice to modify the boundary value conditions (1.14) and obtain in the same manner as above solutions for equation (1.15) that are *almost periodic* in t. Namely, instead of conditions (1.14), we will impose on the solution of equation (1.15) the conditions

$$u(t,0) = 0, \quad u_x(t,\ell) + hu(t,\ell) = 0, \quad h > 0, \tag{1.39}$$

which can be physically interpreted if we assume that equation (1.15) describes the motion of an elastic string fixed at the end $x = 0$ and whose end $x = \ell$ obeys the second relationship in conditions (1.39). In conditions (1.39), $u_x(t,\ell)$ stands for the tension at the end $x = \ell$ and, according to conditions (1.39), should be proportional to the *elongation* $u(t,\ell)$. Following step by step the procedure of separation of variables shown above, we come up with the following equation and boundary conditions for $X(x)$:

$$X''(x) + \lambda X(x) = 0, \quad X(0) = 0, \quad X'(\ell) + hX(\ell) = 0. \tag{1.40}$$

For $T(t)$ such that equation (1.17) will represent a solution of equation (1.15), equation (1.22) remains the same. Of course, different values of λ will lead to the eigenvalues and eigenfunctions of problem (1.40).

We shall now proceed to the investigation of the Sturm–Liouville problem (1.40). Since the general solution of the second-order equation in problem (1.40) is given (using complex functions) by

$$X(x) = c_1 e^{i\sqrt{\lambda}x} + c_2 e^{-i\sqrt{\lambda}x}, \tag{1.41}$$

the first boundary value condition leads to $c_1 + c_2 = 0$, which means

$$X(x) = c_1(e^{i\sqrt{\lambda}x} - e^{-i\sqrt{\lambda}x}), \tag{1.42}$$

with c_1 an arbitrary constant $\neq 0$. Imposing now the second boundary value condition at $x = \ell$, one obtains from equation (1.42), after omitting the factor c_1,

$$i\sqrt{\lambda}(e^{i\sqrt{\lambda}\ell} + e^{-i\sqrt{\lambda}\ell}) + h(e^{i\sqrt{\lambda}\ell} - e^{-i\sqrt{\lambda}\ell}) = 0. \tag{1.43}$$

This relation is equivalent to the trigonometric transcendental equation

$$\frac{\sqrt{\lambda}}{h} + \tan(\sqrt{\lambda}\ell) = 0, \tag{1.44}$$

provided $e^{i\sqrt{\lambda}\ell} + e^{-i\sqrt{\lambda}\ell} \neq 0$. The values of λ to be excluded are given by $e^{2i\sqrt{\lambda}\ell} = -1$, or $2i\sqrt{\lambda}\ell = (2k+1)i\pi$, $k = 0, \pm1, \pm2, \ldots$, or $\sqrt{\lambda} = \left(k + \frac{1}{2}\right)\frac{\pi}{\ell}$. These values of λ cannot be roots of equation (1.43), and $\tan(\sqrt{\lambda}\ell)$ is not defined for these values. Since $\lambda = 0$ is not an eigenvalue (the corresponding eigenfunction should be $X(x) \equiv 0!$), there remains to examine the cases $\lambda = \mu^2 > 0$ and $\lambda = -\mu^2 < 0$.

For $\lambda = \mu^2 > 0$, one obtains from equation (1.44)

$$\frac{\mu}{h} + \tan(\mu\ell) = 0, \tag{1.45}$$

which possesses infinitely many roots. It suffices to look only for positive roots of equation (1.45) because μ and $\tan(\mu\ell)$ are both odd functions. If we draw the graphs of the functions $-\frac{\mu}{h}$ and $\tan(\mu\ell)$, we see that there are infinitely many common points, say μ_m, $\mu_m \longrightarrow \infty$, and $\mu_m \simeq \left(m+\frac{1}{2}\right)\frac{\pi}{\ell}$, $m = 1, 2, \ldots$. The corresponding eigenvalues are

$$\lambda = \lambda_m = \left(m+\frac{1}{2}\right)^2 \frac{\pi^2}{\ell^2}, \quad m = 1, 2, \ldots . \tag{1.46}$$

For $\lambda = -\mu^2 < 0$, equation (1.44) becomes

$$\frac{i\mu}{h} + \tan(i\mu\ell) = 0, \tag{1.47}$$

which is equivalent to

$$-\frac{\mu}{h} + \tan h(\mu\ell) = 0. \tag{1.48}$$

Excepting $\mu = 0$, which has been excluded already, equation (1.48) does not possess any positive root. This can be easily seen by drawing the graph of the two functions involved. Therefore, as seen in the case $\lambda = \mu^2 > 0$, the eigenvalues of problem (1.40) are

$$\lambda_m = \mu_m^2 \simeq \left(m+\frac{1}{2}\right)^2 \frac{\pi^2}{\ell^2}, \quad m \geq 1, \tag{1.49}$$

and the corresponding eigenfunctions are

$$X_m(x) = \sin \mu_m x, \quad m \geq 1, \tag{1.50}$$

excepting a constant factor that is not material for our considerations.

Returning to equation (1.15), with the boundary value conditions (1.39), we can state that any function of the form

$$u_m(t, x) = (A_m \cos \mu_m a t + B_m \sin \mu_m a t) \sin \mu_m x, \quad m \geq 1, \tag{1.51}$$

is a solution. Repeating again the reasoning in the problem (1.14), (1.15), we are led to the conclusion that

$$u(t, x) = \sum_{m=1}^{\infty} u_m(t, x) \tag{1.52}$$

may satisfy equation (1.15), with the initial conditions (1.13) and the boundary value conditions (1.39), if the coefficients A_m and B_m, $m \geq 1$, are conveniently chosen.

The first relationship (1.13) to be satisfied by formula (1.51) leads to

$$u(0, x) = u_0(x) = \sum_{m=1}^{\infty} A_m \sin \mu_m x, \tag{1.53}$$

which appears as a development of $u_0(x)$ on $0 \leq x \leq \ell$ in a series of sines. We must note that an orthogonal relationship holds true for the sequence $\{\sin \mu_m x; \ m \geq 1\}$:

$$\int_0^\ell \sin \mu_m x \sin \mu_k x dx = 0, \quad m \neq k. \tag{1.54}$$

This orthogonality relationship is valid for more general systems of eigenfunctions and its proof can be found in many sources (see, for instance, C. Corduneanu [30]). Taking formula (1.54) into account, multiplying by $\sin \mu_k x$ for given k, one obtains by integration (also assuming the series in (1.53) is uniformly convergent)

$$\int_0^\ell u_0(x) \sin \mu_k x dx = A_k \int_0^\ell \sin^2 \mu_k x dx, \tag{1.55}$$

which uniquely determines A_k ($k \geq 1$). A similar approach leads to formulas for B_k ($k \geq 1$) uniquely determined.

Somewhat more intricate calculations than those shown above in the problem (1.12), (1.13), (1.14) lead to the conclusion that the series (1.52) converges uniformly, and absolutely, together with the series obtained by twice differentiating term by term. In other words, the series (1.52), with $u_m(t,x)$ defined by formula (1.51), represents a solution of the problem (1.15), (1.13), (1.39), if A_m and B_m, $m \geq 1$, are determined as described above.

The most important aspect for us in finding the formula

$$u(t,x) = \sum_{m=1}^\infty (A_m \cos \mu_m at + B_m \sin \mu_m at) \sin \mu_m x \tag{1.56}$$

for the solution of the problem (1.15), (1.13), (1.39) consists in discussing its oscillatory, or wave-like, properties. This discussion will lead to the conclusion that periodic motions are just a rare occurrence in the class of more general wave-like motions that we shall name *almost periodic*.

Indeed, the series (1.29), with the terms given by solution (1.27), represents a periodic motion (of a rather general type because we find in it harmonic terms of arbitrarily large frequency; more precisely, the periods are integer multiples of a basic frequency, $\frac{\pi a}{\ell}$). On the contrary, formula (1.56) represents for each fixed $x \in (0,\ell)$ a function of t that is clearly of the oscillatory type but is not periodic.

As we shall see in Chapter 2, such oscillations belong to a class of *almost periodic* motions (to be more specific, to the class of motions described by almost periodic functions whose Fourier series are absolutely convergent).

In order to get an image of an almost periodic wave, we need to visualize the trajectories of various points x, say in an interval belonging to $(0,\ell)$, as forming a surface. Such a surface provides us with the image of an *almost periodic wave*.

If we take into account the variety of types of boundary value conditions similar to conditions (1.39), then we can get an idea about how seldom the periodic motions occur in the much larger class of almost periodic ones.

We shall briefly return to the nonhomogeneous wave equation (1.12). This equation is also a source of waves that are not periodic but quasi-periodic or almost periodic.

As we have seen above, the homogeneous equation (1.15) possesses periodic as well as almost periodic solutions, depending on the type of boundary value conditions we associate to equation (1.15). Even if we preserve the simplest boundary value conditions (1.14), we can generate quasi-periodic or almost periodic solutions to equation (1.12), by conveniently choosing the term $f(t,x)$ on the right-hand side of equation (1.12).

To give consistency to this statement, we shall note that adding a solution of equation (1.12) to any solution of the homogeneous equation (1.15), one obtains a solution of equation (1.12). Indeed, if $u(t,x)$ is a solution of equation (1.15), while $v(t,x)$ satisfies equation (1.12), then $u(t,x) + v(t,x)$ is a solution of following (1.12). The fact follows easily from $u_{tt} = a^2 u_{xx}$ and $v_{tt} = a^2 v_{xx} + f(t,x)$ by addition: $(u+v)_{tt} = a^2(u+v)_{xx} + f(t,x)$. It remains to deal with the boundary value conditions.

Let us note that

$$u_1(t,x) = \left(A_1 \cos \frac{\pi a}{\ell} t + B_1 \sin \frac{\pi a}{\ell} t \right) \sin \frac{\pi}{\ell} x \qquad (1.57)$$

is a periodic solution, in the variable t, for equation (1.15). It also satisfies the boundary conditions (1.14).

Now, we shall choose the term $f(t,x)$ in equation (1.12) by letting

$$f(t,x) = \sin \frac{at}{\ell}. \qquad (1.58)$$

It is obvious that

$$v(t,x) \equiv v(t) = -\frac{\ell^2}{a^2} \sin \frac{at}{\ell} \qquad (1.59)$$

is a solution of equation (1.12), and therefore $u_1(t,x) + v(t,x)$, from equations (1.57) and (1.59), is also a solution of equation (1.12). It does not satisfy the boundary conditions (1.14), but this feature is not material under our circumstances. It is possible to modify $v(t,x)$ from equation (1.59) such that the conditions (1.14) are satisfied. Now, we can state that

$$u(t,x) = B_1 \sin \frac{\pi x}{\ell} \sin \frac{\pi a}{\ell} t - \frac{\ell^2}{a^2} \sin \frac{a}{t} t \qquad (1.60)$$

is a solution of equation (1.12). We have chosen $A_1 = 0$.

It is obvious that the periods or frequencies of $\sin \frac{\pi a}{\ell} t$ and $\sin \frac{a}{\ell} t$ have an irrational ratio (equivalently, they are incommensurable). As we shall see in the forthcoming chapters, such functions, with obvious oscillatory character, are called *quasi-periodic* when only a finite number of frequencies are involved or *almost periodic* in the general case.

The nearest object of our presentation will be the definition, in rigorous mathematical terms, of various concepts leading to almost periodicity.

2

Metric Spaces and Related Topics

2.1 Metric Spaces

A unified treatment of various classes of almost periodic functions requires some basic theory of *metric spaces* and related topics. The concept of metric spaces was introduced at the beginning of the twentieth century by the French mathematician Maurice Fréchet and has rapidly become one of the most useful tools in building modern analysis (including classical analysis and its development leading to functional analysis, both linear and nonlinear).

It is really a very simple task to define the concept of a metric space. It is likely that this simplicity in formulating the definition has contributed substantially to the diffusion of this concept in modern mathematics.

Definition 2.1. Let E be an abstract set and d a map from $E \times E$ into $R_+ = [0, \infty)$. We shall say that d is a *metric* on E, or a *distance function*, if the following conditions are satisfied:

1. $d(x, y) \geq 0$, for any $x, y \in E$, with $d(x, y) = 0$ only if $x = y$.
2. $d(x, y) = d(y, x)$ for any $x, y \in E$.
3. $d(x, y) \leq d(x, z) + d(z, y)$ for any $x, y, z \in E$.

All these properties of the metric agree with the properties of the usual distance (for instance, in the plane R^2, or more generally in R^n, $n \geq 1$). Property 1 is telling us that the distance between two elements (or *points*) is nonnegative, being zero only when the elements coincide. Property 2 expresses the fact that the distance from x to y is the same as the distance from y to x. Finally, property 3 tells us that the metric obeys the triangle inequality, which in geometry is formulated as follows: In any triangle, the length of any side is smaller than the sum of the lengths of the other two sides.

Definition 2.2. A set E, together with a metric $d : E \times E \to R_+$, satisfying conditions 1, 2, and 3, is called a *metric space.*

C. Corduneanu, *Almost Periodic Oscillations and Waves*,
DOI 10.1007/978-0-387-09819-7_2, © Springer Science+Business Media, LLC 2009

The important consequences of properties 1, 2, and 3 will consist in the fact that they allow us to extend the definition of many classical concepts to the case of metric spaces. These concepts include those of a *limit* (of a sequence or a function at a given point), *continuity* of a function, *convergence* of a function sequence, and many more.

Definition 2.3. A sequence $\{x_n; \ n \geq 1\} \subset E$ is called *convergent* to $x \in E$ if for each $\varepsilon > 0$ there exists a natural number $N = N(\varepsilon)$ such that

$$n \geq N(\varepsilon) \text{ implies } d(x_n, x) < \varepsilon. \tag{2.1}$$

The element $x \in E$ is called the *limit* of the sequence $\{x_n; \ n \geq 1\}$. We also write this property in the form $\lim x_n = x$ as $n \longrightarrow \infty$ or $x_n \longrightarrow x$ as $n \to \infty$.

Proposition 2.1. *The limit of a convergent sequence* $\{x_n; \ n \geq 1\}$ *is unique.*

Indeed, assuming $\lim x_n = x$ as $n \longrightarrow \infty$ and $\lim x_n = \bar{x}$ as $n \to \infty$, we obtain according to property 3

$$d(x, \bar{x}) \leq d(x, x_n) + d(x_n, \bar{x}) \tag{2.2}$$

for every $n \geq 1$. Now we take $\varepsilon > 0$ and $N(\varepsilon)$, $N_1(\varepsilon)$ such that

$$d(x, x_n) < \frac{\varepsilon}{2} \text{ for } n \geq N(\varepsilon), \tag{2.3}$$

$$d(x_n, \bar{x}) < \frac{\varepsilon}{2} \text{ for } n \geq N_1(\varepsilon). \tag{2.4}$$

Combining formulas (2.2), (2.3), and (2.4), we obtain

$$d(x, \bar{x}) < \varepsilon \text{ for } n \geq \max(N, N_1). \tag{2.5}$$

But $d(x, \bar{x}) < \varepsilon$ holds true for any $\varepsilon > 0$, which means $d(x, \bar{x}) = 0$. According to property 1 of the metric, we derive $x = \bar{x}$, i.e., the uniqueness of the limit of a convergent sequence in the metric space (E, d), or simply E.

Definition 2.4. We shall say that the sequence $\{x_n, \ n \geq 1\}$ has the *Cauchy property* if for every $\varepsilon > 0$ there exists a natural number $N = N(\varepsilon)$ such that

$$n, m \geq N(\varepsilon) \text{ imply } d(x_n, x_m) < \varepsilon. \tag{2.6}$$

Proposition 2.2. *Any convergent sequence* $\{x_n; \ n \geq 1\} \subset E$ *has the Cauchy property.*

Indeed, if $\lim x_n = x$ as $n \longrightarrow \infty$, then for any $\varepsilon > 0$ we can find $N(\varepsilon/2)$ such that

$$d(x_n, x) < \frac{\varepsilon}{2} \text{ for } n \geq N(\varepsilon/2) \tag{2.7}$$

and

$$d(x_m, x) < \frac{\varepsilon}{2} \text{ for } m \geq N(\varepsilon/2). \tag{2.8}$$

Since

$$d(x_n, x_m) \leq d(x_n, x) + d(x_m, x) \tag{2.9}$$

for any $n, m \geq 1$, we obtain from formulas (2.7), (2.8), and (2.9)

$$d(x_n, x_m) < \frac{\varepsilon}{2} + \frac{\varepsilon}{2} = \varepsilon \text{ for } n, m \geq N(\varepsilon/2). \tag{2.10}$$

The inequality (2.10) proves the assertion.

Remark 2.1. On the real line R, with the usual metric or distance $d(x, y) = |x - y|$, the Cauchy property is also sufficient for the convergence. This is well known in classical analysis.

Remark 2.2. It is very important to note that the converse of Proposition 2.2 is not true in general. In other words, there are metric spaces in which not every Cauchy sequence is convergent. A simple example is furnished by the set of rational (real) numbers, usually denoted by Q, with the metric $d(x, y) = |x - y|$. It is an elementary exercise to check the validity of properties 1, 2, and 3. We now choose a sequence in Q, say $\{r_n; \ n \geq 1\}$, such that $\lim r_n = \sqrt{2}$, i.e., $|r_n - \sqrt{2}| \longrightarrow 0$ as $n \longrightarrow \infty$. Since $\sqrt{2} = 1.4142\ldots$, we can choose $r_1 = 1$, $r_2 = 1.4$, $r_3 = 1.41$, $r_4 = 1.414, \ldots$. This choice implies $|r_n - \sqrt{2}| < 10^{-n}$, which means that, on the real line R, $r_n \longrightarrow \sqrt{2}$ as $n \to \infty$. It is well known that $\sqrt{2} \notin Q$, being an irrational number. Therefore, the metric space Q, with the metric $d(x, y) = |x - y|$, possesses sequences with Cauchy property ($|r_n - r_m| < 10^{-m+1}$ for $n \geq m$) that are not convergent. The convergence is assured in a larger space $R \supset Q$, a feature suggesting that a metric space in which the Cauchy property does not imply convergence is part of a "larger" space in which the convergence is possible.

Definition 2.5. A metric space (E, d) is called *complete* if any sequence enjoying the Cauchy property is convergent.

As noted in Remark 2.1, the real line R, with the usual metric $d(x, y) = |x - y|$, is a complete metric space.

On the contrary, the set of rationals, with the metric $d(x, y) = |x - y|$, is not complete.

The following result is of utmost significance for the theory of metric spaces and their applications.

Theorem 2.1. *Let (E, d) be an arbitrary metric space. Then, there exists a complete metric space $(\widetilde{E}, \widetilde{d})$ such that $E \subset \widetilde{E}$, while d and $\widetilde{d}$ coincide on E.*

The proof of Theorem 2.1 will be provided after several further concepts related to metric spaces are presented and discussed in this chapter.

Theorem 2.1 constitutes the main tool in constructing various spaces of almost periodic functions by completion of a rather simple set, that of trigonometric polynomials, with respect to various metrics. Let us point out the fact that a *trigonometric polynomial* (in complex form) has the form

$$T(t) = a_1 e^{i\lambda_1 t} + a_2 e^{i\lambda_2 t} + \cdots + a_p e^{i\lambda_p t}, \qquad (2.11)$$

where $\lambda_1, \lambda_2, \ldots, \lambda_p$ are reals and $a_1, a_2, \ldots, a_p$ are complex numbers. One can also assume that a_k are vectors with complex components: $a_k \in \mathcal{C}^r$, $k \in 1, 2, \ldots, p$.

Examples of metrics on the set $\mathcal{T}$ of trigonometric polynomials are

$$d_1(T, \widetilde{T}) = \sum_{k=1}^{p} |a_k - \widetilde{a}_k| \qquad (2.12)$$

with

$$\widetilde{T}(t) = \widetilde{a}_1 e^{i\lambda_1 t} + \cdots + \widetilde{a}_p e^{i\lambda_p t} \qquad (2.13)$$

or

$$d_2(T, \widetilde{T}) = \left(\sum_{k=1}^{p} |a_k - \widetilde{a}_k|^2 \right)^{1/2}, \qquad$$

or even

$$d(T, \widetilde{T}) = \sup_{t \in R} \left| \sum_{k=1}^{p} (a_k - \widetilde{a}_k) e^{i\lambda_k t} \right|. \qquad (2.14)$$

The supremum in equation (2.14) exist because

$$\sup_{t \in R} \left| \sum_{k=1}^{p} (a_k - \widetilde{a}_k) e^{i\lambda_k t} \right| \le d_1(T, \widetilde{T}), \qquad (2.15)$$

which follows easily taking into account

$$|e^{i\lambda_k t}| = |\cos \lambda_k t + i \sin \lambda_k t| = 1, \ k = 1, 2, \ldots, p.$$

We note that in equations (2.11) and (2.13) the exponents $\lambda_1, \lambda_2, \ldots, \lambda_p$ are the same. If this is not the case when dealing with two trigonometric polynomials, we can add other terms in $T(t)$ and $\widetilde{T}(t)$ with zero coefficients.

Each of the metrics d_1, d_2, or d leads to a class of *almost periodic function* by completion of the spaces $(\mathcal{T}, d_1), (\mathcal{T}, d_2)$, or $(\mathcal{T}, d)$, where $\mathcal{T}$ is the set of all trigonometric polynomials of the form (2.11) with arbitrary real λ_k and arbitrary complex $a_k, p \ge 1$.

In this manner, we obtain the spaces of almost periodic functions with absolutely convergent Fourier series, $AP_1(R, \mathcal{C})$, the almost periodic functions in the Besicovitch sense, $AP_2(R, \mathcal{C})$, resp. the space $AP(R, \mathcal{C})$ of Bohr's almost periodic functions – the most usual among the spaces of almost periodic functions (and, historically, the first to be investigated thoroughly).

The details concerning the spaces $AP_1(R,\mathcal{C})$, $AP_2(R,\mathcal{C})$, and $AP(R,\mathcal{C})$ will be presented in Chapter 3. Other spaces of almost periodic functions will also be defined and investigated.

Let us now return to the general theory of metric spaces and discuss further concepts that will be of interest in subsequent chapters.

Definition 2.6. Let $M \subset E$ with (E,d) a metric space. We shall say that $x \in E$ is a *limit point* of M if there exists a sequence $\{x_n; \ n \geq 1\} \subset M$ such that $\lim x_n = x$ as $n \longrightarrow \infty$.

Definition 2.7. The set of limit points of $M \subset E$ is denoted by $\overline{M}$ and is called the *closure* of M.

Definition 2.8. A set $M \subset E$, with (E,d) a metric space, is called *closed* if $M = \overline{M}$.

Remark 2.3. Generally, only the inclusion $M \subset \overline{M}$ is true. Indeed, if $x \in M$, then $x_n \longrightarrow x$ as $n \longrightarrow \infty$ if x_n is chosen in such a way that $d(x, x_n) < n^{-1}$, $n = 1, 2, \ldots$. It may be possible that some x_n coincide.

Definition 2.9. A set $A \subset E$, with (E,d) a metric space, is called *open* if its complementary set in E is closed: $\overline{cA} = cA$.

Proposition 2.3. *The intersection of a finite number of open sets is either empty or a nonempty open set. The union of any number of open sets is open.*

The proof of Proposition 2.3 is elementary. We notice that, according to Definition 2.9, a set $A \subset E$ is open if and only if for any $x \in A$ there exists a positive number $r = r(x)$ such that the *ball* of radius r centered at x belongs to A. In other words, any $y \in E$ with $d(x,y) < r$ satisfies $y \in A$.

Definition 2.10. A subset $M \subset E$, with (E,d) a metric space, is called *relatively compact* if any sequence $\{x_n; \ n \geq 1\} \subset M$ has a convergent subsequence (i.e., there exists a sequence of integers $\{n_k; \ k \geq 1\}$ such that $\lim x_{n_k} = x$ as $k \longrightarrow \infty$).

Remark 2.4. The limit point $x = \lim x_{n_k}$ as $k \longrightarrow \infty$ may or may not belong to M. If any x constructed as above belongs to M, then we shall say that M is *compact*. Hence, any compact subset $M \subset E$ is necessarily closed in E.

Proposition 2.4. *Let (E,d) be a compact metric space. Then (E,d) is also complete.*

The proof is immediate. Indeed, if $\{x_n; \ n \geq 1\} \subset E$ is a Cauchy sequence, then it contains a convergent subsequence, say $\{x_{n_k}; \ k \geq 1\}$: $\lim x_{n_k} = x$ as $k \to \infty$. But $x_n - x = x_n - x_{n_k} + x_{n_k} - x$, and relying on the Cauchy property of $\{x_n; \ n \geq 1\}$, we obtain $x_n \longrightarrow x$ as $n \longrightarrow \infty$.

Proposition 2.5. *The following properties are necessary and sufficient for the relative compactness of the set $M \subset R$, with (E, d) a metric space:*

I.   *From any covering of M by open sets, one can extract a finite covering.*
II. *For any $\varepsilon > 0$, there exists a finite subset $\{x_1, x_2, \ldots, x_p\} \subset E$ such that for each $x \in M$ one can find an integer k, $1 \le k \le p$, such that $d(x_k, x) < \varepsilon$.*

Remark 2.5. We shall omit the proof of Proposition 2.5. It can be found in many sources (see, for instance, A. Friedman [43]).

Definition 2.11. Let $M \subset E$, with (E, d) a metric space. Then $\sup d(x, y)$ for $x, y \in M$ is called the *diameter* of M. It is denoted by $d(M)$ when it is finite.

Proposition 2.6. *Let (E, d) be a metric space whose diameter is infinite. Then we can define another metric $\widetilde{d}$ on E such that $(E, \widetilde{d})$ has finite diameter. Moreover, a sequence converges in (E, d) if and only if it converges in $(E, \widetilde{d})$.*

Proof. We shall denote

$$\widetilde{d}(x, y) = d(x, y)[1 + d(x, y)]^{-1} \tag{2.16}$$

for any $x, y \in E$. From equation (2.16), we derive $\widetilde{d}(x, y) < d(x, y)$ for any $x, y \in E$, $x \ne y$, because $d(x, y) > 0$. We can write, for any sequence $\{x_n;\ n \ge 1\}$ with $x_n \longrightarrow x$ as $n \longrightarrow \infty$, $\widetilde{d}(x_n, x) \le d(x_n, x)$, $n \ge 1$. Therefore, any sequence convergent in (E, d) is also convergent in $(E, \widetilde{d})$. On the other hand, $d(x, y) = \widetilde{d}(x, y)[1 - \widetilde{d}(x, y)]^{-1}$ for $x \ne y$, which leads to $d(x_n, x) \longrightarrow 0$ as $n \longrightarrow \infty$ when $\widetilde{d}(x_n, x) \longrightarrow 0$ as $n \longrightarrow \infty$ (excluding those sequences for which $x_n = x$ for sufficiently large n; for such sequences, the statement is obviously true).

Several examples of metric spaces, as well as the proof of Theorem 2.1 concerning the *completion* of metric spaces, will be considered in the next sections of this chapter.

2.2 Banach Spaces and Hilbert Spaces

Among the metric spaces, there is a very important category known as *linear metric spaces*. Both Banach and Hilbert spaces are special cases of linear metric spaces, and they constitute basic concepts of linear functional analysis (see, for instance, N. Dunford and J. Schwartz [35]).

Besides the metric structure, as defined in the preceding section, the linear metric space possesses one more basic structure, of algebraic nature, namely that of a linear space or linear system. These structures are interrelated, and this feature makes possible the definition and investigation of many useful concepts encountered in classical analysis. Let us define the *linear space*, or *vector space*.

Definition 2.12. A set E is called a *linear space* if the following conditions (or axioms) hold true:

1. There is a map $(x, y) \longrightarrow x + y \in E$, called *addition*, that is commutative and associative: $x + y = y + x$, $(x + y) + z = x + (y + z)$, for any $x, y, z \in E$.
2. There is a map $(\lambda, x) \longrightarrow \lambda x \in E$, called *scalar multiplication*, for $\lambda \in R$ (or $\lambda \in C$), such that $(\lambda + \mu)x = \lambda x + \mu x$, $\lambda(x + y) = \lambda x + \lambda y$, $(\lambda \mu)x = \lambda(\mu x)$, for any $\lambda, \mu \in R$ (or $\lambda, \mu \in C$), and any $x, y \in E$; moreover, there exists an element $\theta \in E$ such that $0x = \theta$ for any $x \in E$, while $1x = x$ for any $x \in E$.

From the axioms above, we derive, for instance, $(1 - 1)x = 0x = \theta = x + (-1)x$. This means that $(-1)x$, denoted simply as $-x$, is the inverse element with respect to addition. Hence, E is an Abelian group with respect to addition. The scalar multiplication is a generalization of the classical operation of scalar multiplication for vectors: If $v = (v_1, v_2, v_3)$, then $\lambda v = (\lambda v_1, \lambda v_2, \lambda v_3)$.

The most common example of a linear (vector) space is $R^n = \underbrace{R \times R \times \cdots \times R}_{n \text{ times}}$, known as the n-dimensional Euclidean space. If $x = (x_1, x_2, \ldots, x_n)$, $y = (y_1, y_2, \ldots, y_n) \in R^n$, then $x + y = (x_1 + y_1, x_2 + y_2, \ldots, x_n + y_n)$ by definition. The scalar multiplication in R^n is defined by $\lambda x = (\lambda x_1, \lambda x_2, \ldots, \lambda x_n)$. The zero (null) element in R^n is $\theta = \underbrace{(0, 0, \ldots, 0)}_{n \text{ times}}$. All the properties listed above can be easily checked in the case of the linear (vector) space R^n.

In a similar manner, one can see that $C^n = \underbrace{C \times C \times \cdots \times C}_{n \text{ times}}$ is a linear space (known as the n-dimensional complex space).

In order to introduce the concept of a Banach space, first we need to define that of a *linear normed space*. At this point, we shall mix up the *algebraic* and the *metric* structures on the set E.

Definition 2.13. Let E be a linear space over the reals E (or C). A *norm* on E is a map from E into R_+, say $x \longrightarrow \|x\|$, such that the following properties are satisfied:

1. $\|x\| \geq 0$ for any $x \in E$, the equal sign occurring only when $x = \theta$.
2. $\|\lambda x\| = |\lambda| \|x\|$ for any $\lambda \in R$ (or C) and any $x \in E$.
3. $\|x + y\| \leq \|x\| + \|y\|$ for $x, y \in E$.

Definition 2.14. A linear space E endowed with a *norm* is called a *linear normed space*.

Proposition 2.7. *Let $(E, \| \cdot \|)$ be a linear normed space. Then, by letting $d(x, y) = \|x - y\|$, (E, d) is a metric space.*

The proof is elementary and is left to the reader. In particular, the triangle inequality for $d(x, y)$ is a consequence of property 3 in Definition 2.13.

Proposition 2.8. *Both addition and scalar multiplication on a linear normed space are continuous operations.*

Indeed, the assertion follows from the inequalities $\|x + y - (u + v)\| \leq \|x - u\| + \|y - v\|$ and $\|\lambda x - \mu y\| \leq |\lambda| \|x - y\| + |\lambda - \mu| \|y\|$.

Definition 2.15. A linear normed space E is called a *Banach space* if (E, d) is complete (with $d(x, y) = \|x - y\|$ for any $x, y \in E$).

The simplest examples of Banach spaces are R, with the absolute value as norm, and C, with the module as norm. We shall examine below the spaces R^n and C^n, $n \geq 1$.

Definition 2.16. Let E be a linear space. A map $(x, y) \longrightarrow \langle x, y \rangle$ from $E \times E$ into C is called an *inner product* if the following conditions are verified:

1. $\langle x, x \rangle \geq 0$ for any $x \in E$ with the equal sign valid only for $x = \theta$.
2. $\langle x, y \rangle = \overline{\langle y, x \rangle}$ for any $x, y \in E$, with the upper bar for the conjugate complex number.
3. $\langle x + y, z \rangle = \langle x, z \rangle + \langle y, z \rangle$ for any $x, y \in E$.
4. $\langle \lambda x, y \rangle = \lambda \langle x, y \rangle$ for $x, y \in E$.

Remark 2.6. If the scalar field is R instead of C, then the second condition in Definition 2.16 expresses the symmetry of the inner product.

Remark 2.7. From Definition 2.16, one can easily derive other properties of the inner product. For instance, $\langle x, y + z \rangle = \langle x, y \rangle + \langle x, z \rangle$ or $\langle x, \lambda y \rangle = \overline{\langle \lambda y, x \rangle} = \overline{\lambda \langle y, x \rangle} = \overline{\lambda} \overline{\langle y, x \rangle} = \overline{\lambda} \langle x, y \rangle$. This shows how to take out a scalar from the second place of an inner product.

Proposition 2.9. *The inner product satisfies the Schwarz inequality*

$$|\langle x, y \rangle|^2 \leq \langle x, x \rangle \langle y, y \rangle \qquad x, y \in E. \qquad (2.17)$$

Indeed, according to the first condition in Definition 2.16, one has $\langle x + \lambda y, x + \lambda y \rangle \geq 0$ for any $\lambda \in C$ and $x, y \in E$. The inequality above can be rewritten in the form

$$|\lambda|^2 \langle y, y \rangle + 2 \operatorname{Re}(\lambda \overline{\langle x, y \rangle}) + \langle x, x \rangle \geq 0. \qquad (2.18)$$

If we choose $\lambda = -\frac{\langle x, y \rangle}{\langle y, y \rangle}$, assuming $y \neq \theta$, then formula (2.18) leads to formula (2.17). When $y = \theta$, formula (2.17) is satisfied because both sides are zero.

Proposition 2.10. *Let E be a linear space endowed with an inner product. Then $(E, \| \cdot \|)$ is a normed space with the norm*

$$\|x\| = (\langle x, x \rangle)^{1/2}. \qquad (2.19)$$

Only the triangle inequality for the norm needs some consideration, the first two conditions for the norm being obvious. Concerning the triangle inequality, we note that

$$\|x+y\|^2 = \langle x+y, x+y \rangle = \langle x, x \rangle + \langle y, y \rangle + 2\,\mathrm{Re}(\langle x, y \rangle)$$
$$\leq \|x\|^2 + \|y\|^2 + 2\|x\|\|y\| = (\|x\| + \|y\|)^2$$

due to the fact that

$$|\,\mathrm{Re}(\langle x, y \rangle)| \leq |\langle x, y \rangle| \leq \|x\|\|y\|$$

in accordance with formula (2.17).

Definition 2.17. Let E be a linear space with the inner product $\langle \cdot, \cdot \rangle$. If $(E, \|\cdot\|)$ is the linear normed space with the norm $\|\cdot\|$ derived from the scalar product according to equation (2.19), then we shall say that E is a *Hilbert space* if it is a complete metric space, the distance being defined by $d(x, y) = \|x - y\|$.

Since it is obvious that a Hilbert space is a Banach space in which the norm is defined by means of an inner product, it is legitimate to ask the question: When is a Banach space a Hilbert space?

Proposition 2.11 gives a rather simple answer to this question.

Proposition 2.11. *Let $(E, \|\cdot\|)$ be a Banach space over the field of reals. Then, a necessary and sufficient condition for E to be a Hilbert space also is the validity of the parallelogram law in E,*

$$\|x+y\|^2 + \|x-y\|^2 = 2(\|x\|^2 + \|y\|^2),$$

for any $x, y \in E$.

The proof of this proposition can be carried out elementarily if one defines the inner product by the formula

$$\langle x, y \rangle = \frac{1}{4}(\|x+y\|^2 - \|x-y\|^2)$$

in terms of the norm of E as a Banach space.

We shall return now to the spaces R^n and C^n, proving that each is a Hilbert space. R^n is a linear space over the field of reals, while C^n is a linear space over the complex field of scalars.

As we may guess, the inner product can be defined by the well-known formula from the vector case, namely

$$\langle x, y \rangle = x_1 y_1 + x_2 y_2 + \cdots + x_n y_n$$

for the space R^n and

$$\langle x, y \rangle = x_1 \bar{y}_1 + x_2 \bar{y}_2 + \cdots + x_n \bar{y}_n$$

in the case of complex space C^n.

These inner products induce the following norms:

$$R^n : \|x\| = (x_1^2 + x_2^2 + \cdots + x_n^2)^{1/2}$$

resp.

$$C^n : \|x\| : (|x_1|^2 + |x_2|^2 + \cdots + |x_n|^2)^{1/2}.$$

The norms for R^n and C^n shown above are known as *Euclidean norms*. There are other norms in R^n or C^n whose definitions and properties can be easily provided. For instance, in the space R^n, one uses norms such as

$$\|x\|_1 = \sum_{j=1}^{n} |x_j|$$

or

$$\|x\|_2 = \max_{1 \leq j \leq n} |x_j|$$

with $x = (x_1, x_2, \ldots, x_n) \in R^n$. Inequalities of the form

$$\|x\|_2 \leq \|x\| \leq \|x\|_1 \leq n\|x\|_2$$

show that the convergence is the same regardless of the kind of norm we are using. Similar considerations are valid in the case of C^n.

Both spaces R^n and C^n are examples of Hilbert spaces of finite dimension. We shall define and investigate now an example of a Hilbert space whose dimension is infinite. Historically, this was the first example of a Hilbert space, and we owe this to Hilbert himself. We shall limit our considerations to the real case and discuss the space $\ell^2(R)$ of sequences of real numbers $x = (x_1, x_2, \ldots, x_k, \ldots)$ such that

$$\sum_{k=1}^{\infty} x_k^2 < \infty. \tag{2.20}$$

Since $(a + b)^2 \leq 2(a^2 + b^2)$ and $(\lambda a)^2 = \lambda^2 a^2$, we see that $\ell^2(R)$ is a linear space. The inner product is defined by

$$\langle x, y \rangle = \sum_{k=1}^{\infty} x_k y_k, \tag{2.21}$$

and the convergence of the series is the result of the inequality $2|x_k y_k| \leq x_k^2 + y_k^2$, $k = 1, 2, \ldots$. The properties of the inner product can be easily checked. $x = \theta = $ the null element is given by $\theta = (0, 0, \ldots, 0, \ldots)$.

A property of Hilbert spaces that requires some consideration is *completeness*. The proof of completeness of $\ell^2(R)$ can be found in various sources (see, for instance, C. Corduneanu [24]). Of course, we deal with the metric $d(x, y) = \|x - y\|$ in proving the completeness of $\ell^2(R)$.

The complex case $\ell^2(C)$ is treated similarly. The inner product is defined by

$$\langle x, y \rangle = \sum_{k=1}^{\infty} x_k \bar{y}_k$$

instead of equation (2.21).

In concluding this section, we shall briefly consider the concept of a *linear metric space*, also known as a *Fréchet space* when it is complete. Basically, a Fréchet space is a linear space E endowed with a metric $d(x, y)$ that is translation invariant,

$$d(x + z, y + z) = d(x, y), \tag{2.22}$$

for any $x, y, z \in E$.

Definition 2.18. A linear space E endowed with a translation-invariant metric d is called a *Fréchet space* if it is complete.

In order to define Fréchet spaces, one often uses the concept of a seminorm.

Definition 2.19. Let E be a linear space. A *seminorm* on E is a map from E to R_+, say $x \longrightarrow |x|$, such that

1. $|x| \geq 0$ for any $x \in E$;
2. $|\lambda x| = |\lambda||x|$ for any $\lambda \in R$ (or C) and $x \in E$; and
3. $|x + y| \leq |x| + |y|$ for any $x, y \in E$.

Remark 2.8. The only difference between a norm and a seminorm is in the fact that a seminorm can vanish for nonzero elements of E. More precisely, based on properties 2 and 3 of Definition 2.19, we see that the set $x \in E$ for which $|x| = 0$ constitutes a linear manifold in E.

Definition 2.20. A sequence of seminorms $\{|x|_k; \ k \geq 1\}$ on the linear space E is called *sufficient* if $|x|_k = 0$, $k \geq 1$, implies $x = \theta$.

In other words, the sequence is sufficient if for each $x \in E$, $x \neq \theta$, there exists a natural number m such that $|x|_m > 0$.

We shall now present a procedure for constructing an invariant metric on a linear space E by means of a sufficient sequence of seminorms.

Proposition 2.12. Let E be a linear space endowed with a countable family of seminorms $\{|x|_k; \ k \geq 1\}$. If this family is sufficient, then (E, d) is a metric space with

$$d(x, y) = \sum_{k=1}^{\infty} \frac{1}{2^k} \frac{|x - y|_k}{1 + |x - y|_k}. \tag{2.23}$$

The proof of Proposition 2.12 is very simple. Obviously, $d(x, y) \geq 0$, the equality being possible only for $|x - y|_k = 0$, $k \geq 1$. This means $x - y = \theta$ because of the sufficiency of the family $\{|x|_k; \ k \geq 1\}$. The triangle inequality for $d(x, y)$ can be proven if we take into account the elementary inequality

$$\frac{|a + b|}{1 + |a + b|} \leq \frac{|a|}{1 + |a|} + \frac{|b|}{1 + |b|}$$

for arbitrary $a, b \in R$.

Remark 2.9. If (E, d) is a complete metric space, then it is a Fréchet space.

A good example of a linear metric space is the space $\mathfrak{s}$ of all sequences $x = \{x_n;\ n \geq 1\}$ with $x_n \in R$, $n \geq 1$. The linearity is an obvious property. The metric is defined by

$$d(x, y) = \sum_{n=1}^{\infty} \frac{1}{2^n} \frac{|x_n - y_n|}{1 + |x_n - y_n|}.$$

The translation invariance is easily checked. A sequence $\{x^k;\ k \geq 1\} \subset \mathfrak{s}$ is convergent if and only if the sequences (of real numbers) $\{x_n^k;\ n \geq 1\}$ converge for each $k \geq 1$. It can be shown that $\mathfrak{s} = \mathfrak{s}(R)$ and $\mathfrak{s} = \mathfrak{s}(\mathcal{C})$ are Fréchet spaces, i.e., they are complete.

Many important sequence spaces are subspaces of $\mathfrak{s}$. We have already seen $\ell^2(R)$, which is a Hilbert space.

Another important sequence space is the space of bounded sequences $x = \{x_n;\ n \geq 1\}$, i.e., such that for each x there exists $M_x > 0$ with $|x_n| \leq M_x$, $n \geq 1$. The norm in this space, denoted by $b(R)$, is given by

$$\|x\| = \sup_{n \geq 1} |x_n|. \tag{2.24}$$

$b(R)$ is a Banach space. An important subspace of $b(R)$ is that consisting of all convergent sequences. It is denoted by $c(R)$, and the norm is that defined by equation (2.24). $c(R)$ is also a Banach space. Finally, $c_0(R)$ denotes the Banach space of all sequences convergent to zero. The norm is that given by equation (2.24). Further examples will be considered in subsequent chapters.

Finally, we shall briefly deal with the concept of *factor space* (sometimes also called quotient space). Let E be a linear space (in the algebraic sense). Assume L is a linear manifold (or subspace) of E. We shall introduce in E an equivalence relation as follows: $u \simeq v$, $u, v \in E$, if $u - v \in L$. It is easy to check that this relation, symbolized by $\simeq$ or $\equiv$ (congruent), is indeed an equivalence relation. Indeed, $u - u = \theta \in L$ (reflexivity), if $u \simeq v$, then $v \simeq u$ because $v - u = -(u - v) \in L$ (symmetry); and from $u - v \in L$ and $v - w \in L$ we obtain $u - w = (u - v) + (v - w) \in L$ (transitivity).

It is well known that such a relation of equivalence defines the disjoint classes of equivalent elements of E. Such a class is determined by a single representative $u \in E$ and consists of all elements in E that belong to $u + L = \{u + v;\ v \in L\}$. It is obvious that any element of $u + L$ characterizes this class (because if we consider $v + L$ with $v \in u + L$, we have $v + L \in u + L + L = u + L$). We shall denote as the equivalence class $u + L = \{u\}$.

Definition 2.21. The *factor space* E/L is the set of equivalence classes with respect to the relation $u \simeq v$ if $u - v \in L$, organized in accordance with the operations $\{u\} + \{v\} = \{u + v\}$, $u, v \in E$, and $\lambda\{u\} = \{\lambda u\}$ for $u \in E$ and any scalar $\lambda \in R$ (or $\mathcal{C}$).

Thus, E/L is a linear space. The relations $(u + L) + (v + L) = u + v + L$ (because $L + L = L$) and $\lambda(u + L) = \lambda u + L$ (because $\lambda L = L$) motivate the assertion.

More interesting, and significant for our purpose, is the case where E is a normed space (or even a Banach space). In this case, we can introduce a seminorm, or even a norm, in the factor space E/L.

Proposition 2.13. *Let E be a normed space with norm $\| \cdot \|$ and L a linear manifold in E. If we let*

$$\|u + L\| = \inf\{\|v\|; \ v \in u + L\}, \tag{2.25}$$

then equation (2.25) defines a seminorm on the factor space E/L. When L is a closed set in E, equation (2.25) defines a norm on E/L. If E is a Banach space, so is E/L.

Proof. Since $\|u + L\| \geq 0$ for any $u \in E$, we must show that the equal sign implies $u \in L$, and hence $\|u + L\| = \|L\| = 0$ because L contains the null element in E. If $\|u + L\| = 0$ and L is closed, then $u + L$ is *closed* in E and from equation (2.25) we see that $\theta \in u + L$. This means $u \in L$. It is an elementary task to show that $\|\lambda(u + L)\| = \|\lambda u + L\|$ and $\|(u + L) + (v + L)\| \leq \|u + L\| + \|v + L\|$. In other words, we have shown that equation (2.25) defines a seminorm in general and a norm on E/L when L is a closed subspace in E.

There remains to prove that E/L is a Banach space if E belongs to this class.

Assume now that $\{u_n + L; \ n \geq 1\}$ is a Cauchy sequence in E/L. There exists then a subsequence $\{u_{n_k} + L; \ k \geq 1\}$ such that

$$\|u_{n_{k+1}} - u_{n_k} + L\| < 2^{-k-2}, \quad k \geq 1. \tag{2.26}$$

If we take into account the definition of the norm in E/L by formula (2.25), we get that each element $\{u_{n_{k+1}} - u_{n_k} + L; \ k \geq 1\} \in E/L$ contains a v_k such that

$$\|v_k\| < \|u_{n_{k+1}} - u_{n_k}\| + 2^{-k-2} < 2^{-k-1}, \tag{2.27}$$

relying also on formula (2.26). Choose now $x_{n_1} \in \{u_{n_1} + L\}$, and consider the series (in E)

$$x_{n_1} + v_1 + v_2 + \cdots + v_k + \cdots . \tag{2.28}$$

Based on formula (2.27), we can infer the convergence of the series (2.28) in E. Denote by x the sum of the series (2.28) and $s_k = x_{n_1} + v_1 + v_2 + \cdots + v_k$. Let $\{x + L\}$ be the corresponding element of E/L. We will show that

$$\|x + L - (u_{n_{k+1}} + L)\| \leq \|x - s_k\| \longrightarrow 0 \text{ as } k \to \infty, \tag{2.29}$$

which means that $u_{n_k} + L \longrightarrow x + L$ in E/L.

We have to justify formula (2.29). Indeed, from $x_{n_1} \in \{u_{n_1} + L\}$ and $v_k \in \{u_{n_{k+1}} - u_{n_k} + L\}$, one obtains $s_k \in \{u_{n_{k+1}} + L\}$, $k \geq 1$. Therefore,

$$\|x + L - (u_{n_k} + L)\| \le \|x - s_k\|, \quad k \ge 1, \tag{2.30}$$

which is the first part in formula (2.29).

In other words, the subsequence $\{u_{n_k} + L; \ k \ge 1\}$ is convergent in E/L. But we can write the inequality

$$\|x + L - (u_n + L)\| \le \|x + L - (u_{n_k} + L)\| + \|(u_{n_k} + L) - (u_n + L)\|, \tag{2.31}$$

which leads immediately to the conclusion

$$x + L = \lim\{u_n + L\} \ \text{ as } n \to \infty.$$

This ends the proof of Proposition 2.13.

Remark 2.10. If E is endowed with a seminorm instead of a norm and is complete in this seminorm, then Proposition 2.13 remains true.

Remark 2.11. The map $u \longrightarrow u + L$ from E into E/L is called the *canonical* map. If E is a normed linear space and L is closed in E, then this map is linear, continuous and surjective (i.e., the image of this map coincides with E/L).

We shall proceed now to the concept of *function space*. This is fundamental for our development of the theory of almost periodic functions.

2.3 Function Spaces: Continuous Case

The metric spaces we shall define and investigate in this section consist of functions or maps between different spaces. They are usually encountered in many chapters of books on classical or modern analysis, and their applications embrace a large number of domains, such as sequences or series of functions, differential equations, and other kinds of functional equations (integral, discrete, etc.).

Unlike Section 2.2, in which only finite-dimensional spaces or spaces of sequences have been dealt with, the function spaces are somewhat more complex. They can be normed spaces, Banach spaces, or Hilbert spaces, and even more general, linear metric spaces. Several spaces of almost periodic functions will belong to these categories, and we shall present several examples here.

The space $CB(R, C)$ consists of all continuous functions on the real line R with complex values in C such that

$$\sup(|f(t)|; \ t \in R) = M_f < \infty. \tag{2.32}$$

$|\cdot|$ is the absolute value in C (i.e., for $z = x + iy$, $x, y \in R$, $|z| = (x^2 + y^2)^{1/2}$).

It is a simple fact to prove that $CB(R, C)$ is a Banach space over the complex scalars. Indeed, if $f(t)$ and $g(t)$ are continuous on R, then $\alpha f(t) + \beta g(t)$ is also continuous on R for any $\alpha, \beta \in C$. The elementary properties of continuous functions are implied here: The sum of two continuous functions is also continuous, and the product of a continuous function and a constant is also continuous. Hence, the linearity of the space $CB(R, \mathcal{C})$ will follow if we notice that the sum of two functions, say f and g, both satisfying the condition (2.32), will satisfy a similar estimate. Or, for each $t \in R$, $|f(t) + g(t)| \leq |f(t)| + |g(t)|$, which implies $\sup |f(t) + g(t)| \leq \sup |f(t)| + \sup |g(t)|$ for $t \in R$. This means

$$\sup |f(t) + g(t)| \leq M_f + M_g, \ t \in R, \tag{2.33}$$

which proves the assertion.

Let us now introduce the norm

$$\|f(t)\| = \|f\| = \sup(|f(t)|; \ t \in R). \tag{2.34}$$

Since $\|f\| \geq 0$ for any $f \in CB(R, \mathcal{C})$ according to formula (2.31), it remains to note that $\|f\| = 0$ implies $f(t) \equiv 0$ on R. Formula (2.33) means $\|f + g\| \leq \|f\| + \|g\|$, while $\|\lambda f\| = |\lambda| \|f\|$ is obvious.

The only property that we must check for $CB(R, \mathcal{C})$ in order to conclude that this is a Banach space is the completeness. Assume that the sequence $\{f_n(t); \ n \geq 1\} \subset CB(R, \mathcal{C})$ is a Cauchy sequence. This means that given $\varepsilon > 0$, there exists $N(\varepsilon) > 0$ such that

$$\|f_n - f_{n+m}\| < \varepsilon, \ n \geq N(\varepsilon), \ m \geq 1. \tag{2.35}$$

Formula (2.35) implies

$$|f_n(t) - f_{n+m}(t)| < \varepsilon, \ n \geq N(\varepsilon), \ m \geq 1, \tag{2.36}$$

for all $t \in R$. If we fix $t \in R$, then formula (2.36) reduces to the usual Cauchy property in $\mathcal{C}$. Therefore, for each $t \in R$, the limit of the sequence $\{f_n(t); \ n \geq 1\} \subset \mathcal{C}$ does exist. Denote

$$\lim f_n(t) = f(t), \ n \longrightarrow \infty, \tag{2.37}$$

for each fixed t. Returning to formula (2.36), we obtain as $m \longrightarrow \infty$

$$|f_n(t) - f(t)| \leq \varepsilon, \ t \in R, \ n \geq N(\varepsilon). \tag{2.38}$$

But formula (2.38) means that $f_n(t) \longrightarrow f(t)$ as $n \longrightarrow \infty$ uniformly on the whole R. As we know from classical analysis, $f(t)$ is continuous on R, being the uniform limit of a sequence of continuous functions. Obviously, formula (2.38) leads to

$$\|f_n - f\| \leq \varepsilon, \ n \geq N(\varepsilon). \tag{2.39}$$

From formula (2.39), we also derive

$$\|f\| \leq \|f_n\| + \|f - f_n\| = M_{f_n} + \varepsilon, \tag{2.40}$$

which shows that $f \in CB(R, \mathcal{C})$.

The following statement holds true.

Proposition 2.14. *The set* $\mathrm{CB}(R, \mathcal{C})$ *of all continuous and bounded functions from* R *into* $\mathcal{C}$ *endowed with the usual addition and scalar multiplication and the norm* (2.34) *is a Banach space over the complex field.*

The proof of Proposition 2.14 has been conducted above.

Remark 2.12. Spaces similar to $\mathrm{CB}(R, \mathcal{C})$ can be defined by substituting other spaces to $\mathcal{C}$. For instance, $\mathrm{CB}(R, R^n)$ will denote the space of all continuous and bounded functions from R into R^n. These functions are also known as n-vector functions.

Remark 2.13. The space $\mathrm{CB}(R, \mathcal{C})$ has an important (for our purpose) subspace, that of *almost periodic functions in Bohr's sense*, $\mathrm{AP}(R, \mathcal{C})$. This space can be defined as follows. We consider all complex trigonometric polynomials $\mathcal{T}$, as defined by equation (2.11),

$$T(t) = a_1 e^{i\lambda_1 t} + a_2 e^{i\lambda_2 t} + \cdots + a_k e^{i\lambda_k t}, \tag{2.41}$$

with real λ's and complex a's. It is obvious that $\mathcal{T} \subset \mathrm{CB}(R, \mathcal{C})$. The space $\mathrm{AP}(R, \mathcal{C})$ is the closure $\overline{\mathcal{T}}$ of $\mathcal{T}$ in the sense of convergence in the norm (2.34). In other words, $f \in \mathrm{CB}(R, \mathcal{C})$ belongs to $\overline{\mathcal{T}} = \mathrm{AP}(R, \mathcal{C})$ if for any $\varepsilon > 0$ there exists a trigonometric polynomial of the form (2.41) such that

$$\|f - T\| < \varepsilon. \tag{2.42}$$

Equivalently, $f \in \mathrm{AP}(R, \mathcal{C})$ if and only if there exists a sequence of trigonometric polynomials $\{T_n(t)\}$ of the form (2.41) such that

$$\lim T_n(t) = f(t) \text{ as } n \longrightarrow \infty, \tag{2.43}$$

uniformly on the real line R. Indeed, equation (2.43) means that, for each $\varepsilon > 0$, one can find $N(\varepsilon) > 0$ such that

$$|f(t) - T_n(t)| < \varepsilon, \ t \in R, \ n \geq N(\varepsilon),$$

which is the same thing as

$$\|f - T_n\| \longrightarrow 0 \text{ as } n \longrightarrow \infty. \tag{2.44}$$

In subsequent chapters, we shall consider in detail the properties of almost periodic functions. Here, we invite the reader to check the validity of the following basic properties:

I.   Any almost periodic function in Bohr's sense is bounded on R.
II. Any almost periodic function is uniformly continuous on R.

It is useful to notice first that each trigonometric polynomial of the form (2.41) is uniformly continuous on R.

III. Any sequence $\{f_n(t);\ n \geq 1\} \subset AP(R, \mathcal{C})$ that converges in the norm (and therefore also in $AP(R, \mathcal{C})$) has as its limit an almost periodic function in Bohr's sense.

IV. If $f, g \in AP(R, \mathcal{C})$, then λf, with $\lambda \in \mathcal{C}$, $f + g$, $\bar{f}$, fg, are also in $AP(R, \mathcal{C})$.

Actually, property IV shows that $AP(R, \mathcal{C})$ is more complex than a Banach space. $AP(R, \mathcal{C})$ is in fact a *Banach algebra*, a structure allowing the multiplication of its elements.

Further properties of almost periodic functions in the sense of Bohr will be discussed in subsequent chapters, among them the original definition of these functions as formulated by H. Bohr:

V. If $f \in AP(R, \mathcal{C})$, then for each $\varepsilon > 0$ there exists a positive $\ell = \ell(\varepsilon)$ such that any interval $(a, a + \ell) \subset R$ contains a number τ with the property

$$|f(t + \tau) - f(t)| < \varepsilon, \ t \in R. \tag{2.45}$$

This property is very suggestive in regard to the concept of almost periodicity. The number τ, as described above, is called the *ε-almost period* of f. They constitute a *relatively dense* set on the real line.

The proof of equivalence between property V and the definition given to $AP(R, \mathcal{C})$ in Remark 2.13 and Proposition 2.14 will be presented in Chapter 3.

We shall now define another function space that consists of all functions from R into $\mathcal{C}$ representable in the form

$$f(t) = \sum_{k=1}^{\infty} a_k e^{i\lambda_k t}, \ t \in R, \tag{2.46}$$

where $\lambda_k \in R$, $k \geq 1$, $a_k \in \mathcal{C}$, $k \geq 1$, and

$$\sum_{k=1}^{\infty} |a_k| < \infty. \tag{2.47}$$

Definition 2.22. The set of functions defined by formulas (2.46) and (2.47) constitutes a Banach space, denoted by $AP_1(R, \mathcal{C})$, with the norm

$$\|f\|_1 = \sum_{k=1}^{\infty} |a_k|. \tag{2.48}$$

Remark 2.14. Condition (2.47) expresses the absolute convergence of the series appearing on the right-hand side of equation (2.46). Since

$$\left| f(t) - \sum_{k=1}^{n} a_k e^{i\lambda_k t} \right| \leq \sum_{k=n+1}^{\infty} |a_k|, \tag{2.49}$$

Formula (2.47) proves that $f(t)$ can be uniformly approximated on R by trigonometric polynomials. Therefore, $AP_1(R, \mathcal{C}) \subset AP(R, \mathcal{C})$.

In order to motivate Definition 2.22, we must show that $AP_1(R, \mathcal{C})$ is a linear normed space with norm (2.40). Then we must show the completeness. We leave to the reader the task of proving that $AP_1(R, \mathcal{C})$ is a linear normed space over the complex field, and we shall dwell on the proof of the completeness of this space with respect to the norm (2.48).

Let $\{f^j(t); \ j \geq 1\} \subset AP_1(R, \mathcal{C})$ be a Cauchy sequence. We must show that $\{f^j(t); \ j \geq 1\}$ is convergent in the norm (2.46). Assume

$$f^j(t) = \sum_{k=1}^{\infty} a_k^j e^{i\lambda_k t}, \ j \geq 1, \tag{2.50}$$

with each series on the right-hand side absolutely convergent:

$$\sum_{k=1}^{\infty} |a_k^j| < \infty, \ j \geq 1. \tag{2.51}$$

The assumption that λ_k's are the same for each f^j is not a restriction. For all f^j, $j \geq 1$, we have a countable set of λ_k's. Their union is also countable. Including terms with zero a_k's if necessary, we obtain representation (2.50).

The Cauchy property means

$$\|f^j - f^m\|_1 < \varepsilon, \ j, m \geq N(\varepsilon). \tag{2.52}$$

But according to norm (2.48), formula (2.52) translates into

$$\sum_{k=1}^{\infty} |a_k^j - a_k^m| < \varepsilon, \ j, m \geq N(\varepsilon). \tag{2.53}$$

For each n, $n \geq 1$, we derive from formula (2.53)

$$|a_n^j - a_n^m| < \varepsilon, \ j, m \geq N(\varepsilon). \tag{2.54}$$

Formula (2.54) is the Cauchy property (in the complex field $\mathcal{C}$) for the sequence $\{a_n^k; \ k \geq 1\}$, which implies the existence of

$$\lim a_n^k = a_n \ \text{as} \ k \longrightarrow \infty. \tag{2.55}$$

It is our aim now to show that

$$f(t) = \sum_{n=1}^{\infty} a_n e^{i\lambda_n t}, \ t \in R, \tag{2.56}$$

is an element of $AP_1(R, \mathcal{C})$, and

$$f^j(t) \longrightarrow f(t) \ \text{as} \ j \longrightarrow \infty \tag{2.57}$$

in $AP_1(R, \mathcal{C})$. Indeed, from formula (2.53) and equation (2.55), we derive for fixed $p > 0$

$$\sum_{k=1}^{p} |a_k^j - a_k^m| < \varepsilon, \ j, m \geq N(\varepsilon),$$

which implies (for $m \longrightarrow \infty$)

$$\sum_{k=1}^{p} |a_k^j - a_k| \leq \varepsilon, \ j \geq N(\varepsilon),$$

and letting $p \longrightarrow \infty$, we obtain

$$\sum_{k=1}^{\infty} |a_k^j - a_k| \leq \varepsilon, \ j \geq N(\varepsilon). \tag{2.58}$$

Since $a_k = (a_k - a_k^j) + a_k^j$, from $|a_k| \leq |a_k - a_k^j| + |a_k^j|$, for any $k \geq 1$ and a fixed j, we derive the absolute convergence of the series (2.56) and the fact that $f \in AP_1(R, \mathcal{C})$. The inequality (2.58) expresses the validity of formula (2.57). This ends the proof of the assertion that $AP_1(R, \mathcal{C})$ is a Banach space.

We shall see in the subsequent chapters that $AP_1(R, \mathcal{C})$ is the space of almost periodic functions whose *Fourier series* are absolutely convergent. The concept of Fourier series attached to an almost periodic function (in classes even larger than $AP_1(R, \mathcal{C})$ or $AP(R, \mathcal{C})$) will be introduced and investigated in Chapter 4.

Let us note, in concluding the discussion of properties of $AP_1(R, \mathcal{C})$, that each function in this space describes a complex oscillation, the result of compounding infinitely many harmonic oscillations (of frequencies $2\pi/\lambda_k$, $k \geq 1$). We did summarily refer to such oscillations in the introduction.

2.4 Completion of Metric Spaces

We shall now consider the problem of *completion* of a metric space. In Section 2.1, we stated Theorem 2.1, which guarantees the possibility of completing an arbitrary metric space. We are not repeating Theorem 2.1 here, but we shall use the same notation as in Section 2.1. Once this result is established, we shall be able to interpret the spaces $AP(R, \mathcal{C})$ and $AP_1(R, \mathcal{C})$ as the completed spaces of the set of trigonometric polynomials $\mathcal{T}$ with respect to the norms $\| \cdot \|$ of $CB(R, \mathcal{C})$, resp. $\| \cdot \|_1$, given by $\|T\|_1 = \sum_{j=1}^{k} |a_j|$.

The proof of Theorem 2.1 requires some preparation, and the concept of a *function space* will prove to be the right tool to achieve the objective of a (constructing the completion of a metric space).

Proposition 2.15. *Let (E, d) be a metric space. Then the map $d : E \times E \longrightarrow R_+$ is continuous.*

From the triangle inequality we can write $d(x,y) \leq d(x,u)+d(u,v)+d(v,y)$ and $d(u,v) \leq d(u,x)+d(x,y)+d(y,v)$. Also taking into account the symmetry of the distance function, we derive from the inequalities above

$$|d(x,y) - d(u,v)| \leq d(x,u) + d(y,v), \tag{2.59}$$

which is valid for any $x, y, u, v \in E$. From formula (2.59), we obtain the proof of Proposition 2.15.

This proposition is useful in what follows but also has intrinsic interest because it shows that the distance between the points of a metric space varies continuously (with respect to the points involved).

The next proposition shows how to construct a function space, which will be denoted by $\mathrm{CB}(E, R)$, consisting of all continuous and bounded maps from a metric space E into the real line R. The space $\mathrm{CB}(E, R)$ will itself be a metric space with the metric

$$\rho(f,g) = \sup\{|f(x) - g(x)|;\ x \in E\} \tag{2.60}$$

for each $f, g \in \mathrm{CB}(E, R)$.

Proposition 2.16. *Let (E, d) be a metric space, and consider the set of maps $\mathrm{CB}(E, R)$ as described above. With the usual addition (i.e., pointwise) for functions and multiplication by (real) scalars, $\mathrm{CB}(E, R)$ becomes a linear space. Endowed with the distance/norm (2.60), it is a Banach space.*

Proof. First, the linearity is a simple consequence of the following definitions: For $f, g \in \mathrm{CB}(E, R)$, one denotes $(f + g)(x) = f(x) + g(x)$, and for λ real, $(\lambda f)(x) = \lambda f(x)$ for any $x \in E$.

Then, equation (2.60) makes sense for each $f, g \in \mathrm{CB}(E, R)$ because of the boundedness of these maps on E. The conditions required by the distance function can be verified elementarily.

Finally, there remains to show that $\mathrm{CB}(E, R)$ is complete with respect to the distance ρ, given by equation (2.60). Consider a Cauchy sequence $\{f_n;\ n \geq 1\} \subset \mathrm{CB}(E, R)$. This implies $\rho(f_n, f_m) < \varepsilon$, for any $\varepsilon > 0$, provided $n, m \geq N(\varepsilon)$. Taking equation (2.60) into account, we can write, for any $x \in E$,

$$|f_n(x) - f_m(x)| < \varepsilon,\ n, m \geq N(\varepsilon). \tag{2.61}$$

But formula (2.61) shows that $\{f_n(x);\ n \geq 1\}$ is a Cauchy sequence in R for a given x. Hence, $\lim f_n(x) = f(x)$ as $n \longrightarrow \infty$. If in formula (2.61) we let $m \longrightarrow \infty$ for fixed $x \in E$, we obtain

$$|f_n(x) - f(x)| \leq \varepsilon,\ n \geq N(\varepsilon). \tag{2.62}$$

This means

$$\rho(f_n, f) \leq \varepsilon,\ n \geq N(\varepsilon), \tag{2.63}$$

which proves the completeness of the space $\mathrm{CB}(E, R)$.

Remark 2.15. The proof can be carried out to the more general case of the space $CB(E, F)$, where F stands for a Banach space.

Proof of Theorem 2.1. We can now move to provide the proof of Theorem 2.1, which constitutes a basic tool in our presentation.

More precisely, we shall prove the following statement: *Let (E, d) be a metric space. Then, there exists a subset of the Banach space $CB(E, R)$ that can be identified with (E, d).*

The term *identified* should be understood in the sense that there exists a one-to-one correspondence between (E, d) and the subset of $CB(E, R)$ with distance preservation.

Let us consider the metric space (E, d) and the associated Banach space $CB(E, R)$, which is endowed with the metric ρ defined by equation (2.60). Let us fix a point $a \in E$ and for each $u \in E$ consider the real-valued function

$$f_u(x) = d(x, u) - d(x, a), \quad x \in E. \tag{2.64}$$

From the triangle inequality for the metric d, we obtain $|f_u(x)| \leq d(u, a)$. Hence f_u is bounded on E. Moreover, f_u is continuous on E if one takes into account equation (2.64) and Proposition 2.15. It is easy to see that the map $u \longrightarrow f_u$ from E into $CB(E, R)$ is one-to-one. Indeed, if we assume $f_u = f_v$, this amounts to $d(x, u) = d(x, v)$ for any $x \in E$. But for $x = u$ one has $d(u, u) = 0 = d(u, v)$, which implies $u = v$. We can also prove that

$$\rho(f_u, f_v) = d(u, v) \tag{2.65}$$

for any $u, v \in E$. This will mean that the map $u \longrightarrow f_u$ preserves the distance (one also says that the map $u \longrightarrow f_u$ is an *isometry*). In order to prove equation (2.65), we shall note that, for each $x \in E$,

$$\begin{aligned} f_u(x) - f_v(x) &= d(x, u) - d(x, a) - d(x, v) + d(x, a) \\ &= d(x, u) - d(x, v) \leq d(u, v). \end{aligned}$$

Changing the roles of u and v, we obtain a similar inequality. Both inequalities lead to

$$|f_u(x) - f_v(x)| \leq d(u, v), \quad x \in E, \tag{2.66}$$

which implies

$$\rho(f_u, f_v) \leq d(u, v), \quad u, v \in E. \tag{2.67}$$

On the other hand, $f_u(v) - f_v(v) = d(v, u) - d(v, a) + d(v, a) = d(u, v)$. Therefore,

$$d(u, v) \leq \sup_{x \in E} |f_u(x) - f_v(x)| = \rho(f_u, f_v),$$

which, together with formula (2.67), leads to equation (2.65).

At this stage in the proof of Theorem 2.1, there remains to note that the space $(\widetilde{E}, \widetilde{d})$ in the statement of the theorem is nothing but the closure in

CB(E, R) of the range of the map $u \longrightarrow f_u$. Indeed, this closure is a closed subset of the complete metric space CB(E, R) with respect to the metric $\rho = \tilde{d}$. As a closed subset of a complete metric space, it is itself complete. Indeed, if a Cauchy sequence is in $(\widetilde{E}, \rho)$, then it is convergent in CB(E, R). But $(\widetilde{E}, \rho)$ is closed in CB(E, R), which means that the limit belongs to $(\widetilde{E}, \rho)$. The relation (2.65) shows that $d = \tilde{d} = \rho$ for any $u, v \in E$.

Theorem 2.1 is thus proven.

Remark 2.16. Since each subset M of a metric space E is *dense* in its closure $\overline{M} \subset E$, there results from Definition 2.9 that, for any element $\bar{x} \in \overline{M}$, one can find a sequence $\{x_n;\ n \geq 1\} \subset M$ such that $x_n \longrightarrow \bar{x}$ as $n \longrightarrow \infty$.

Definition 2.23. The space $(\widetilde{E}, \tilde{d})$ constructed in the proof of Theorem 2.1 is called the *completion* of the space (E, d).

The following result is a natural complement to Theorem 2.1.

Proposition 2.17. *For any metric space E, there exists, up to an isometry, a unique complete metric space $\widetilde{E}$ containing a dense subset isometric to E.*

Proof. In order to prove Proposition 2.17, it is obviously sufficient to prove the more general assertion: If F and G are complete metric spaces and A and B are subsets dense in F resp. G (i.e., $\overline{A} = F$, $\overline{B} = G$), then for every isometric mapping $f : A \longrightarrow B$, there exists an isometry $\tilde{f} : F \longrightarrow G$ such that $\tilde{f}(x) = f(x)$ for each $x \in A$. Let $x \in F$ and $\{x_n;\ n \geq 1\} \subset A$ such that $x_n \longrightarrow x$ as $n \longrightarrow \infty$. $\{x_n;\ n \geq 1\}$ satisfies the Cauchy condition, which means that the sequence $\{y_n = f(x_n);\ n \geq 1\}$ also satisfies the Cauchy property. This implies that $y_n \longrightarrow y \in G$. Let us point out that y does not depend on the particular sequence $\{x_n;\ n \geq 1\} \subset A$, $x_n \longrightarrow x$ as $n \longrightarrow \infty$. If $\{\tilde{x}_n;\ n \geq 1\} \subset A$ and $\tilde{x}_n \longrightarrow x$ as $n \longrightarrow \infty$, then considering $\tilde{y}_n = f(\tilde{x}_n)$, $n \geq 1$, we notice that the sequence $f(x_1), f(\tilde{x}_1), f(x_2), f(\tilde{x}_2), \ldots, f(x_n), f(\tilde{x}_n), \ldots$ is convergent in G, which implies $\lim y_n = \lim \tilde{y}_n = y$ as $n \longrightarrow \infty$. We define $\tilde{f} : F \longrightarrow G$ by letting $\tilde{f}(x) = y$, where y has been determined as described above. When $x \in A$, we obviously have $\tilde{f}(x) = f(x) = y$.

This ends the proof of uniqueness of the completion of a metric space.

Remark 2.17. If the metric space E is linear, then the completion of E with respect to the metric given by the norm in E is a Banach space. This can be easily seen by defining in the completion the operations of addition of elements and the scalar multiplication.

Indeed, let $x, y \in \widetilde{E} =$ the completion of E. Then, there exist two Cauchy sequences $\{x_n\}, \{y_n\} \in E$, such that $x_n \to x$, $y_n \to y$ as $n \to \infty$. But $\{x_n + y_n\}$ is also a Cauchy sequence in E, which means there exists $z \in \widetilde{E}$ with $x_n + y_n \to z$ as $n \to \infty$. By definition, $x + y = z$. Obviously, z does not depend on the choice of the sequences $\{x_n\}, \{y_n\}$, provided $x_n \to x$ and $y_n \to y$ as $n \to \infty$. Similarly, for $\lambda \in C$ and $x \in \widetilde{E}$, one defines $\lambda x \in \widetilde{E}$ as the (unique) limit of any sequence $\{\lambda x_n\} \subset E$, with $x_n \to x$ as $n \to \infty$. It is an elementary exercise to check the basic properties of these operations.

The uniqueness of the completion of a metric space, as stipulated by Proposition 2.17, allows us to conclude that the spaces $AP(R, \mathcal{C})$ or $AP_1(R, \mathcal{C})$ are uniquely determined by the set of trigonometric polynomials $\mathcal{T}$, of the form (2.41), and the norms (or metrics) indicated by $\sup\{|T(t)|;\ t \in R\}$, resp. $\|T\|_1 = |a_1| + |a_2| + \cdots + |a_k|$.

Several other spaces consisting of almost periodic functions can be constructed taking the spaces $AP_1(R, \mathcal{C})$ or $AP(R, \mathcal{C})$ as basic elements. For instance, the space $AP(R, \mathcal{C}^n)$ is defined as the direct product of n factors, each identical to $AP(R, \mathcal{C})$. In a similar manner, one defines $AP(R, R^n)$, the starting element being $AP(R, R)$; i.e., the space of real-valued almost periodic functions. This space is the completion of the set $\mathcal{T}_1$ of *real trigonometric polynomials* with respect to sup-norm. A real trigonometric polynomial is a function of the form $T_0(t) = \alpha_0 + \alpha_1 \cos \lambda_1 t + \beta_1 \sin \lambda_1 t + \cdots + \alpha_k \cos \lambda_k t + \beta_k \sin \lambda_k t$, with $\lambda_1, \lambda_2, \ldots, \lambda_k$ arbitrary real numbers, as well as α_j, $1 \leq j \leq k$, β_j, $1 \leq j \leq k$. The connection between the concept of almost periodicity and that of a *general* trigonometric series is not yet investigated in depth. The term general stands for the fact that λ_j's are arbitrary and not the multiples of a given positive number. In the latter case, we have a *Fourier series* that corresponds to the case of periodic functions.

Remark 2.18. When we introduced the space $AP_1(R, \mathcal{C})$, we in fact considered the completion of the set $\mathcal{T}$ of trigonometric polynomials with respect to the norm

$$\|T(t)\|_1 = \sum_{j=1}^{k} |a_j|. \tag{2.68}$$

We can generalize this type of norm on $\mathcal{T}$ by considering

$$\|T\|_r = \left(\sum_{j=1}^{k} |a_j|^r\right)^{1/r}, \qquad 1 \leq r \leq 2. \tag{2.69}$$

For $r = 1$, we reobtain the norm $\| \cdot \|_1$, while $r = 2$ corresponds to the Euclidean norms. It is easy to see that $\| \cdot \|_2 \leq \| \cdot \|_1$. In general, $\| \cdot \|_r$ is decreasing when r is increasing, $1 \leq r \leq 2$.

Since each $\| \cdot \|_r$ is a norm on $\mathcal{T}$, the completion of $\mathcal{T}$ with respect to the norm $\| \cdot \|_r$ leads to a class of almost periodic functions that we shall denote by $AP_r(R, \mathcal{C})$.

The complete spaces $AP_r(R, \mathcal{C})$, $1 < r \leq 2$, may not consist of functions from R into $\mathcal{C}$, but rather of classes of functions (equivalent in a precise sense). We shall illustrate this fact in detail in the case $r = 2$ when we deal with almost periodic functions in the sense of Besicovitch. The space $AP_2(R, \mathcal{C})$ is also denoted $B^2(R, \mathcal{C})$.

2.5 Function Spaces: Measurable Case

The concept of almost periodicity is not restricted to continuous functions. In the case of measurable (Lebesgue) functions, one can define the almost periodicity by using adequate metrics, distances, and norms in the linear space $\mathcal{T}$ of trigonometric polynomials.

In order to define the class of almost periodic functions in the sense of V. Stepanov, we shall first deal with the space $M = M(R, \mathcal{C})$ of *locally integrable* maps from R into $\mathcal{C}$ satisfying a certain condition of boundedness. This space has been considered in the literature by N. Wiener and V. Stepanov. A more recent treatment of its definition and properties can be found in the book by J. L. Massera and J. J. Schäffer [68].

Definition 2.24. The space M can be defined by the condition

$$M = \left\{ x : R \to \mathcal{C}, \sup \int_t^{t+1} |x(s)| ds, \ t \in R \right\} < +\infty. \qquad (2.70)$$

One says that the elements of M are *bounded in the mean*.

Obviously, $BC(R, \mathcal{C}) \subset M(R, \mathcal{C})$ because $x \in BC$ implies local integrability and $|x(t)| \leq A$, $t \in R$. Hence

$$\sup \int_t^{t+1} |x(s)| ds \leq A, \ t \in R, \qquad (2.71)$$

which proves the fact that $x \in M$.

To get a more precise idea of what a function in M represents, we observe that it can be obtained by taking a bounded sequence in the space $L = L^1([0,1], \mathcal{C})$, say $\{x_n; \ n \in Z\}$, and then letting

$$x(t) = x_n(t - n), \ n < t \leq n + 1, \ n \in Z, \qquad (2.72)$$

Z standing for the ring of integers.

Conversely, each element $x \in M$ is generating a sequence in $L^1([0,1], \mathcal{C})$. Formula (2.72) explains how this happens.

Returning now to the formula (2.70), which is the definition of the space $M(R, \mathcal{C})$, it is easy to see that $M(R, \mathcal{C})$ is a linear space over the field $\mathcal{C}$ of complex numbers. Of course, as stipulated above in this section, each element of $M(R, \mathcal{C})$ is a locally integrable function. Therefore, by definition, $M(R, \mathcal{C}) \subset L_{loc}(R, \mathcal{C})$.

The norm in the space M is defined by

$$|x|_M = \sup \left\{ \int_t^{t+1} |x(s)| ds; \ t \in R \right\}. \qquad (2.73)$$

It is an elementary exercise to show that the map $x \to |x|_M$ is a norm on M. In order to show that M is a Banach space endowed with the norm (2.73), one has to show that *any sequence in M satisfying the Cauchy condition is convergent in the norm* $|\cdot|_M$ (to an element of M). This requires some details, and we are providing the proof here (it can also be found elsewhere; see, for instance, J. L. Massera and J. J. Schäffer [68]).

Indeed, let $\{x_m\} \subset M$ be a sequence satisfying Cauchy's condition; i.e., for each $\varepsilon > 0$, there exists a natural number $N = N(\varepsilon) > 0$ such that

$$|x_m - x_p|_M < \varepsilon, \ m, p \geq N(\varepsilon). \tag{2.74}$$

We need to show that there exists $x \in M$ such that

$$|x_m - x|_M \longrightarrow 0 \ \text{as} \ m \longrightarrow \infty. \tag{2.75}$$

From formula (2.74), we derive

$$\sup \int_t^{t+1} |x_m(s) - x_p(s)| ds < \varepsilon, \ m, p \geq N(\varepsilon). \tag{2.76}$$

Let us now fix an arbitrary finite interval $(a, b) \subset R$ and note that for some integers $r, s \in Z$ one has $(a, b) \subset (r, s)$. Then, formula (2.76) leads to

$$\int_a^b |x_m(u) - x_p(u)| du \leq (s - r)|x_m - x_p|_M < (s - r)\varepsilon \tag{2.77}$$

for $m, p \geq N(\varepsilon)$. In other words, the restrictions of the functions of $\{x_m\}$ to (a, b) form a Cauchy sequence in $L^1((a, b), C)$. Hence, one can define

$$x(t) = \lim x_p(t) \ \text{as} \ p \longrightarrow \infty \tag{2.78}$$

on each finite interval $(a, b) \subset R$. The limit function $x(t)$ is thus defined on R, and it is locally integrable. It remains to show that $x(t)$ given by equation (2.78) belongs to the space M. But $x = (x - x_m) + x_m$ with $x_m \in M$. On the other hand, letting $p \to \infty$ in formula (2.76), one obtains

$$|x_m - x|_M \leq \varepsilon \ \text{for} \ m \geq N(\varepsilon).$$

This means that $x - x_m \in M$. Therefore, $x \in M$, which ends the proof of the assertion that M endowed with the norm $|\cdot|_M$ is a Banach space.

The space M is a fairly large function space, and it does contain as subspaces all the spaces $L^p(R, C)$, $1 \leq p \leq \infty$. The inequality

$$\int_t^{t+1} |x(s)| ds \leq \left(\int_t^{t+1} |x(s)|^p ds \right)^{1/p}, \ 1 \leq p < \infty,$$

leads immediately to the inclusions $L^p \subset M$, $1 \leq p < \infty$. The inclusion $L^\infty \subset M$ is obvious. Moreover, one has $|x|_M \leq |x|_{L^p}$, $1 \leq p \leq \infty$. These inequalities imply that convergence in L^p always has as a consequence convergence in M.

Another significant property of the space M is related to the *convolution product*

$$(f * g)(t) = \int_R f(t - s)g(s)ds, \qquad (2.79)$$

which is known to make sense for $f \in L^1$ and $g \in L^p$, $1 \le p \le \infty$. One has $f * g \in L^p$, and this is a consequence of the inequality

$$|f * g|_{L^p} \le |f|_{L^1}|g|_{L^p}. \qquad (2.80)$$

For the validity of formula (2.80), see for instance the author's book [23], where the case of the space M is also discussed. It is shown that, for $f \in L^1$ and $g \in M$,

$$|f * g|_M \le |f|_{L^1}|g|_M. \qquad (2.81)$$

Inequalities like (2.80) or (2.81) will be needed in subsequent chapters when dealing with operators on function spaces and associated functional equations.

We are now able to define the almost periodic functions in the sense of Stepanov. The space of these functions is actually a subspace of the space $M(R, \mathcal{C})$.

Definition 2.25. The space $\mathcal{S} = S(R, \mathcal{C})$ of almost periodic functions in the Stepanov sense is the closure in $M = M(R, \mathcal{C})$ of the set $\mathcal{T}$ of (complex) trigonometric polynomials of the form (2.41).

The definition above implies that for any $x \in \mathcal{S}(R, \mathcal{C})$ and $\varepsilon > 0$ there exists a trigonometric polynomial $T(t)$ such that

$$|x - T|_M < \varepsilon. \qquad (2.82)$$

We also say that any almost periodic function in the sense of Stepanov can be approximated with any degree of accuracy by trigonometric polynomials in the norm $|\cdot|_M$.

From Definition 2.25, we know that S is a Banach space over the complex field $\mathcal{C}$. It has properties similar to those of $AP(R, \mathcal{C})$, but only to a certain level. For instance, S is not closed with respect to pointwise multiplication of its elements. Its basic properties, similar to property V of $AP(R, \mathcal{C})$, will be discussed in subsequent chapters and used in connection with almost periodicity of solutions of functional equations.

So far, we know about the following inclusions of the spaces of almost periodic functions:

$$AP_1(R, \mathcal{C}) \subset AP(R, \mathcal{C}) \subset \mathcal{S}(R, \mathcal{C}). \qquad (2.83)$$

To the sequence of inclusions (2.83), we shall add one more after introducing a new class of almost periodic functions known as Besicovitch class $B^2(R, \mathcal{C})$. We made a brief reference to the space $B^2(R, \mathcal{C})$ at the end of Section 2.4.

We shall conclude the discussion of Stepanov's almost periodic functions (in this section) by noting that several other spaces can be defined similarly. The spaces $S(R, R)$, $S(R, R^n)$, $S(R, C^n)$ are obviously defined, starting in the first case with real trigonometric polynomials instead of T (to be more specific, with $\operatorname{Re} T$) and then proceeding by completion.

There are many references in the literature to the spaces $S^p(R, C)$, $1 \leq p$, of almost periodic functions in Stepanov's sense. They are defined in a manner similar to that used above for the space $S = S^1(R, C)$. Namely, the norm in $S^p(R, C)$ is given by

$$\|x\|_{S^p} = \sup_{t \in R} \left(\int_t^{t+1} |x(s)|^p ds \right)^{1/p},$$

and the relationship with the norm of S (or M) follows from the inequality mentioned above, namely

$$\int_t^{t+1} |x(s)| ds \leq \left(\int_t^{t+1} |x(s)|^p ds \right)^{1/p}, \qquad 1 \leq p < \infty.$$

This inequality shows that $S^p \subset S$ for $p > 1$, which tells us that the space S is the richest space of almost periodic functions in Stepanov's sense.

The space $S^2(R, C)$ will be considered later.

We shall now prepare to introduce *Besicovitch almost periodic functions* by following the same procedure as for the spaces $AP(R, C)$ or $S(R, C)$. In other words, we will first define a Banach function space that will naturally contain the Besicovitch space. The latter will appear as a (closed) subspace of the space $\mathcal{M}_2(R, C)$, which will be immediately defined and is known as a *Marcinkiewicz function space*.

As a suggestion for a possible norm in both Marcinkiewicz and Besicovitch spaces, we first note the following connection between the norm $\| \cdot \|_2$ on T (i.e., the Euclidean norm) and the so-called *mean value* for trigonometric polynomials:

$$\sum_{j=1}^k |a_j|^2 = \|T\|_2^2 = \lim_{\ell \to \infty} (2\ell)^{-1} \int_{-\ell}^\ell |T(t)|^2 dt. \qquad (2.84)$$

In the representation (2.41) of the trigonometric polynomial $T(t)$, we assume that the λ_j's are distinct reals, while a_j's are complex. Since

$$|T(t)|^2 = T(t)\overline{T}(t) = \sum_{j=1}^k |a_j|^2 + \sum_{j=1}^k \sum_{m=1}^k a_j \bar{a}_m e^{i(\lambda_j - \lambda_m)t},$$

with $j \neq m$ in the double sum, equation (2.84) follows immediately from the equation above if we take into account that

$$\lim_{\ell \to \infty} (2\ell)^{-1} \int_{-\ell}^{\ell} e^{i\lambda t} dt = 0, \qquad \lambda \in R, \ \lambda \neq 0. \tag{2.85}$$

The norm suggested by equation (2.84) is

$$x \longrightarrow \left[\lim_{\ell \to \infty} (2\ell)^{-1} \int_{-\ell}^{\ell} |x(t)|^2 dt \right]^{1/2},$$

which is finite for some $x \in L^2_{loc}(R, C)$ = the space of all measurable functions from R into C that are locally square integrable. Slightly modified, the considerations above lead to the possible norm (Marcinkiewicz)

$$\|x\|_{\mathcal{M}}^2 = \limsup_{\ell \to \infty} (2\ell)^{-1} \int_{-\ell}^{\ell} |x(t)|^2 dt \tag{2.86}$$

whenever the right-hand side in equation (2.86) is finite.

It is an elementary exercise to see that the set $\mathcal{M}$ of all $x \in L^2_{loc}(R, C)$ for which equation (2.86) is finite is a linear space over C. For instance, to check that $x + y \in \mathcal{M}$ if $x, y \in \mathcal{M}$, we rely on the well-known inequality $(a + b)^2 \leq 2(a^2 + b^2)$ for positive a and b, which implies

$$\int_{-\ell}^{\ell} [x(t) + y(t)]^2 dt \leq 2 \int_{-\ell}^{\ell} |x(t)|^2 dt + 2 \int_{-\ell}^{\ell} |y(t)|^2 dt.$$

In fact, the map $x \to \|x\|_{\mathcal{M}}$ from $\mathcal{M}$ into $R_+ = [0, \infty)$ is not a norm but a *seminorm*. Indeed, the properties of the seminorm, $\|x\|_{\mathcal{M}} \geq 0$ and $\|\lambda x\|_{\mathcal{M}} = |\lambda| \|x\|_{\mathcal{M}}$, are obviously satisfied. The triangle inequality $\|x + y\|_{\mathcal{M}} \leq \|x\|_{\mathcal{M}} + \|y\|_{\mathcal{M}}$ follows from the well-known inequality

$$\left(\int_{-\ell}^{\ell} |x(t) + y(t)|^2 dt \right)^{1/2} \leq \left(\int_{-\ell}^{\ell} |x(t)|^2 dt \right)^{1/2} + \left(\int_{-\ell}^{\ell} |y(t)|^2 dt \right)^{1/2},$$

also taking into account $\limsup(u + v) \leq \limsup u + \limsup v$.

On the other hand, $\|x\|_{\mathcal{M}} = 0$ means

$$\lim_{\ell \to \infty} (2\ell)^{-1} \int_{-\ell}^{\ell} |x(t)|^2 dt = 0. \tag{2.87}$$

It is easy to see that x satisfying equation (2.87) need not necessarily be zero (a.e.). For instance, $x(t) = \exp\{-|t|\}$, $t \in R$, is such a function, nowhere vanishing on R. Indeed, one easily finds

$$(2\ell)^{-1} \int_{-\ell}^{\ell} e^{-|t|} dt = \ell^{-1}(1 - e^{-\ell}),$$

which means that equation (2.87) is satisfied.

Since $\| \cdot \|_{\mathcal{M}}$ is only a seminorm on $\mathcal{M}$, in order to obtain *Marcinkiewicz space* we will consider the factor space $\mathcal{M}/\mathcal{L}$, where $\mathcal{L}$ stands for the linear manifold consisting of all $y \in \mathcal{M}$ satisfying equation (2.87). Therefore, the elements of the Marcinkiewicz space $\mathcal{M}_2(R,\mathcal{C}) = \mathcal{M}/\mathcal{L}$ are constituted by the classes of equivalence in $\mathcal{M}$ according to the relation $x \sim y$ iff $x - y \in \mathcal{L}$. In other words, an element of $\mathcal{M}_2$ is determined by a set of elements in $\mathcal{M}$ of the form $\{x + x_0\}$ with $x \in \mathcal{M}$ and $x_0 \in \mathcal{L}$. Because the $\mathcal{M}$-norm of each element in an equivalence class is the same (as noted above), it is natural to take it as a norm in $\mathcal{M}_2$ for that class (as an element of $\mathcal{M}_2$).

We can now formulate the definition of a Marcinkiewicz function space $\mathcal{M}_2(R,\mathcal{C})$ just by summarizing the considerations above.

Definition 2.26. The factor space $\mathcal{M}/\mathcal{L}$ with $\mathcal{M}(R,\mathcal{C})$ defined by equation (2.86) and $\mathcal{L}$ defined by equation (2.87) is called a Marcinkiewicz function space and is denoted by $\mathcal{M}_2(R,\mathcal{C})$.

In a similar manner, one can define the Marcinkiewicz function spaces $\mathcal{M}_2(R,R)$, $\mathcal{M}_2(R,R^n)$, and others.

Of course, the discussion above does not imply that $\mathcal{M}_2(R,\mathcal{C})$ is a Banach space. Unfortunately, Proposition 2.13 cannot be applied here directly and can only serve as a model (for instance, $\mathcal{M}$ is not a Banach space).

In order to provide full consistency to Definition 2.26, we need to proceed in a manner similar to the proof of Proposition 2.13 but keeping in mind the special situation since $\mathcal{M}$ is not a Banach space. Fortunately, $\mathcal{L}$ *is a closed linear manifold* in $\mathcal{M}$. This condition is requested in Proposition 2.13.

Indeed, let $x \in \mathcal{M}$ be the limit of the sequence $\{x_n; \; n \geq 1\} \subset \mathcal{L}$. This means

$$\limsup_{n \longrightarrow \infty} (2\ell)^{-1} \int_{-\ell}^{\ell} |x_n(t) - x(t)|^2 dt = 0. \tag{2.88}$$

But $x = (x - x_n) + x_n$, and therefore

$$\limsup_{\ell \longrightarrow \infty} (2\ell)^{-1} \int_{-\ell}^{\ell} |x(t)|^2 dt \leq \limsup_{\ell \longrightarrow \infty} (2\ell)^{-1} \int_{-\ell}^{\ell} |x(t) - x_n(t)|^2 dt$$

because $x_n \in \mathcal{L}$. Taking equation (2.88) into account, we obtain

$$\limsup_{\ell \longrightarrow \infty} (2\ell)^{-1} \int_{-\ell}^{\ell} |x(t)|^2 dt = 0,$$

which proves that $x \in \mathcal{L}$. This shows that $\mathcal{L}$ is closed in $\mathcal{M}$ in the topology induced by the seminorm $\| \cdot \|_{\mathcal{M}}$.

The next challenge is to prove that $\mathcal{M}$ *is complete* with respect to the convergence generated by $\| \cdot \|_{\mathcal{M}}$. In other words, if we consider a Cauchy sequence in $\mathcal{M}$, one must prove its convergence in the norm $\| \cdot \|_{\mathcal{M}}$.

Let $\{x_n;\ n \geq 1\} \subset \mathcal{M}$ be such that, given $\varepsilon > 0$, one has

$$\|x_n - x_m\|_{\mathcal{M}} < \sqrt{\varepsilon} \quad \text{for } n, m \geq N(\varepsilon). \tag{2.89}$$

This means that

$$\limsup_{\ell \to \infty} (2\ell)^{-1} \int_{-\ell}^{\ell} |x_n(t) - x_m(t)|^2 dt < \varepsilon \quad \text{for } n, m \geq N(\varepsilon). \tag{2.90}$$

From formula (2.90), we derive (in accordance with the definition of $\limsup$) that there exists a positive number $L = L(\varepsilon)$ such that $\ell \geq L(\varepsilon)$ implies

$$(2\ell)^{-1} \int_{-\ell}^{\ell} |x_n(t) - x_m(t)|^2 dt < 2\varepsilon \quad \text{for } n, m \geq N_1(\varepsilon). \tag{2.91}$$

Let us now fix ℓ in formula (2.91), $\ell \geq L$, and rewrite this inequality in the form

$$\int_{-\ell}^{\ell} |x_n(t) - x_m(t)|^2 dt < 4\ell\varepsilon \quad \text{for } n, m \geq N_1(\varepsilon). \tag{2.92}$$

The inequality (2.92) shows that the restrictions of the functions $x_n(t)$, $n \geq 1$, to the interval $[-\ell, \ell]$ form a Cauchy sequence in the space $L^2([-\ell, \ell], \mathcal{C})$. Therefore, based on the completeness of the space $L^2([-\ell, \ell], \mathcal{C})$, we can state that the sequence $\{x_n;\ n \geq 1\}$, restricted to the interval $[-\ell, \ell]$, is L^2-convergent: $\lim x_n(t) = x(t)$ as $n \to \infty$. Since the limit is unique (a.e.) in any space $L^2([-\ell, \ell], \mathcal{C})$, $\ell \geq L$, we conclude that there exists an element $x \in L^2_{\text{loc}}(R, \mathcal{C})$ such that $\lim x_n(t) = x(t)$ in $L^2_{\text{loc}}(R, \mathcal{C})$. But convergence in $L^2_{\text{loc}}(R, \mathcal{C})$ does not imply convergence in $\mathcal{M}(R, \mathcal{C})$. As seen above, the converse is true. Also, from the considerations above, it does not follow that $x \in \mathcal{M}(R, \mathcal{C})$. These facts must now be clarified.

First, let us note that $x(t)$ defined above satisfies, for each $\ell \geq L(\varepsilon)$, the inequality

$$\int_{-\ell}^{\ell} |x_n(t) - x(t)|^2 dt \leq 4\ell\varepsilon \quad \text{for } n \geq N_1(\varepsilon) \tag{2.93}$$

because in the L^2-convergence it is allowed to let $m \to \infty$ in formula (2.92).

Second, formula (2.93) is equivalent to

$$(2\ell)^{-1} \int_{-\ell}^{\ell} |x_n(t) - x(t)|^2 dt \leq 2\varepsilon \quad \text{for } n \geq N_1(\varepsilon). \tag{2.94}$$

Hence, keeping in mind that $\ell \geq L(\varepsilon)$ is arbitrary,

$$\|x_n - x\|_{\mathcal{M}}^2 \leq 2\varepsilon \quad \text{for } n \geq N_1(\varepsilon). \tag{2.95}$$

Because $x = (x - x_n) + x_n$ and both terms on the right-hand side are in $\mathcal{M}$, one derives $x \in \mathcal{M}$. The inequality (2.95) also shows that $\lim x_n = x$ as $n \to \infty$ in the sense of the convergence in $\mathcal{M}$ (i.e., as defined by the seminorm $\|\cdot\|_{\mathcal{M}}$).

Let us summarize the discussion above by stating the following result.

Proposition 2.18. *The space* $\mathcal{M} = \mathcal{M}(R,\mathcal{C})$ *of locally square integrable functions for which* $\|\cdot\|_{\mathcal{M}}$ *given by equation* (2.86) *is finite is complete in the sense of convergence induced by* $\|\cdot\|_{\mathcal{M}}$. *Moreover, the linear manifold* $\mathcal{L} \subset \mathcal{M}$, *defined by equation* (2.87), *is closed in* $\mathcal{M}$.

Remark 2.19. Actually, the closedness of $\mathcal{L}$ in $\mathcal{M}$ is a consequence of the fact that $\|x\|_{\mathcal{M}}$ is continuous in x (each norm or seminorm is continuous in the topology it generates!). The fact that $\|x\|_{\mathcal{M}} = 0$ defines a closed linear manifold is then obvious.

In Definition 2.26, we introduced the Marcinkiewicz space $\mathcal{M}_2(R,\mathcal{C})$ as the factor space $\mathcal{M}(R,\mathcal{C})/\mathcal{L}(R,\mathcal{C})$. It is now our aim to prove that $\mathcal{M}_2(R,\mathcal{C})$ is a Banach space over the complex field. This fact is basically a consequence of Remark 2.10 to Proposition 2.13. Nevertheless, we prefer to get into some details, given the fact that Proposition 2.13 cannot be directly applied to derive the result we have in mind.

Proposition 2.19. *The Marcinkiewicz function space* $\mathcal{M}_2 = \mathcal{M}/\mathcal{L}$ *is a Banach space with the norm defined by* $\|x + \mathcal{L}\|_{\mathcal{M}} = \|x\|_{\mathcal{M}}$, $x \in \mathcal{M}$.

Remark 2.20. In Section 2.2, the norm $\|x + \mathcal{L}\|_{\mathcal{M}} = \inf\{\|y\|_{\mathcal{M}}; y \in x + \mathcal{L}\}$ was defined. It is easy to see that this definition is equivalent to that indicated in Proposition 2.19. Indeed, taking into account that $\mathcal{L} = \{x;\ x \in \mathcal{M},\ \|x\|_{\mathcal{M}} = 0\}$, we have, for any $y \in x + \mathcal{L}$, $\|y\|_{\mathcal{M}} \leq \|x\|_{\mathcal{M}} + \|x_0\|_{\mathcal{M}}$ with $\|x_0\|_{\mathcal{M}} = 0$ because $x_0 \in \mathcal{L}$. Hence, $\|y\|_{\mathcal{M}} \leq \|x\|_{\mathcal{M}}$. But $y = x + x_0$ gives $x = y - x_0$, $x_0 \in \mathcal{L}$, which means $\|x\|_{\mathcal{M}} \leq \|y\|_{\mathcal{M}}$. Therefore, $\|x + \mathcal{L}\|_{\mathcal{M}} = \inf\{\|y\|_{\mathcal{M}}; y \in x + \mathcal{L}\} = \|x\|_{\mathcal{M}}$, and we see that all the elements in $\mathcal{M}$ that define an element in $\mathcal{M}_2 = \mathcal{M}/\mathcal{L}$ have the same $\mathcal{M}$-seminorm. This common value is hence taken as the norm in $\mathcal{M}_2$.

Proof of Proposition 2.19. From the proof of Proposition 2.13, we know that $\mathcal{M}_2$ is a linear space and $x + \mathcal{L} \longrightarrow \|x + \mathcal{L}\|_{\mathcal{M}}$ is a seminorm on $\mathcal{M}_2$. Moreover, according to Proposition 2.18, the seminormed space $\mathcal{M}$ is complete with respect to $\|\cdot\|_{\mathcal{M}}$.

There remains to prove that the map $x + \mathcal{L} \longrightarrow \|x + \mathcal{L}\|_{\mathcal{M}} = \|x\|_{\mathcal{M}}$ is actually a norm on the factor space and that this space is also complete.

First, if $\|x + \mathcal{L}\|_{\mathcal{M}} = 0$, this means $\|x\|_{\mathcal{M}} = 0$. Therefore, $x \in \mathcal{L} =$ the equivalence class in the factor space that is the zero element of this space. This shows that $x + \mathcal{L} \longrightarrow \|x + \mathcal{L}\|_{\mathcal{M}}$ is actually a norm.

Second, in order to prove the completeness of $\mathcal{M}_2$, we will show that a Cauchy sequence $\{x_n + \mathcal{L}; n \geq 1\} \subset \mathcal{M}_2$ does converge in this space. Indeed, for each $\varepsilon > 0$, one can find an integer $N = N(\varepsilon)$ such that

$$\|(x_n + \mathcal{L}) - (x_m + \mathcal{L})\|_{\mathcal{M}} < \varepsilon \text{ for } n \geq N. \tag{2.96}$$

But $(x_n + \mathcal{L}) - (x_m + \mathcal{L}) = x_n - x_m + \mathcal{L}$ because $\mathcal{L}$ is a linear manifold. Therefore, formula (2.96) implies

$$\|x_n - x_m + \mathcal{L}\|_{\mathcal{M}} = \|x_n - x_m\|_{\mathcal{M}} < \varepsilon \text{ for } n, m \geq N(\varepsilon),$$

which means that $\{x_n; \ n \geq 1\} \subset \mathcal{M}$ is a Cauchy sequence. Hence, there exists $x \in \mathcal{M}$ such that

$$\lim_{n \to \infty} \|x_n - x\|_{\mathcal{M}} = 0. \tag{2.97}$$

We now note that equation (2.97) implies

$$\|x_n - x + \mathcal{L}\|_{\mathcal{M}} = \|x_n - x\|_{\mathcal{M}} \longrightarrow 0 \text{ as } n \longrightarrow \infty, \tag{2.98}$$

which is the same as saying

$$\lim_{n \to \infty} \|(x_n + \mathcal{L}) - (x + \mathcal{L})\|_{\mathcal{M}} = 0. \tag{2.99}$$

In other words, the sequence $\{x_n + \mathcal{L}; \ n \geq 1\} \subset \mathcal{M}_2$ is convergent to $x + \mathcal{L} \subset \mathcal{M}_2$.

This ends the proof of Proposition 2.19, proving that the Marcinkiewicz function space $\mathcal{M}_2 = \mathcal{M}_2(R, \mathcal{C})$ is a Banach space over the complex field $\mathcal{C}$.

We now have all the elements to introduce the Besicovitch space of almost periodic functions, denoted by $\mathrm{AP}_2(R, \mathcal{C})$ or $B^2(R, \mathcal{C})$. We have seen at the end of Section 2.4 that $\mathrm{AP}_2(R, \mathcal{C})$ is uniquely determined as the completion of $\mathcal{T}$ with respect to the Euclidean norm $\|\cdot\|_2$ for trigonometric polynomials of the form (2.41). What we now want to point out is the fact that $\mathrm{AP}_2(R, \mathcal{C})$ is a (closed) subspace of the space $\mathcal{M}_2(R, \mathcal{C})$. Hence, the elements of $\mathrm{AP}_2(R, \mathcal{C})$ will be classes of equivalent functions, each class containing all functions in $\mathcal{M}(R, \mathcal{C})$ of equal norm, or, equivalently, $x, y \in \mathcal{M}$ belong to the same element of $\mathcal{M}_2(R, \mathcal{C})$ iff $x - y \in \mathcal{L}$.

In the space $\mathcal{M}_2(R, \mathcal{C})$, we consider the linear manifold $\mathcal{T} + \mathcal{L}$, where $\mathcal{T}$ denotes the set of polynomials of the form (2.41). Since both terms $\mathcal{T}$ and $\mathcal{L}$ are linear manifolds, $\mathcal{T} + \mathcal{L}$ is also a linear manifold in $\mathcal{M}_2(R, \mathcal{C})$.

Definition 2.27. The space $\mathrm{AP}_2(R, \mathcal{C})$ of almost periodic functions in the Besicovitch sense is the closure in $\mathcal{M}_2(R, \mathcal{C})$ of the linear manifold $\mathcal{T} + \mathcal{L}$, with $\mathcal{T}$ standing for the set of trigonometric polynomials of the form (2.41) and $\mathcal{L}$ the subspace of $\mathcal{M}$ of the elements with zero $\mathcal{M}$-seminorm.

Proposition 2.20. *The space* $\mathrm{AP}_2(R, \mathcal{C})$ *is a Banach space over* $\mathcal{C}$ *with the norm induced by* $\|\cdot\|_{\mathcal{M}}$, *namely*

$$\|x + \mathcal{L}\|_{\mathcal{M}} = \|x\|_{\mathcal{M}}, \ x \in \mathrm{AP}_2(R, \mathcal{C}). \tag{2.100}$$

The proof of Proposition 2.20 is the result of the discussion above and Definition 2.24.

Remark 2.21. Since the almost periodic functions in the Besicovitch sense form a Banach space, it is obvious that the sum of two such functions is in $\mathrm{AP}_2(R, \mathcal{C})$ as well as the product of a function in $\mathrm{AP}_2(R, \mathcal{C})$ by a constant.

Moreover, the limit of a sequence of functions in $AP_2(R, C)$ in the sense of an $\mathcal{M}$-seminorm, i.e., such that

$$\lim_{n \to \infty} \left\{ \limsup_{\ell \to \infty} (2\ell)^{-1} \int_{-\ell}^{\ell} |x_n(t) - x(t)|^2 dt \right\} = 0,$$

is also almost periodic in the sense of Besicovitch. Further properties will be established in the following chapters.

In concluding this section, we will dwell on the sequence of inclusions for spaces of almost periodic functions considered in this chapter. In Section 2.4, we established (see formula (2.83)) the inclusions between the spaces $AP_1(R, C)$, $AP(R, C)$, and $S(R, C)$. Now we have to find out the place of $AP_2(R, C)$ in the sequence of inclusions. One can easily see that, similar to formula (2.83), we have the inclusions

$$AP_1(R, C) \subset AP(R, C) \subset AP_2(R, C). \tag{2.101}$$

Indeed, if $x \in AP(R, C)$, it is bounded on R, and the right-hand side in equation (2.86) is finite. This justifies the second inclusion (2.101). Unfortunately, we cannot interpose $S(R, C)$ between $AP(R, C)$ and $AP_2(R, C)$. This is because the functions in $S(R, C)$ are only locally integrable, while those in $AP_2(R, C)$ are locally square integrable – a stronger property than local integrability. Later on, when we develop the theory of almost periodic functions in various senses, we shall be able to produce concrete examples of functions in $S(R, C)$ that do not belong to $AP_2(R, C)$. In particular, a periodic function in $S(R, C)$ that is locally integrable but not locally square integrable constitutes such an example.

If we consider the Stepanov space $S^2(R, C)$, however, then the sequence of inclusions of spaces of almost periodic functions looks like

$$AP_1(R, C) \subset AP(R, C) \subset S^2(R, C) \subset AP_2(R, C), \tag{2.102}$$

which shows that each space in the sequence is richer than the preceding one. Of course, moving from one space in the sequence to the next, one loses some of the properties related to almost periodicity.

In order to conclude our discussion, we need to prove the validity of the last inclusion in the sequence (2.102). Let us consider a function $x \in S^2(R, C)$ and show that for each function the seminorm $\|x\|_{\mathcal{M}}$, as defined by equation (2.86), is finite. Indeed, one easily obtains

$$\limsup_{\ell \to \infty} (2\ell)^{-1} \int_{-\ell}^{\ell} |x(t)|^2 dt \leq \limsup_{\ell \to \infty} (2\ell)^{-1} \int_{-[\ell]-1}^{[\ell]+1} |x(t)|^2 dt$$

$$\leq \limsup_{\ell \to \infty} \frac{[\ell]+1}{\ell} \left[\sup_{t \in R} \int_{t}^{t+1} |x(s)|^2 ds \right] \leq \|x\|_{S^2}^2$$

because $\ell^{-1}([\ell] + 1)$, with $[\ell]$ = greatest integer less than or equal to $\ell > 0$, tends to 1 as $\ell \longrightarrow \infty$. The inequalities above tell us that

$$\|x\|_{\mathcal{M}_2} \leq \|x\|_{S^2}^2, \tag{2.103}$$

which justifies the last inclusion in (2.102).

We shall later indicate other classes of almost periodic functions. For a recent survey paper examining the classification of almost periodic functions, see J. Andres et al. [4].

3

Basic Properties of Almost Periodic Functions

3.1 The Space $AP_1(R, C)$

We defined in Section 2.3 the space $AP_1(R, C)$, where elements can be represented by

$$f(t) = \sum_{k=1}^{\infty} a_k e^{i\lambda_k t}, \qquad t \in R, \tag{3.1}$$

where $a_k \in C$, $\lambda_k \in R$, $k = 1, 2, \ldots$, such that

$$\sum_{k=1}^{\infty} |a_k| < \infty. \tag{3.2}$$

Condition (3.2) implies the absolute and uniform convergence of the series in equation (3.1) on the whole real line R. The norm of f is the sum of the series in formula (3.2).

Taking into account the formula

$$e^{i\lambda t} = \cos \lambda t + i \sin \lambda t, \qquad \lambda, t \in R,$$

one obtains from equation (3.1), with $a_k = \alpha_k + i\beta_k$,

$$f_1(t) = \operatorname{Re} f(t) = \sum_{k=1}^{\infty} (\alpha_k \cos \lambda_k t - \beta_k \sin \lambda_k t),$$

$$f_2(t) = \operatorname{Im} f(t) = \sum_{k=1}^{\infty} (\alpha_k \sin \lambda_k t + \beta_k \cos \lambda_k t),$$

formulas showing that both $f_1(t)$ and $f_2(t)$ can be represented by means of absolutely and uniformly convergent series whose terms describe simple harmonic motions (oscillations). In other words, the motion described by the function $f(t)$, given by equation (3.1), is the result of compounding infinitely

C. Corduneanu, *Almost Periodic Oscillations and Waves*,
DOI 10.1007/978-0-387-09819-7_3, © Springer Science+Business Media, LLC 2009

many simple harmonics. This shows the oscillatory property of the function $f(t)$ or the motion described by this function.

Since $AP_1(R,\mathcal{C})$ is the "smallest" space of almost periodic functions we are investigating, any property of functions in the spaces $AP(R,\mathcal{C})$ or $S(R,\mathcal{C})$ is also valid for functions in $AP_1(R,\mathcal{C})$. In the next section of this chapter, we shall get acquainted with several of these properties. For instance, properties I, II, and V, stated in Section 2.3 for the functions in $AP(R,\mathcal{C})$, are valid for any function in $AP_1(R,\mathcal{C})$.

It is our aim in this section to point out some features of functions in $AP_1(R,\mathcal{C})$ that do not hold in general for almost periodic functions in larger spaces of almost periodic functions.

First, let us start by clarifying the relationship between the function $f(t)$ in equation (3.1) and the coefficients a_k in the representation of $f(t)$ as a trigonometric series. The formulas we shall obtain also have historical significance since they were obtained by H. Poincaré long before the theory of almost periodic functions was developed by H. Bohr. Following Poincaré, let us multiply both sides of equation (3.1) by $e^{-i\lambda_j t}$. We obtain

$$f(t)e^{-i\lambda_j t} = \sum_{k=1}^{\infty} a_k e^{i(\lambda_k - \lambda_j)t}, \tag{3.3}$$

with the series on the right-hand side of equation (3.3) absolutely and uniformly convergent on R. Unfortunately, we cannot integrate both sides of equation (3.3) on R because the left-hand side may not be in $L^1(R,\mathcal{C})$. But we can integrate both sides of equation (3.3) in a valid manner on any interval $(-\ell, \ell)$, $\ell > 0$:

$$\int_{-\ell}^{\ell} f(t)e^{-i\lambda_j t}dt = \sum_{k=1}^{\infty} a_k \int_{-\ell}^{\ell} e^{i(\lambda_k - \lambda_j)t}dt.$$

We now multiply both sides above by $(2\ell)^{-1}$ and then let $\ell \to \infty$. Keeping in mind that (see Section 2.3)

$$\lim_{\ell \to \infty} (2\ell)^{-1} \int_{-\ell}^{\ell} e^{i\lambda t}dt = \begin{cases} 0 \ \lambda \neq 0 \\ 1 \ \lambda = 0 \end{cases},$$

one obtains

$$\lim_{\ell \to \infty} (2\ell)^{-1} \int_{-\ell}^{\ell} f(t)e^{-i\lambda_j t}dt = a_j, \tag{3.4}$$

which provides the coefficients a_j, $j = 1, 2, \ldots$, in terms of $f(t)$ and λ_j.

Let us note that in the representation (3.1) for $f(t)$, we assumed λ_k's to be distinct real numbers.

In Chapter 4, when dealing with Fourier series attached to almost periodic functions (in various spaces), we shall reobtain formula (3.4) in another context, which will be written as $a_j = M\left\{f(t)e^{-\lambda_j t}\right\}$, $j = 1, 2, \ldots$, where M

denotes the *mean value* (of an almost periodic function; it will be shown that this value exists for any almost periodic function).

Let us consider the problem of *differentiability* of the function $f(t)$ given by equation (3.1). It is obvious that the condition

$$\sum_{k=1}^{\infty} |a_k| \, |\lambda_k| < \infty \tag{3.5}$$

will assure the existence of the derivative $f'(t)$ as well as its representation in the form

$$f'(t) = i \sum_{k=1}^{\infty} a_k \lambda_k e^{i\lambda_k t}, \qquad t \in R. \tag{3.6}$$

Indeed, the series on the right-hand side of equation (3.6) is absolutely and uniformly convergent in R due to assumption (3.5). It only remains to apply a well-known result from classical analysis in order to derive the validity of equation (3.6).

Proposition 3.1. *Let $f(t)$ be given by equation (3.1) under condition (3.2). If condition (3.5) is satisfied, then $f'(t)$ exists on R, and equation (3.6) holds true.*

Corollary 3.1. *If $f(t)$ given by equation (3.1) is such that condition (3.2) holds true while the set $\{\lambda_k; \ k \geq 1\}$ is bounded on R, then formula (3.6) is valid.*

This is a consequence of the fact that Condition (3.2) and the boundedness of the set $\{\lambda_k; \ k \geq 1\}$ imply assumption (3.5). One observes that in the case of periodic functions, when representation (3.1) is possible, Corollary 3.1 is not applicable. Indeed, $\lambda_k = k\pi/\omega$, $\omega > 0$, $k = 0, \pm 1, \pm 2, \ldots$, is unbounded on R.

Corollary 3.2. *If $f(t)$ given by equation (3.1) is such that*

$$\sum_{k=1}^{\infty} |a_k| \, |\lambda_k|^m < \infty, \qquad m = 1, 2, \ldots, p, \tag{3.7}$$

then $f(t)$ is differentiable up to the order p, and for each $m = 1, 2, \ldots, p$, one has

$$f^{(m)}(t) = (i)^m \sum_{k=1}^{\infty} a_k \lambda_k^m e^{i\lambda_k t}, \qquad t \in R. \tag{3.8}$$

Remark 3.1. If the series on the right-hand side of equation (3.6) is uniformly convergent, then formula (3.6) is still valid. Unlike the case discussed in Proposition 3.1, $f'(t)$ may not belong to $AP_1(R, C)$.

The elementary considerations above concerning the derivatives of functions representable in the form (3.1) under condition (3.2) will serve in forthcoming chapters when dealing with applications to the theory of oscillations and waves.

Considerations similar to those related to the derivatives of functions in $AP_1(R,\mathcal{C})$ can be made in regard to the (indefinite) integral of functions in $AP_1(R,\mathcal{C})$.

We leave it to the reader to prove the following assertion.

Let $f(t)$ be given by equation (3.1), with a_k's satisfying formula (3.2). If there exists $\delta > 0$ such that $|\lambda_k| \geq \delta$, $k \geq 1$, then

$$\int^t f(s)ds = C - i\sum_{k=1}^{\infty} \frac{a_k}{\lambda_k} e^{i\lambda_k t}, \tag{3.9}$$

with C an arbitrary constant. Moreover,

$$\int^t f(s)ds \in AP_1(R,\mathcal{C}).$$

A remarkable property of the space $AP_1(R,\mathcal{C})$ is the following proposition.

Proposition 3.2. *The space $AP_1(R,\mathcal{C})$ is a Banach algebra, the operation of multiplication being the usual pointwise multiplication.*

Proof. As mentioned in Section 3.1, a Banach space that is also endowed with an operation of multiplication such that $|fg| \leq |f|\,|g|$ is called a Banach algebra. In order to prove Proposition 3.2, we have to show that the pointwise product of two elements of $AP_1(R,\mathcal{C})$ can be represented in the form (3.1) with coefficients satisfying condition (3.2).

Assume now that $f(t) \in AP_1(R,\mathcal{C})$ is given by equation (3.1), while

$$g(t) = \sum_{j=1}^{\infty} b_j e^{i\mu_j t}, \qquad t \in R, \tag{3.10}$$

with $\mu_j \in R, j \geq 1$, and

$$\sum_{j=1}^{\infty} |b_j| < \infty. \tag{3.11}$$

The Cauchy product of the functions f and g can be represented in the form

$$f(t)g(t) = a_1 b_1 e^{i(\lambda_1+\mu_1)t} + a_1 b_2 e^{i(\lambda_1+\mu_2)t} + a_2 b_1 e^{i(\lambda_2+\mu_1)t}$$
$$+ a_1 b_3 e^{i(\lambda_1+\mu_3)t} + a_2 b_2 e^{i(\lambda_2+\mu_2)t} + a_3 b_1 e^{\lambda_3+\mu_1)t} + \cdots. \tag{3.12}$$

Based on a classical result concerning the multiplication of absolutely convergent numerical series, we can infer that the series on the right-hand side of equation (3.12) is absolutely and uniformly convergent. Indeed, taking into account formulas (3.2) and (3.11), one obtains the convergence of the series

$$|a_1|\,|b_1| + |a_1|\,|b_2| + |a_2|\,|b_1| + |a_1|\,|b_3| + |a_2|\,|b_2| + |a_3|\,|b_1| \\ + \cdots + |a_1|\,|b_k| + |a_2|\,|b_{k-1}| + \cdots + |a_k|\,|b_1| + \cdots. \tag{3.13}$$

Moreover, this classical result in the theory of series states that the sum of series (3.13) is equal to the product

$$(|a_1| + |a_2| + \cdots + |a_k| + \cdots)\,(|b_1| + |b_2| + \cdots + |b_k| + \cdots). \tag{3.14}$$

Taking into account the definition of the norm in $AP_1(R,C)$, one obtains the equality $|fg| = |f|\,|g|$.

Remark 3.2. In the representation (3.12) of the product $f(t)g(t)$, it is possible to have terms with equal complex exponentials (for instance, when $\lambda_k + \mu_j = 2\pi + \lambda_m + \mu_n$). Due to the absolute convergence expressed by conditions (3.2) and (3.11), the repetition of equal exponentials does not matter. In an absolutely convergent series, one can arbitrarily change the order of terms without changing the sum of the series.

Remark 3.3. Since $AP_1(R,C)$ can be organized as a Banach algebra, many properties of the latter can be used to derive properties for the space $AP_1(R,C)$. A simple example is given by compounding a polynomial P over C with an arbitrary function in $AP_1(R,C)$: $p(t) = P(f(t)) \in AP_1(R,C)$. An interesting problem consists in finding those elements of $AP_1(R,C)$ that are invertible in this algebra. Let's point out that the unit element in $AP_1(R,C)$ is $1 = e^{0 \cdot t}$. We leave to the reader the task of exploiting this property of AP_1 (R,C) in order to derive more information about this space of almost periodic functions. Let us mention that the periodic case has been thoroughly investigated (see, for instance, J. P. Kahane [53]).

3.2 The Space AP(R, C)

In Section 2.3, we defined the almost periodic functions in Bohr's sense and stated some properties that follow more or less easily from their definition (see Remark 2.13 to Proposition 2.14 of Section 2.3). We shall briefly review those properties and provide some details in cases requiring clarification.

I. Any almost periodic function in Bohr's sense is bounded on R.

Indeed, the elements of $AP(R,C)$ are by definition elements of $BC(R,C)$ and hence bounded on R.

II. Any function in $AP(R, \mathcal{C})$ is uniformly continuous on R.

Let $f \in AP(R, \mathcal{C})$ and $\{T_n(t); \ n \geq 1\}$ be a sequence of trigonometric polynomials such that

$$\lim_{n \to \infty} T_n(t) = f(t) \text{ uniformly on } R.$$

We notice that any trigonometric polynomial $T(t) = T_1(t) + iT_2(t)$ is uniformly continuous on R. Indeed, $T_1(t)$ is a real trigonometric polynomial, it is differentiable on R, and its first derivative (also a trigonometric polynomial!) is bounded on R. By the Lagrange formula, $T_1(t) - T_1(s) = (t - s)T_1'(u)$, $u \in (s, t)$, which leads to the inequality $|T_1(t) - T_1(s)| \leq M|t - s|$ with $M = \sup\{|T_1'(t)|; \ t \in R\}$. The last inequality proves the uniform continuity on R of $T_1(t)$. The same conclusion is valid for $T_2(t)$, which leads to the uniform continuity of $T(t)$ on R. The remaining part of the proof follows easily from the inequality

$$|f(t) - f(s)| \leq |f(t) - T_n(t)| + |T_n(t) - T_n(s)| + |T_n(s) - f(s)|,$$

in which each term on the right-hand side can be made less than $\varepsilon/3$, provided n is chosen sufficiently large and $|t - s|$ sufficiently small ($|t - s| \leq \delta(\varepsilon)$). This means $|f(t) - f(s)| < \varepsilon$ for $|t - s| < \delta(\varepsilon)$, which means the uniform continuity of f on R.

Remark 3.4. There exists a function space that contains $AP(R, \mathcal{C})$ and is contained in $BC(R, \mathcal{C})$. This is the space $BC_u(R, \mathcal{C})$, which consists of all elements of $BC(R, \mathcal{C})$ that are uniformly continuous on R:

$$AP(R, \mathcal{C}) \subset BC_u(R, \mathcal{C}) \subset BC(R, \mathcal{C}).$$

III. Any sequence $\{f_n(t); \ n \geq 1\} \subset AP(R, \mathcal{C})$ that converges uniformly on R has as its limit a function in $AP(R, \mathcal{C})$.

This property is obvious because $AP(R, \mathcal{C})$ is the closure of the set $\mathcal{T}$ of trigonometric polynomials in $BC(R, \mathcal{C})$ and hence is closed in $BC(R, \mathcal{C})$.

IV. Among the properties listed under IV, the only one requiring some consideration is that stating $f, g \in AP(R, \mathcal{C})$ implies $fg \in AP(R, \mathcal{C})$. In other words, $AP(R, \mathcal{C})$ is closed with respect to pointwise multiplication.

Indeed, if $f(t) = \lim T_n(t)$ and $g(t) = \lim S_n(t)$ as $n \to \infty$ uniformly on R, with $\{T_n(t); \ n \geq 1\}, \{S_n(t); \ n \geq 1\} \subset \mathcal{T}$, then noting that $T_n(t)S_n(t) = \widetilde{T}_n(t)$ is also a trigonometric polynomial, one easily obtains

$$f(t)g(t) = \lim_{n \to \infty} \widetilde{T}_n(t) \text{ uniformly on } R,$$

which proves our assertion.

Since

$$\sup_{t \in R} |f(t)g(t)| \leq \sup_{t \in R} |f(t)| \cdot \sup_{t \in R} |g(t)|,$$

one concludes that AP($R,\mathcal{C}$) is a Banach algebra with a pointwise product of functions as the operation of multiplication.

An immediate consequence of the fact that AP($R,\mathcal{C}$) can be organized as a Banach algebra is the following property: If $P(z_1, z_2, \ldots, z_n)$ is a complex polynomial and $f_1(t), \ldots, f_n(t) \in$ AP($R,\mathcal{C}$), then

$$p(t) = P(f_1(t), f_2(t), \ldots, f_n(t)) \in \text{AP}(R,\mathcal{C}).$$

The property above can be generalized to the case where $P(z_1, z_2, \ldots, z_n)$ is substituted by a continuous function $F(z_1, z_2, \ldots, z_n)$. More precisely, the result can be stated as follows.

Proposition 3.3. *Let* $F(z_1, z_2, \ldots, z_n)$ *be a continuous, complex-valued function defined on a compact set* $\mathcal{M} \subset \mathcal{C}^n$. *If* $f_1(t), f_2(t), \ldots, f_n(t) \in$ AP($R,\mathcal{C}$) *and, for each* $t \in R$, $(f_1(t), f_2(t), \ldots, f_n(t)) \in \mathcal{M}$, *then*

$$f(t) = F(f_1(t), f_2(t), \ldots, f_n(t)) \in \text{AP}(R,\mathcal{C}).$$

Proof. The proof is immediate if one takes into account the Weierstrass approximation theorem for a continuous function (in the complex domain) by polynomials. This theorem (see, for instance, the author's book [23]) states that for each $\varepsilon > 0$ there exists a polynomial $P_\varepsilon(z_1, z_2, \ldots, z_n; \bar{z}_1, \bar{z}_2 \ldots, \bar{z}_n)$ such that

$$|F(z_1, z_2, \ldots, z_n) - P_\varepsilon(z_1, z_2, \ldots, z_n; \bar{z}_1, \bar{z}_2, \ldots, \bar{z}_n)| < \varepsilon \text{ for } (z_1, z_2, \ldots, z_n) \in \mathcal{M}.$$

Taking into account that the complex conjugate of a function in AP($R,\mathcal{C}$) is also in AP($R,\mathcal{C}$), one derives the fact that $f(t) \in$ AP($R,\mathcal{C}$), being uniformly approximable on R by functions from this space.

Corollary 3.3. *Let* $f, g \in$ AP($R,\mathcal{C}$) *with* g *such that* $0 < m \le |g(t)|$, $t \in R$. *Then* $f(t)/g(t) \in$ AP($R,\mathcal{C}$).

Indeed, it suffices to show that $1/g(t)$ is in AP($R,\mathcal{C}$). We note that $F(z) = 1/z$ is continuous in the crown $m \le |z| \le M$ with $M = \sup |g(t)|$, $t \in R$. Proposition 3.3 applies, concluding the inclusion $1/g(t) \in$ AP($R,\mathcal{C}$). Property IV leads to the conclusion that $f(t)/g(t) \in$ AP($R,\mathcal{C}$).

Concerning property V, formulated in Section 2.3, we shall provide its proof after discussing other basic properties of functions in AP($R,\mathcal{C}$).

Proposition 3.4. *Let* $f \in$ AP($R,\mathcal{C}$) *be differentiable on* R. *Then* $f' \in$ AP($R,\mathcal{C}$) *if, and only if,* f' *is uniformly continuous on* R.

Proof. The uniform continuity of f' is a necessary condition for its almost periodicity (in Bohr's sense) according to property II.

In order to prove that this condition is also sufficient, we shall consider the real and imaginary parts of $f(t) = f_1(t) + if_2(t)$. Since $f \in$ AP($R,\mathcal{C}$) if and only if $f_1, f_2 \in AP(R, R)$, we can restrict to the case of the space $AP(R, R)$. For $f \in AP(R, R)$ differentiable on R, we consider the sequence

$$\varphi_n(t) = n \left[f \left(t + \frac{1}{n} \right) - f(t) \right], \qquad n \geq 1,$$

obviously in $AP(R, R)$. But $\varphi_n(t) = f' \left(t + \frac{\theta_n}{n} \right)$, $0 < \theta_n < 1$, $n \geq 1$. The uniform continuity of $f'(t)$ allows us to write $\lim_{n \to \infty} \varphi_n(t) = f'(t)$ uniformly on R. Hence, based on property III, $f'(t) \in AP(R, \mathcal{C})$.

This ends the proof of Proposition 3.4.

For the primitive (indefinite integral) of a function $f \in AP(R, \mathcal{C})$, the following result is valid.

Proposition 3.5. *Let $F(t)$ be such that $F'(t) = f(t) \in AP(R, \mathcal{C})$. The necessary and sufficient condition for having $F \in AP(R, \mathcal{C})$ is $F \in BC(R, \mathcal{C})$.*

The proof of Proposition 3.5 will be given later, after establishing the equivalence of property V with the definition of the space $AP(R, \mathcal{C})$. This will also mean the discussion of almost periodicity of the solutions of the simplest differential equation $x'(t) = f(t) \in AP(R, \mathcal{C})$.

Let us point out the fact that the properties of functions in $AP(R, \mathcal{C})$ emphasized earlier in this section keep their validity in the spaces $AP(R, \mathcal{C}^n)$, $n \geq 1$. This is the result of the fact that $AP(R, \mathcal{C}^n) = [AP(R, \mathcal{C})]^n$; i.e., $AP(R, \mathcal{C}^n)$ is the Cartesian product of n factors, each being identical to $AP(R, \mathcal{C})$. Similar remarks are valid for $AP(R, R^n)$.

The concept of almost periodicity in Bohr's sense is intimately related to the concept of compactness in the space $BC(R, \mathcal{C})$. More precisely, the following result, due to S. Bochner, is valid.

Proposition 3.6. *Let $f \in AP(R, \mathcal{C})$. Then the family of translates*

$$\mathcal{F} = \{f(t + h); \ h \in R\} \tag{3.15}$$

is relatively compact in $BC(R, \mathcal{C})$.

Proof. The meaning of Proposition 3.6, if one takes into account Definition 2.10, is as follows: If $\{h_k; \ k \geq 1\} \subset R$ is an arbitrary sequence, from the sequence $\{f(t + h_k); \ k \geq 1\} \subset AP(R, \mathcal{C}) \subset BC(R, \mathcal{C})$, one can extract a subsequence, say $\{f(t + h_{1k}); \ k \geq 1\}$, that converges in $BC(R, \mathcal{C})$ (i.e., uniformly on R) and hence also in $AP(R, \mathcal{C})$.

Assume first that $f(t) = e^{i\lambda t}$ for some $\lambda \in R$. If $\{h_k; \ k \geq 1\}$ is an arbitrary sequence in R, then $f(t + h_k) = e^{i\lambda t} \cdot e^{i\lambda h_k}$, $k \geq 1$. One must show that $\{e^{i\lambda t} \cdot e^{i\lambda h_k}; \ k \geq 1\}$ contains a subsequence that is uniformly convergent on R. But $\left| e^{i\lambda h_k} \right| = 1$, and this guarantees the existence of a subsequence $\{h_{1k}; \ k \geq 1\}$ such that $\{e^{i\lambda h_{1k}}; \ k \geq 1\}$ is convergent. Hence,

$$|f(t + h_{1j}) - f(t + h_{1k})| = \left| e^{i\lambda h_{1j}} - e^{i\lambda h_{1k}} \right|,$$

which proves the uniform convergence on R of the sequence of functions $\{f(t + h_{1k}); \ k \geq 1\}$.

Consider now the case where $f(t)$ is a trigonometric polynomial

$$f(t) = \sum_{k=1}^{n} a_k e^{i\lambda_k t} \tag{3.16}$$

with $a_k \in C$, $\lambda_k \in R$, $k = 1, 2, \ldots, n$. As seen above, if $\{h_k; \ k \geq 1\} \subset R$, then $\{a_1 e^{i\lambda_1 (t + h_k)}; \ k \geq 1\}$ contains a subsequence $\{a_1 e^{i\lambda_1 (t + h_{1k})}; \ k \geq 1\}$ that converges uniformly in R. Next, from $\{h_{1k}; \ k \geq 1\}$, we extract a subsequence $\{h_{2k}; \ k \geq 1\}$ such that $\{a_2 e^{i\lambda_2 (t + h_{2k})}; \ k \geq 1\}$ converges uniformly on R. We continue this procedure until we obtain a sequence $\{h_{nk}; \ k \geq 1\}$ such that $\{a_n e^{i\lambda_n (t + h_{nk})}; \ k \geq 1\}$ converges uniformly on R. Since h_{nk} is a subsequence of any $\{h_{mk}; \ k \geq 1\}$ with $1 \leq m < n$, there results that $\{f(t + h_k); \ k \geq 1\}$ contains a subsequence $\{f(t + h_{nk}); \ k \geq 1\}$ that converges uniformly on R. This means that any trigonometric polynomial (3.16) has the property of relative compactness in $BC(R,C)$ of the family of its translates.

The third step in the proof will consist in proving that the property in the statement of Proposition 3.6 is valid in general; i.e., for any $f \in AP(R,C)$. Let $\{T_n(t); \ n \geq 1\} \subset T$ be a sequence of trigonometric polynomials of the form (3.16) such that $T_n(t) \to f(t)$ uniformly on R as $n \to \infty$. If $\{h_n; \ n \geq 1\} \subset R$ is an arbitrary sequence, then from $\{T_1(t + h_n); \ n \geq 1\}$ we can extract a subsequence $\{T_1(t + h_{1n}); \ n \geq 1\}$ that converges uniformly on R. Then we move to the sequence $\{T_2(t + h_{1n}); \ n \geq 1\}$, and we know there exists a subsequence $\{T_2(t + h_{2n}); \ n \geq 1\}$ that converges uniformly on R. We further proceed in the same way as above, and for any integer p we obtain a sequence of reals $\{h_{pn}; \ n \geq 1\}$ such that $\{T_q(t + h_{pn}); \ n \geq 1\}$ is uniformly convergent on R for $q = 1, 2, \ldots, p$. Let us consider now (using the so-called diagonal procedure) the sequence $\{h_{nn}; \ n \geq 1\}$, whose terms belong to any $\{h_{pn}; \ n \geq 1\}$ except perhaps a finite number of them. This means that $\{T_n(t + h_{mm}); \ m \geq 1\}$ with fixed n is uniformly convergent on R. Now let $\varepsilon > 0$ and choose n large enough that

$$|f(t) - T_n(t)| < \frac{\varepsilon}{3}, \qquad t \in R. \tag{3.17}$$

There exists $N = N(\varepsilon) > 0$ such that

$$|T_n(t + h_{mm}) - T_n(t + h_{pp})| < \frac{\varepsilon}{3}, \qquad t \in R, \tag{3.18}$$

for $m, p \geq N(\varepsilon)$. From formulas (3.17) and (3.18), we derive

$$|f(t + h_{mm}) - f(t + h_{pp})| \leq |f(t + h_{mm}) - T_n(t + h_{mm})|$$
$$+ |T_n(t + h_{mm}) - T_n(t + h_{pp})| + |T_n(t + h_{pp}) - f(t + h_{pp})| < \varepsilon,$$

for $t \in R$, provided $m, p \geq N(\varepsilon)$. From the last inequality, we conclude that the sequence $\{f(t + h_{mm}); \ m \geq 1\}$ is uniformly convergent on R, which ends the proof of Proposition 3.6.

The next proposition establishes a connection between the compactness property (Bochner) and property V of functions in $AP(R,\mathcal{C})$. This property is, as mentioned above, the one taken by H. Bohr as the definition for almost periodic functions in $AP(R,\mathcal{C})$.

Proposition 3.7. *If $f \in AP(R,\mathcal{C})$, then for every $\varepsilon > 0$ there exists $\ell = \ell(\varepsilon) > 0$ such that in each interval $(a, a + \ell)$ there is a number τ with the property*

$$|f(t + \tau) - f(t)| < \varepsilon, \qquad t \in R. \tag{3.19}$$

Proof. According to Proposition 3.6, it suffices to show that the compactness property (Bochner) implies Bohr's property (or property V).

Let us assume, on the contrary, that $f(t)$ does not possess Bohr's property. Then we can find at least one $\varepsilon > 0$ for which $\ell(\varepsilon)$ does not exist. In other words, for each $\ell > 0$, one can find an interval of length ℓ that contains no points τ with property (3.19). Now choose an arbitrary number $h_1 \in R$ and an interval $(a_1, b_1) \subset R$ of length larger than $2|h_1|$ such that (a_1, b_1) does not contain any number τ with property (3.19). Denote $h_2 = (a_1 + b_1)/2$. Then $h_2 - h_1 \in (a_1, b_1)$, and therefore $h_2 - h_1$ cannot be taken as τ, satisfying (3.19). Let us take an interval $(a_2, b_2) \in R$ of length larger than $2(|h_1| + |h_2|)$ that does not contain any number τ with property (3.19). We continue the process, letting $h_3 = (a_2 + b_2)/2$. Then we have $h_3 - h_2, h_3 - h_1 \in (a_2, b_2)$, and this means that $h_3 - h_2$ and $h_3 - h_1$ cannot be taken as τ in formula (3.19). Proceeding similarly, we construct the numbers $h_4, h_5, \ldots$, with the property that none of the differences $h_i - h_j$, $i > j$, could be taken as number τ in formula (3.19). Hence, for $i > j$, $\sup |f(t + h_i) - f(t + h_j)| = \sup |f(t + h_i - h_j) - f(t)| \geq \varepsilon$, $t \in R$. This inequality contradicts the property of relative compactness of the family $\mathcal{F} = \{f(t + h); h \in R\}$. This ends the proof of Proposition 3.7.

Remark 3.5. Any number τ satisfying formula (3.19) is called an ε-*translation number* of f. If f is almost periodic, then according to Proposition 3.7, for each $\varepsilon > 0$ there exist ε-translation numbers in each interval of length $\ell = \ell(\varepsilon)$. A number τ satisfying formula (3.19) is then called an ε-*almost period* of f.

Remark 3.6. Bohr's property says that, for each $\varepsilon > 0$, there exists $\ell = \ell(\varepsilon) > 0$ such that each interval of length ℓ contains an ε-almost period of $f \in AP(R,\mathcal{C})$. This property of the set of all ε-almost periods is also known as *relative density*.

In order to end the proof of the equivalence of the properties:

(1) $f \in AP(R,\mathcal{C})$,

(2) $\mathcal{F} = \{f(t + h); h \in R\}$ is relatively compact in $BC(R,\mathcal{C})$,

(3) f satisfies Bohr's property,

we need to show that Bohr's property implies the approximation property (by trigonometric polynomials); i.e., $f \in \mathrm{AP}(R,\mathcal{C})$.

The logical scheme corresponding to this equivalence looks as follows: (1) $\longrightarrow$ (2) $\longrightarrow$ (3) $\longrightarrow$ (1). Only the last implication has to be proven. We shall complete the proof in Chapter 4, which is dedicated to the concept of Fourier series associated to almost periodic functions.

We shall now discuss a method of constructing functions in $\mathrm{AP}(R,\mathcal{C})$ based on the operation of convolution. More precisely, we shall prove that the operator

$$f \longrightarrow \int_R K(s)f(t-s)ds \qquad (3.20)$$

takes the space $\mathrm{AP}(R,\mathcal{C})$ into itself if K is in $L^1(R,\mathcal{C})$; i.e.,

$$\int_R |K(s)|ds < \infty. \qquad (3.21)$$

Let us note that the operator (3.20) takes the linear manifold $\mathcal{T}$ of trigonometric polynomials into itself.

Indeed, if $T(t) = \sum_{k=1}^n a_k e^{i\lambda_k t}$, with complex a_k and real $\lambda_k, k = 1, 2, \ldots, n$, then

$$\int_R K(s)T(t-s)ds = \sum_{k=1}^n \left[a_k \int_R K(s)e^{-i\lambda_k s}ds \right] e^{i\lambda_k t}. \qquad (3.22)$$

Proposition 3.8. *If $f \in \mathrm{AP}(R,\mathcal{C})$ and $K \in L^1(R,\mathcal{C})$, then*

$$(K * f)(t) = \int_R K(s)f(t-s)ds \in \mathrm{AP}(R,\mathcal{C}). \qquad (3.23)$$

Proof. Let $f \in \mathrm{AP}(R,\mathcal{C})$ and $\{T_n(t); \ n \geq 1\} \subset \mathcal{T}$ a sequence of trigonometric polynomials such that $T_n(t) \longrightarrow f(t)$ as $n \longrightarrow \infty$ uniformly on R. As seen above, $K * T_n$ are also trigonometric polynomials from $\mathcal{T}$. Since

$$\left| \int_R K(s)f(t-s)ds - \int_R K(s)T_n(t-s)ds \right|$$

$$= \left| \int_R K(s)[f(t-s) - T_n(t-s)]ds \right| \qquad (3.24)$$

$$\leq \int_R |K(s)|\,|f(t-s) - T_n(t-s)|ds,$$

the last integral in formula (3.24) can be estimated in the following manner (see formula (2.80) in Section 2.4):

$$\int_R |K(s)|\,|f(t-s) - T_n(t-s)|ds \leq |K|_{L^1(R,\mathcal{C})}|f - T_n|_{\mathrm{AP}(R,\mathcal{C})}. \qquad (3.25)$$

But the second factor in formula (3.25) tends to zero as $n \longrightarrow \infty$, while the first factor is finite from formula (3.21).

Taking into account formulas (3.21), (3.24), and (3.25), we obtain equation (3.23), which proves Proposition 3.8.

Several applications related to convolution will appear in forthcoming chapters when dealing with almost periodic solutions of certain functional equations. For instance, a possible choice for the kernel $K(t)$ is

$$K(t) = \begin{cases} 0 & t < 0, \\ e^{-\alpha t} & t \geq 0, \end{cases} \qquad \alpha > 0,$$

which leads to the formula

$$(K * f)(t) = e^{-\alpha t} \int_{-\infty}^{t} e^{\alpha s} f(s) ds. \qquad (3.26)$$

It is easy to check that the right-hand side in equation (3.26) represents the (only) almost periodic solution (i.e., in $AP(R, \mathcal{C})$) of the differential equation $x'(t) + \alpha x(t) = f(t)$.

Let us note the fact that Proposition 3.8 could be regarded as stating the invariance of the space $AP(R, \mathcal{C})$ with respect to the linear operator (3.20), with $K(t)$ satisfying formula (3.21). This property of invariance will also be encountered in the case of the space $S(R, \mathcal{C})$.

3.3 The Space $S(R, \mathcal{C})$

We introduced the space $S(R, \mathcal{C})$ in Section 2.5 as the closure in $M(R, \mathcal{C})$ of the set $\mathcal{T}$ of trigonometric polynomials. The norm in $M(R, \mathcal{C})$, which is also taken as the norm in $S(R, \mathcal{C})$, has been defined by

$$\|x\|_M = \sup \left\{ \int_t^{t+1} |x(s)| ds; \ t \in R \right\}. \qquad (3.27)$$

It can be easily seen that, for fixed $\ell > 0$,

$$\sup \left\{ \ell^{-1} \int_t^{t+\ell} |x(s)| ds; \ t \in R \right\} \qquad (3.28)$$

defines a norm equivalent to $\|x\|_M$ on $M(R, \mathcal{C})$ or $S(R, \mathcal{C})$. For instance, when $\ell > 1$, we can write $\ell = p + \theta$, with p a natural number and $\theta \in (0, 1)$. Hence

$$\sup_{t \in R} \ell^{-1} \int_t^{t+\ell} |x(s)| ds = \sup_{t \in R} \frac{1}{p+\theta} \int_t^{t+p+\theta} |x(s)| ds$$

$$= \sup_{t \in R} (p+\theta)^{-1} \left[\int_t^{t+1} |x(s)| ds + \cdots + \int_{t+p-1}^{t+p} |x(s)| ds + \int_{t+p}^{t+p+\theta} |x(s)| ds \right]$$

$$\leq \frac{p+1}{p} \sup_{t \in R} \int_t^{t+1} |x(s)| ds = \frac{p+1}{p} \|x\|_M.$$

On the other hand, we obtain from above that

$$\sup \ell^{-1} \int_t^{t+\ell} |x(s)|ds \geq \frac{1}{p+1} \|x\|_M,$$

which shows the equivalence of formulas (3.27) and (3.28).

In the following, the norm in $S(R,\mathcal{C})$, also defined by formula (3.27) or (3.28), will be denoted by $\|x\|_S$.

From the definition of Stepanov's almost periodic functions, we easily derive some properties of these functions.

I. Each function $f \in S(R,\mathcal{C})$ is bounded in the mean; i.e., $\|f\|_S \leq A_f$ for some $A_f > 0$.

II. Each function $f \in S(R,\mathcal{C})$ is uniformly continuous in the S-norm.

If one denotes $f_h(t) = f(t+h)$, $t, h \in R$, then for each $\varepsilon > 0$ there exists $\delta = \delta(\varepsilon) > 0$ such that

$$\|f_h - f\|_S < \varepsilon \qquad \text{for } |h| < \delta. \tag{3.29}$$

When f is a trigonometric polynomial, formula (3.29) is obvious. Taking into account the fact that $S(R,\mathcal{C})$ is the closure of $\mathcal{T}$ with respect to the S-norm, one obtains formula (3.29).

III. Any convergent sequence $\{f_n; \; n \geq 1\} \subset S(R,\mathcal{C})$ (i.e., with respect to the S-norm) has its limit in $S(R,\mathcal{C})$.

IV. If $f, g \in S(R,\mathcal{C})$ and $\lambda \in \mathcal{C}$, then $\bar{f}$, λf and $f + g$ are also in $S(R,\mathcal{C})$.

Remark 3.7. Unlike in the case of Bohr's almost periodic functions, the product fg does not necessarily belong to $S(R,\mathcal{C})$.

We leave to the reader the task of proving that $fg \in S(R,\mathcal{C})$ when $f \in S(R,\mathcal{C})$ and $g \in \text{AP}(R,\mathcal{C})$.

V. If $f \in S(R,\mathcal{C})$, then for each $\varepsilon > 0$ there exists $\ell = \ell(\varepsilon) > 0$ such that each interval $(a, a + \ell) \subset R$ contains a point τ with the property

$$|f_\tau - f|_S < \varepsilon. \tag{3.30}$$

We shall see later that this property can serve as the definition of a function belonging to $S(R,\mathcal{C})$, being equivalent to Definition 2.25 of Chapter 2. Further properties of the functions in $S(R,\mathcal{C})$ will be established later on.

Another property, similar to the property established in Section 2.2 for the case $\text{AP}(R,\mathcal{C})$, is related to convolution; namely, if $f \in S(R,\mathcal{C})$ and $k \in L^1(R,\mathcal{C})$, then

$$k * f \in S(R,\mathcal{C}). \tag{3.31}$$

This fact results from the inequality (2.81) of Section 2.5. The detailed proof follows the same lines as in Section 3.2, where the case of the space $AP(R, \mathcal{C})$ is dealt with instead of the space $S(R, \mathcal{C})$.

One problem arising in generalizing the classical almost periodic functions (i.e., those in $AP(R, \mathcal{C})$) is when $f \in S(R, \mathcal{C})$ also satisfies $f \in AP(R, \mathcal{C})$.

An answer to this question has been given by S. Bochner, and it can be stated as follows.

Proposition 3.9. *Let* $f \in S(R, \mathcal{C})$. *A necessary and sufficient condition for having* $f \in AP(R, \mathcal{C})$ *is the uniform continuity of* f.

Proof. The condition is obviously necessary, according to property II from Section 2.3.

In order to prove the sufficiency, let us note that for each $h > 0$

$$\varphi_h(t) = h^{-1} \int_0^h f(t+s)ds, \ t \in R, \tag{3.32}$$

is in $AP(R, \mathcal{C})$. Indeed, for $\tau \in R$, one has for $t \in R$

$$|\varphi_h(t+\tau) - \varphi_h(t)| \leq h^{-1} \int_t^{t+h} |f(s+\tau) - f(s)|ds. \tag{3.33}$$

The uniform continuity of f on R implies, as seen from formula (3.33), the uniform continuity of $\varphi_h(t)$. On the other hand, $f \in S(R, \mathcal{C})$ shows that any ε-almost period of f is an (ε/h)-almost period of $\varphi_h(t)$. Hence, $\varphi_h \in AP(R, \mathcal{C})$. We further obtain

$$|\varphi_h(t) - f(t)| = h^{-1} \left| \int_0^h [f(t+s) - f(t)]ds \right|$$
$$\leq h^{-1} \int_0^h |f(t+s) - f(t)|ds. \tag{3.34}$$

The uniform continuity of f allows us to write $|f(t+s) - f(t)| < \varepsilon$ as soon as $0 \leq s \leq h < \delta(\varepsilon)$. Therefore, formula (3.34) leads to $|\varphi_h(t) - f(t)| < \varepsilon$ for $h < \delta(\varepsilon)$, $t \in R$. This means that $f(t)$ can be uniformly approximated on R by functions from $AP(R, \mathcal{C})$. According to property III from Section 2.3, there results that $f \in AP(R, \mathcal{C})$.

Remark 3.8. In the proof above, we have made use of the fact that Bohr's definition of an almost periodic function is equivalent to the approximation property (by trigonometric polynomials), which has been taken as the definition in our exposition. As mentioned already, the equivalence will be proven later.

We shall now consider a property similar to Proposition 3.6; namely, if we consider the family $\mathcal{F} = \{f(t+h); \ h \in R\}$ with $f \in S(R, \mathcal{C})$, is this family relatively compact in $S(R, \mathcal{C})$?

Proposition 3.10. *Let $f \in S(R,\mathcal{C})$, and consider the family $\mathcal{F} = \{f(t+h);$ $h \in R\} \subset S(R,\mathcal{C})$. Then $\mathcal{F}$ is relatively compact in $S(R,\mathcal{C})$ [or $M(R,\mathcal{C})$].*

Proof. What we have to prove is that any sequence $\{f(t+h_n); n \geq 1\} \subset S(R,\mathcal{C})$ contains a subsequence that converges in $S(R,\mathcal{C})$.

The proof relies partly on the proof of Proposition 3.6. As shown there, if $f \in \mathcal{T}$, from $\{f(t+h_n); n \geq 1\}$ we can extract a subsequence that converges uniformly on R (i.e., in $AP(R\mathcal{C})$). Based on this finding and taking into account that uniform convergence on R implies the convergence in $S(R,\mathcal{C})$ because

$$\sup_{t \in R} \int_t^{t+1} |f_n(s) - f(s)|ds \leq \sup_{t \in R} |f_n(t) - f(t)|, \tag{3.35}$$

any time the sup on the right-hand side exists, there remains to be shown that the statement is valid in general (i.e., when $f \in S(R,\mathcal{C})$ is arbitrary).

Now let $f \in S(R,\mathcal{C})$ be any function, and consider a sequence $\{T_n(t); n \geq 1\} \subset \mathcal{T}$ such that $T_n \longrightarrow f$ in $S(R,\mathcal{C})$. This means

$$|T_n - f|_S = \sup_{t \in R} \int_t^{t+1} |T_n(s) - f(s)|ds \longrightarrow 0 \tag{3.36}$$

as $n \longrightarrow \infty$. Further, we will apply the diagonal procedure; namely, if $\{h_n; n \geq 1\} \subset R$ is a sequence, then it contains a subsequence $\{h_{1n}; n \geq 1\} \subset R$ such that $\{T_1(t+h_{1n}); n \geq 1\} \subset AP \subset S$ is uniformly convergent on R. From the sequence $\{h_{1n}; n \geq 1\}$, we extract a subsequence $\{h_{2n}; n \geq 1\}$ with the property that $\{T_2(t+h_{2n}), n \geq 1\}$ does converge uniformly on R, and so on. We have seen that the sequence $\{T_n(t+h_{pp}); p \geq 1\}, n \geq 1$, is uniformly convergent on R and hence in $S(R,\mathcal{C})$.

Let us now take $\varepsilon > 0$ arbitrarily and choose n large enough that

$$|f - T_n|_M < \frac{\varepsilon}{3}. \tag{3.37}$$

There exists $N(\varepsilon) > 0$ for which

$$|T_n(t+h_{mm}) - T_n(t+h_{pp})| < \frac{\varepsilon}{3}, \qquad t \in R, \tag{3.38}$$

as soon as $m, p \geq N(\varepsilon)$. From formulas (3.37) and (3.38), we obtain

$$|f(t+h_{mm}) - f(t+h_{pp})|_M \leq |f(t+h_{mm}) - T_n(t+h_{mm})|_M$$
$$+ |T_n(t+h_{mm}) - T_n(t+h_{pp})|_M + |T_n(t+h_{pp}) - f(t+h_{pp})|_M$$
$$< \frac{\varepsilon}{3} + \frac{\varepsilon}{3} + \frac{\varepsilon}{3} = \varepsilon, \qquad m, p \geq N(\varepsilon).$$

For the first and third terms, we rely on formula (3.37). For the middle term of the second part in the last inequality, we keep in mind formula (3.35); i.e.,

$$|T_n(t+h_{mm}) - T_n(t+h_{pp})|_M \leq \sup_{t \in R} |T_n(t+h_{mm}) - T_n(t+h_{pp})|.$$

In conclusion, we have shown that the sequence $\{f(t+h_n);\ n \geq 1\} \subset S(R,\mathcal{C})$ contains a subsequence $\{f(t+h_{nn});\ n \geq 1\}$ that is a Cauchy sequence in $S(R,\mathcal{C})$. Therefore, $\{f(t+h_{mm});\ n \geq 1\}$ is convergent in $S(R,\mathcal{C})$.

This ends the proof of Proposition 3.10.

In concluding this section, we shall prove that a property similar to Bohr's property in the case where $AP(R,\mathcal{C})$ holds true in the case of the space $S(R,\mathcal{C})$.

Proposition 3.11. *Let* $f \in S(R,\mathcal{C})$ *and* $\varepsilon > 0$ *arbitrary. Then there exists* $\ell = \ell(\varepsilon) > 0$ *such that any interval* $(a, a + \ell) \subset R$ *contains a number* τ *with the property*

$$|f(t+\tau) - f(t)|_M < \varepsilon. \tag{3.39}$$

The proof follows the same lines as in the case of Proposition 3.7 in the preceding section. We rely on the result stated in Proposition 3.10; i.e., the compactness of the family $\mathcal{F} = \{f(t+h);\ h \in R\} \subset S(R,\mathcal{C})$. One constructs the sequence $\{h_k;\ h \geq 1\} \subset R$ such that all differences $h_i - h_j$, $i > j$, cannot serve as ε-translation numbers to f. This easily implies formula (3.39), and exactly as in the proof of Proposition 3.7, the condition (3.19) has been obtained.

The equivalence of the three basic properties of functions almost periodic in Stepanov's sense (approximation by polynomials from $\mathcal{T}$, compactness of the family $\mathcal{F}$ of translates, and Bohr's type property) will be proven when we show that any function satisfying the property stipulated in Proposition 3.11 also enjoys the approximation property.

In the next section, we shall see that Stepanov's almost periodic functions (i.e., in $S(R,\mathcal{C})$ or even $S^p(R,\mathcal{C})$, $p > 1$) can be obtained as special cases of almost periodic functions in $AP(R,X)$, where X stands for a convenient Banach space.

3.4 Besicovitch Spaces; the Mean Value

We defined in Section 2.5 the space $B^2(R,\mathcal{C})$, or $AP_2(R,\mathcal{C})$, consisting of almost periodic functions in the sense of Besicovitch (see Besicovitch [10]). From that definition, we derive some basic properties of these almost periodic functions, such as the sum of a finite number of functions almost periodic in the Besicovitch sense is also almost periodic; the product of an almost periodic function in Besicovitch's sense by a constant (from $\mathcal{C}$) is also almost periodic (i.e., in B^2); the limit of a sequence $\{x_n(t);\ n \geq 1\} \subset B^2(R,\mathcal{C})$, convergent in the Besicovitch–Marcinkiewicz norm, also belongs to the space $B^2(R,\mathcal{C})$; if $x \in B^2(R,\mathcal{C})$, then $\bar{x} = $ the complex conjugate of x, and the translates $x_h(t) = x(t+h)$, $h \in R$, also belong to $B^2(R,\mathcal{C})$.

Another elementary property easily derived from the definition is that each $x \in B^2(R,\mathcal{C})$ can be represented as a sum

$$x(t) = T(t) + y(t), \qquad t \in R, \tag{3.40}$$

with $T \in \mathcal{T}$ a trigonometric polynomial and $y \in B^2(R, \mathcal{C})$ arbitrarily small in the norm of this space. This means that for each $\varepsilon > 0$ there exists a trigonometric polynomial $T(t)$, and $y \in B^2(R, \mathcal{C})$, such that equation (3.40) takes place and

$$|y|_{\mathcal{M}} < \varepsilon, \tag{3.41}$$

which is the same as

$$\limsup_{\ell \longrightarrow +\infty} (2\ell)^{-1} \int_{-\ell}^{\ell} |y(t)|^2 dt < \varepsilon^2. \tag{3.42}$$

This property of Besicovitch's almost periodic functions will be used later. It is also useful to note that a representation of the form (3.40) holds true for any almost periodic function (Bohr's, Stepanov's) with $y(t)$ arbitrarily "small" in the corresponding norm.

We can now proceed to establish properties of $B^2(R, \mathcal{C})$ similar to Propositions 3.10 and 3.11 of the preceding section.

Proposition 3.12. *Let* $f \in B^2(R, \mathcal{C})$ *be arbitrary. Then the family* $\mathcal{F} = \{f\ (t+h);\ h \in R\}$ *is relatively compact in the topology of* $B^2(R, \mathcal{C})$, *which is the same as the topology (convergence!) in the Marcinkiewicz space* $\mathcal{M}_2(R, \mathcal{C})$.

Proof. We again follow the procedure (diagonal) used in the proofs of Proposition 3.6 and Proposition 3.10.

As seen in the proof of Proposition 3.6, each trigonometric polynomial has the property that the family of its translates is relatively compact in the topology of $BC(R, \mathcal{C})$. Given $f \in B^2(R, \mathcal{C})$, let us consider a sequence of trigonometric polynomials $\{T_n(t);\ n \geq 1\} \subset \mathcal{T}$ such that $T_n \longrightarrow f$ in $B^2(R, \mathcal{C})$ or, what is the same, in $\mathcal{M}_2(R, \mathcal{C})$. For sufficiently large $n = n(\varepsilon)$, we will have the inequality

$$|T_n - f|_{\mathcal{M}} < \frac{\varepsilon}{3}, \tag{3.43}$$

where $\varepsilon > 0$ is chosen arbitrarily. If $\{h_k;\ k \geq 1\} \subset R$ is an arbitrary sequence, then we know there exists a subsequence $\{h_{mm};\ m \geq 1\}$, obtained by means of the diagonal procedure, such that $\{T_n(t + h_{mm});\ m \geq 1\}$ is uniformly convergent on R for each fixed n; i.e., in $BC(R, \mathcal{C})$. We can obviously write for n chosen as in formula (3.43)

$$|f(t + h_{mm}) - f(t + h_{pp})|_{\mathcal{M}} \leq |f(t + h_{mm}) - T_n(t + h_{mm})|_{\mathcal{M}}$$
$$+ |T_n(t + h_{mm}) - T_n(t + h_{pp})|_{\mathcal{M}} + |T_n(t + h_{pp}) - f(t + h_{pp})|_{\mathcal{M}}. \tag{3.44}$$

The first and third terms in formula (3.44) are dominated by $\varepsilon/3$, as shown by formula (3.43). In regard to the middle term, we note that for any $x \in BC(R, \mathcal{C})$ one has

$$|x|_{\mathcal{M}} \leq \sup_{t \in R} |x(t)|, \tag{3.45}$$

an inequality that follows immediately from

$$|x|_{\mathcal{M}}^2 = \limsup_{\ell \longrightarrow \infty} (2\ell)^{-1} \int_{-\ell}^{\ell} |x(t)|^2 dt.$$

Since $\{T_n(t+h_{mm}); \ m \geq 1\}$ is convergent in $BC(R,\mathcal{C})$, we can write according to formula (3.45)

$$|T_n(t + h_{mm}) - T_n(t + h_{pp})|_{\mathcal{M}} \leq \sup |T_n(t + h_{mm}) - T_n(t + h_{pp})| < \frac{\varepsilon}{3}$$

for $m, p \geq N(\varepsilon)$. We now obtain from formulas (3.44) and (3.45) that

$$|f(t + h_{mm}) - f(t + h_{pp})|_{\mathcal{M}} < 3\frac{\varepsilon}{3} = \varepsilon, \tag{3.46}$$

provided $m, p \geq N(\varepsilon)$. From formula (3.46), we derive the convergence in $B^2(R,\mathcal{C})$ of the sequence $\{f(t + h_{mm}); \ m \geq 1\}$.

This ends the proof of Proposition 3.12.

The next proposition deals with Bohr's property, this time extended to the case of Besicovitch almost periodic functions.

Proposition 3.13. *If $f \in B^2(R,\mathcal{C})$, then to each $\varepsilon > 0$ there corresponds $\ell = \ell(\varepsilon) > 0$ such that in any interval $(a, a+\ell) \subset R$ one can find a number τ with the property*

$$|f(t + \tau) - f(t)|_{\mathcal{M}} < \varepsilon. \tag{3.47}$$

The proof of Proposition 3.13 follows, step by step, the proof of Proposition 3.7 above. One has to substitute for the norm of $BC(R,\mathcal{C})$ that of $B^2(R,\mathcal{C})$ (i.e., the Marcinkiewicz norm $|\cdot|_{\mathcal{M}}$).

Further properties of functions from $B^2(R,\mathcal{C})$ can be obtained starting from their definition, but we prefer to postpone such topics until we create new tools.

As in the case of Stepanov almost periodic functions, when instead of the norm $|\cdot|_{\mathcal{M}}$ one uses the norms

$$\sup_{t \in R} \left\{ \int_t^{t+1} |x(s)|^p ds \right\}^{1/p}, \quad . \quad 1 \leq p < \infty, \tag{3.48}$$

to construct the spaces $S^p(R,\mathcal{C})$ with $S^1 = S$, the Besicovitch spaces $B^p(R,\mathcal{C})$ are also obtained by the completion of $\mathcal{T}$ with respect to the seminorms

$$\left\{ \limsup_{\ell \longrightarrow \infty} (2\ell)^{-1} \int_{-\ell}^{\ell} |x(s)|^p ds \right\}^{1/p}. \tag{3.49}$$

In particular, the space $B^1(R,\mathcal{C})$ is denoted by $B(R,\mathcal{C})$, and it is constructed in the same manner as $B^2(R,\mathcal{C})$ by using the seminorm

$$\limsup_{\ell \to \infty} (2\ell)^{-1} \int_{-\ell}^{\ell} |x(s)| ds. \qquad (3.50)$$

It is important to point out the fact that

$$B^2(R, \mathcal{C}) \subset B(R, \mathcal{C}). \qquad (3.51)$$

The conclusion (3.51) follows easily if we keep in mind that

$$(2\ell)^{-1} \int_{-\ell}^{\ell} |x(s)| ds \le (2\ell)^{-1} \left(\int_{-\ell}^{\ell} |x(s)|^2 ds \right)^{1/2} \left(\int_{-\ell}^{\ell} ds \right)^{1/2},$$

which amounts to

$$(2\ell)^{-1} \int_{-\ell}^{\ell} |x(s)| ds \le \left[(2\ell)^{-1} \int_{-\ell}^{\ell} |x(s)|^2 ds \right]^{1/2}. \qquad (3.52)$$

Inequality (3.52) states that

$$|x|_B \le |x|_{\mathcal{M}} = |x|_{B^2}, \qquad (3.53)$$

which obviously implies formula (3.51).

As in the case of the space $B^2(R, \mathcal{C})$, the elements of the space $B(R, \mathcal{C})$ consist of classes of equivalent functions such that $f \simeq g$ iff

$$\lim_{\ell \to \infty} (2\ell)^{-1} \int_{-\ell}^{\ell} |f(t) - g(t)| dt = 0.$$

We noticed in Chapter 2 that the space $S(R, \mathcal{C})$ does not belong to $B^2(R, \mathcal{C})$. The converse is also true; i.e., $B^2(R, \mathcal{C})$ is not part of $S(R, \mathcal{C})$. It turns out that

$$S(R, \mathcal{C}) \subset B(R, \mathcal{C}), \qquad (3.54)$$

which means that $B(R, \mathcal{C})$ contains all spaces of almost periodic functions we have discussed so far.

The proof of inclusion (3.54) is very similar to the proof of the inclusion $S^2(R, \mathcal{C}) \subset B^2(R, \mathcal{C})$ we established in Section 2.5. Instead of the norms of S^2 and B^2, one needs to use those of S and B. We leave the details to the reader.

We shall now establish a basic result, valid for all classes of almost periodic functions, that plays an important role in Chapter 4 when the Fourier analysis of almost periodic functions will be developed.

Theorem 3.1. *Let $f \in B(R, \mathcal{C})$. Then there exists the mean value*

$$M\{f\} = \lim_{\ell \to \infty} (2\ell)^{-1} \int_{-\ell}^{\ell} f(t) dt. \qquad (3.55)$$

Corollary 3.4. *The mean value exists for any type of almost periodic function; i.e., for any function in* $\mathrm{AP}_1(R,\mathcal{C})$, $\mathrm{AP}(R,\mathcal{C})$, $S(R,\mathcal{C})$, $\mathrm{AP}_2(R,\mathcal{C}) = B^2(R,\mathcal{C})$.

Proof of Theorem 3.1. Since the limit in equation (3.55) exists if and only if similar limits exist for Re f and Im f, we can assume without loss of generality that $f \in B(R,R)$. Then, according to the definition of this space, we can write (see formula (3.40) above) for any $\varepsilon > 0$

$$f(t) = T(t) + r(t), \qquad t \in R, \tag{3.56}$$

with $T(t) \in \operatorname{Re}\mathcal{T}(t)$ and $r(t) \in B(R,R)$ such that

$$|r|_B = \limsup_{\ell \longrightarrow \infty} (2\ell)^{-1} \int_{-\ell}^{\ell} |r(s)| ds < \frac{\varepsilon}{4}. \tag{3.57}$$

But equation (3.57) implies that, for any $\ell \geq \ell_0 = \ell_0(\varepsilon)$,

$$(2\ell)^{-1} \int_{-\ell}^{\ell} |r(s)| ds < \frac{\varepsilon}{4}. \tag{3.58}$$

On the other hand, $M\{T(t)\}$ does exist for any $T \in \operatorname{Re}\mathcal{T}$, as noted in Section 2.5. Therefore, one can write, for $\ell \geq \ell_1(\varepsilon) > 0$,

$$\left| (2\ell)^{-1} \int_{-\ell}^{\ell} T(t) dt - M\{T\} \right| < \frac{\varepsilon}{4}. \tag{3.59}$$

Taking into account formulas (3.56)–(3.59), we can write, for $\ell \geq \max\{\ell_0, \ell_1\} = \ell_2$,

$$\left| (2\ell)^{-1} \int_{-\ell}^{\ell} f(t) dt - M\{T\} \right| < \frac{\varepsilon}{2}. \tag{3.60}$$

From formula (3.60), one derives, for $\ell, \bar{\ell} > \ell_2(\varepsilon)$,

$$\left| (2\ell)^{-1} \int_{-\ell}^{\ell} f(t) dt - (2\bar{\ell})^{-1} \int_{-\bar{\ell}}^{\bar{\ell}} f(t) dt \right| < \varepsilon, \tag{3.61}$$

which is exactly Cauchy's criterion for existence of the limit in equation (3.55), defining $M\{f\}$.

Theorem 3.1 is thereby proven.

The *mean value* is in fact a linear functional from the space $B(R,\mathcal{C})$ into $\mathcal{C}$. It has some basic properties, which we shall now emphasize.

Proposition 3.14. *The mean value, as defined by equation (3.55) for any* $f \in B(R,\mathcal{C})$, *possesses the following properties:*

(a) $M\{f + g\} = M\{f\} + M\{g\}$.

(b) $M\{\lambda f\} = \lambda M\{f\}$, $\lambda \in C$.

(c) $M\{\bar{f}\} = \overline{M\{f\}}$.

(d) If $f(t) \geq 0$ on R, $M\{f\} \geq 0$.

(e) $|M\{f\}| \leq M\{|f|\} = |f|_B$.

The proof of Proposition 3.14 is immediate. From property (e), we derive the boundedness on $B(R, C)$ of the linear functional $f \longrightarrow M\{f\}$.

A consequence of the boundedness of the mean value functional $M\{f\}$ is that $f_n \longrightarrow f$ in $B(R, C)$, as $n \longrightarrow \infty$, implies

$$M\{f\} = \lim_{n \to \infty} M\{f_n\}. \tag{3.62}$$

Equation (3.62) follows from property (e) applied to $f_n - f$. One has

$$|M\{f_n\} - M\{f\}| = |M\{f_n - f\}| \leq |f_n - f|_B,$$

which justifies equation (3.62).

From the inequalities

$$|f|_B \leq |f|_S \leq |f|_{AP}, \tag{3.63}$$

one easily derives that equation (3.62) takes place when we deal with the spaces $S(R, C)$ or $AP(R, C)$ using their type of convergence; in other words, when $f_n \longrightarrow f$ either in $S(R, C)$ or in $AP(R, C)$. In the last case, we have uniform convergence on R.

From the properties of the mean value $M\{f\}$, there results immediately that the map $f \longrightarrow M\{|f|\}$ is a seminorm on $B(R, C)$. This seminorm is actually a norm on the factor space B/B_0, where $B_0(R, C)$ consists of these locally integrable functions on R such that

$$\lim_{\ell \to \infty} (2\ell)^{-1} \int_{-\ell}^{\ell} |x(s)| ds = 0. \tag{3.64}$$

The elements of the factor space B/B_0, which we continue to denote by $B(R, C)$ because $|x + B_0|_B = |x|_B$, consist of all classes (of equivalence) $\{x + B_0;\ x \in B\}$. B_0 stands for the null element in B/B_0, and we can say that equation (3.64) implies $x = \theta =$ the null element in B/B_0 but not in $B(R, C)$. This is because there exist locally integrable functions on R such that equation (3.64) holds but $x(t) \not\equiv 0$. We can choose in equation (3.64), for instance, $x(t) = \exp\{-|t|\}$.

The following result plays an important role in the forthcoming developments of the theory of Bohr's almost periodic functions.

Proposition 3.15. *On the space* $AP(R, C)$, *the map* $f \longrightarrow M\{|f|\}$ *is a norm. The same is true for the map* $f \longrightarrow (M\{|f|^2\})^{1/2}$.

Remark 3.9. We know that the norm on $AP(R, \mathcal{C})$ is $f \longrightarrow \sup\{|f(t)|; \, t \in R\}$. With respect to this norm, $AP(R, \mathcal{C})$ is a Banach space (i.e., it is complete). The norms indicated in Proposition 3.15 induce in $AP(R, \mathcal{C})$ different types of convergence (not the uniform one!). In this way, we obtain only normed spaces (i.e., not complete). By completion, we reobtain the spaces $B(R, \mathcal{C})$ and $B^2(R, \mathcal{C})$. We do not elaborate on this topic here.

Proof of Proposition 3.15. This will follow from the following statement: Let $f \in AP(R, R)$ be such that $f(t) \geq 0$, $t \in R$. If $M\{f\} = 0$, then $f(t) \equiv 0$ on R.

In order to prove the last assertion, we shall rely on Bohr's definition of almost periodic functions. As seen in Section 3.2, any function that is almost periodic in the sense of the definition from Section 2.3 (i.e., uniformly approximable on R by trigonometric polynomials) satisfies Bohr's property. This fact was established in Proposition 3.7.

Now, assume to the contrary that $M\{f\} = 0$ but $f \not\equiv 0$. Since $f(t) \geq 0$, $t \in R$, we can find $t_0 \in R$ such that $f(t_0) = \alpha > 0$. We shall now define two positive numbers ℓ and δ such that any interval of length ℓ on R contains a subinterval of length 2δ whose points must be $(\alpha/3)$-translation numbers for f. This is possible due to the uniform continuity of f on R (property II in Section 3.2). For $\alpha > 0$, we can find $\delta_1 = \delta_1(\alpha/6)$ such that $|f(t + h) - f(t)| < \alpha/6$, $t \in R$, for $|h| < \delta_1$, due to uniform continuity. Further, based on Bohr's property for f, one can find $\ell_1 > 0$ such that every interval $(a, a + \ell_1) \ni \tau$ with $|f(t + \tau) - f(t)| < \alpha/6$, $t \in R$. Since $|h| < \delta_1$, $\tau + h \in (a - \delta_1, a + \ell_1 + \delta_1)$. Therefore,

$$|f(t + \tau + h) - f(t)| \leq |f(t + \tau + h) - f(t + h)|$$
$$+ |f(t + h) - f(t)| < \frac{\alpha}{6} + \frac{\alpha}{6} = \frac{\alpha}{3}, \qquad t \in R.$$

In other words, one can take $\ell = \ell + \delta_1$ and $\delta = 2\delta_1$ and conclude that each interval $(a, a + \ell) \subset R$ contains a subinterval of length δ with the property that any number in this subinterval is an $(\alpha/3)$-translation number for f. This will help to show that each interval of length ℓ contains a subinterval of length 2δ on which $f(t) \geq \alpha/3$. Indeed, with $a \in R$ arbitrary, consider the interval $(a - \delta - t_0, a + \ell + \delta - t_0)$. There exists an $(\alpha/3)$-translation number in this interval. But $t + \tau \in (a - \delta, a + \ell - \delta)$, and taking t such that $|t - t_0| < \delta$, $t + \tau$ will cover an interval of length 2δ. We obtain

$$f(t + \tau) = f(t_0) + [f(t) - f(t_0)] + [f(t + \tau) - f(t)] > \alpha - \frac{\alpha}{3} - \frac{\alpha}{3} = \frac{\alpha}{3}.$$

With this property of $f(t)$ established, and keeping in mind that $f(t) \geq 0$ on R, one obtains

$$\frac{1}{2n\ell} \int_{-n\ell}^{n\ell} f(t)dt = \frac{1}{2n\ell} \sum_{k=-n+1}^{n} \int_{(k-1)\ell}^{k\ell} f(t)dt > \frac{1}{2n\ell}(2n)(2\delta)\frac{\alpha}{3} = \frac{2\alpha\delta}{3\ell} > 0.$$

Letting $n \to \infty$, there results $M\{f\} > 0$, which contradicts our hypothesis.

Obviously, the argument above in the proof of Proposition 3.15 is applicable in proving that the map $f \longrightarrow (M\{|f|^2\})^{1/2}$ is also a norm on AP(R, R). Proposition 3.15 is thereby proven.

3.5 The Space AP(R, X), X-Banach Space

Most features of almost periodic functions in various senses have been illustrated in preceding sections. The case where the values taken by such functions belong to a Banach space and we have in mind not R^n or C^n but the infinite-dimensional space X deserves special attention. We shall briefly present the basic properties of AP(R, X), with X an arbitrary Banach space over R or C. This presentation will be helpful when we discuss certain applications of the almost periodic functions to such phenomena as wave propagation.

Of course, it is possible to build up a theory of almost periodicity in cases like AP$_1(R, X)$, $S(R, X)$, or even $B^2(R, X) = $ AP$_2(R, X)$. We shall not proceed on this path.

We shall present the theory of AP(R, X) spaces with X a (real or complex) Banach space, starting with Bohr's property as the definition. As we shall see later on, this property is equivalent to the relative compactness of the set of translates as well as the approximation property (by trigonometric polynomials, this time with coefficients being elements of the Banach space).

Definition 3.1. Let X be a Banach space and $f : R \longrightarrow X$ a continuous function. Then $f \in $ AP(R, X), and it is called *almost periodic* if for each $\varepsilon > 0$ there exists $\ell = \ell(\varepsilon) > 0$ such that in any interval of length ℓ of R one can find a number $\tau \in (a, a + \ell)$ with the property

$$\|f(t + \tau) - f(t)\| < \varepsilon, \qquad t \in R. \tag{3.65}$$

The norm in AP(R, C) will be

$$|f|_{\text{AP}} = \sup\{\|f(t)\|; \ t \in R\} < \infty. \tag{3.66}$$

Remark 3.10. It is obvious that any continuous periodic function is almost periodic.

Proposition 3.16. *If $f \in $ AP(R, X), then f is bounded and uniformly continuous.*

Proof. Let $f \in $ AP(R, X), and let $\ell = \ell(1)$ be the number corresponding to $\varepsilon = 1$ according to Definition 3.1 above. The function $f(t)$, restricted to the interval $[0, \ell]$, is bounded. This means there exists $M > 0$ such that $\|f(t)\| \leq M$, $t \in [0, \ell]$. For arbitrary $t \in R$, consider the interval $[-t, -t + \ell]$, which must contain τ (with $\varepsilon = 1$) such that

$$\|f(t + \tau) - f(t)\| < 1, \qquad t \in R. \tag{3.67}$$

Therefore, based on formula (3.67),

$$\|f(t)\| \le \|f(t) - f(t+\tau)\| + \|f(t+\tau)\| < 1 + M, \qquad t \in R,$$

which proves the boundedness of f on R.

Next, to prove the uniform continuity of f on R, take $\ell = \ell(\varepsilon/3)$ for an arbitrary $\varepsilon > 0$ according to Definition 3.1, and note that f is uniformly continuous on the interval $[-1, 1+\ell]$. Let $\delta = \delta(\varepsilon/3)$, $\delta < 1$, from uniform continuity; i.e., $\|f(t) - f(s)\| < \varepsilon/3$ when $t, s \in [-1, 1+\ell]$, $|t - s| < \delta$. Then, for $|t - s| < \delta$ with $t, s \in R$, we obtain for $\tau \in [-t, -t+\ell]$ an $(\varepsilon/3)$-translation number of f,

$$\|f(t) - f(s)\| \le \|f(t) - f(t+\tau)\| + \|f(t+\tau) - f(s+\tau)\|$$
$$+ \|f(s+\tau) - f(s)\| < \frac{\varepsilon}{3} + \frac{\varepsilon}{3} + \frac{\varepsilon}{3} = \varepsilon,$$

because $t+\tau \in [0, \ell] \subset [-1, 1+\ell]$, while $s+\tau \in [-1, 1+\ell]$ since $|t-s| < \delta < 1$. Therefore, the first and third terms on the right-hand side are less than $\varepsilon/3$ from almost periodicity, while the middle term is also less than $\varepsilon/3$ from uniform continuity of f on $[-1, 1+\ell]$.

This ends the proof of Proposition 3.16.

The following is an elementary statement.

Proposition 3.17. *If $f \in AP(R, X)$, then λf, with $\lambda \in R$ or $\lambda \in C$, $f(t+h)$, $h \in R$, and $\|f(t)\|$ are almost periodic (the last one is in $AP(R, R)$).*

It suffices to note that

$$\big| \|f(t+\tau)\| - \|f(t)\| \big| \le \|f(t+\tau) - f(t)\|$$

for $t, \tau \in R$.

Let us point out that, according to the definition of the norm in $AP(R, X)$, as shown in equation (3.66), the convergence in $AP(R, X)$ is the uniform convergence on R. In other words, if $f_n, f \in AP(R, X)$, then $f_n \longrightarrow f$ in this space, as $n \to \infty$, if to any $\varepsilon > 0$ there corresponds a positive integer $N = N(\varepsilon)$ such that

$$\|f_n(t) - f(t)\| < \varepsilon, \qquad t \in R,$$

as soon as $n \ge N$.

The completeness of $AP(R, X)$ with respect to the norm (3.66) is easily obtained if we keep in mind the inequality

$$\|f(t+\tau) - f(t)\| \le \|f(t+\tau) - f_n(t+\tau)\|$$
$$+ \|f_n(t+\tau) - f_n(t)\| + \|f_n(t) - f(t)\|,$$

where $\{f_n;\ n \ge 1\}$ is a sequence in $AP(R, X)$ uniformly convergent to f.

Proposition 3.18. *Let $f \in \mathrm{AP}(R, X)$. Then the set $\{f(t); \ t \in R\} \subset X$ is relatively compact.*

Proof. According to Proposition 2.5 from Chapter 2, it suffices to show that for each $\varepsilon > 0$ there exists a finite number of balls of radius ε in X such that their union covers the set $\{f(t); \ t \in R\}$. Indeed, let $\ell = \ell(\varepsilon/2)$ be the number corresponding to Bohr's definition of almost periodic functions. Due to the continuity of $f(t), \ t \in [0, \ell]$, one can find because of the compactness of the set $\{f(t); \ t \in [0, \ell]\} \subset X$ a finite number of points $x_k \in X$, $k = 1, 2, \ldots, m$, such that the balls $\Sigma(x_k; \varepsilon/2)$, $k = 1, 2, \ldots, m$, cover this set. For each $t \in R$, let $\tau \in [-t, -t + \ell]$ be an $(\varepsilon/2)$-translation number of f. But $t + \tau \in [0, \ell]$, and consequently there exists p, $1 \leq p \leq m$, such that $f(t + \tau) \in \Sigma(x_p; \varepsilon/2)$. Hence,

$$\|f(t) - x_p\| \leq \|f(t) - f(t + \tau)\| + \|f(t + \tau) - x_p\| < \frac{\varepsilon}{2} + \frac{\varepsilon}{2} = \varepsilon,$$

which proves that $\bigcup_{p=1}^{m} \Sigma(x_p; \varepsilon) \supset \{f(t); \ t \in R\}$.
Proposition 3.18 is thus proven. $\quad\blacksquare$

Remark 3.11. The compactness of the set $\{f(t); \ t \in [0, \ell]\} \subset X$ is assured by the continuity of f. Indeed, if $\{f(t_j); \ j \geq 1, \ t_j \in [0, \ell]\}$ is a sequence in $\{f(t); \ t \in [0, \ell]\}$, then $\{t_j; \ j \geq 1\}$ contains a convergent subsequence, say $\{t_{j_k}; \ k \geq 1\}$, with $t_{j_k} \longrightarrow \bar{t} \in [0, \ell]$ as $k \longrightarrow \infty$. But $f(t_{j_k}) \longrightarrow f(\bar{t})$ from the continuity of f, and since $f(\bar{t}) \in \{f(t); \ t \in [0, \ell]\}$, the compactness of this set is established.

Remark 3.12. From the relative compactness (in X) of the set $\{f(t); t \in R\}$, when f is in $\mathrm{AP}(R, X)$; there results that from any sequence $\{f(t_j); \ t_j \geq 1, \ t_j \in R\}$ one can extract a convergent subsequence.

We shall now prove a result that represents a characterization of almost periodicity.

Theorem 3.2. *Bohr's definition of almost periodicity (i.e., $f \in \mathrm{AP}(R, X)$) is equivalent to the property of relative compactness for the family $\mathcal{F} = \{f(t+h); \ h \in R\}$ of translates in $\mathrm{BC}(R, X)$.*

Proof. As we have seen in Section 2.2, the relative compactness of $\mathcal{F}$ implies Bohr's property. So, relative compactness is a sufficient condition for Bohr's property.

In order to prove the necessity, we shall use again a diagonal procedure of extraction. If $\{f(t + h_n); \ n \geq 1\} \subset \mathcal{F}$ is a sequence of translates of f and $S = \{s_n\} \subset R$ a dense subset (for instance, the set Q of rational numbers), then the sequence $\{f(s_1+h_n); \ n \geq 1\} \subset X$ consists of values of $f(t)$, and as seen in Remark 3.12, it contains a convergent subsequence $\{f(s_1+h_{1n}); \ n \geq 1\}$. Then we consider the sequence $\{f(s_2+h_{1n}); \ n \geq 1\}$ and note that it contains a subsequence $\{f(s_2 + h_{2n}); \ n \geq 1\}$ that converges. We then move to

$\{f(s_3 + h_{2n}); \ n \geq 1\}$, and so on. The diagonal sequence $\{f(t + h_{nn}); \ n \geq 1\}$ is convergent for each $t \in S$. Actually, we will show that $\{f(t + h_{nn}); \ n \geq 1\}$ is uniformly convergent in R.

Indeed, let $\varepsilon > 0$ and $\ell = \ell(\varepsilon/5)$ be the number corresponding to $\varepsilon/5$ according to Bohr's definition of almost periodicity. Let $\delta = \delta(\varepsilon/5)$ be the number associated to $\varepsilon/5$ according to the property of uniform continuity of $f(t)$ on R. We take a subdivision of the interval $[0, \ell]$, say $\Delta_0 = \{0 = \alpha_0, \alpha_1, \alpha_2, \ldots, \alpha_p = \ell\}$, such that $\max(\alpha_k - \alpha_{k-1}) < \delta$, $k = 1, 2, \ldots, p$.

In each interval of Δ_0, we choose a point in S, thus obtaining a finite set of points $S_0 = \{r_1, r_2, \ldots, r_p\} \subset S$. Since S_0 is a finite set, $\{f(t + h_{nn}); \ n \geq 1\}$ is uniformly convergent with respect to $t \in S_0$. Hence, we can determine $N = N(\varepsilon/5)$ such that for $n, m \geq N$ one has

$$\|f(r_k + h_{nn}) - f(r_k + h_{mm})\| < \varepsilon/5, \ k = 1, 2, \ldots, p.$$

Now take an arbitrary $t \in R$, and let τ be an $(\varepsilon/5)$-translation number of f such that $\tau \in [-t, -t + \ell]$, which means $t + \tau \in [0, \ell]$. Denote by r_j the number in S_0 such that $|t + \tau - r_j| < \delta$. If one takes $n, m \geq N$, then there results

$$
\begin{aligned}
\|f(t + h_{nn}) - f(t + h_{mm})\| &\leq \|f(t + h_{nn}) - f(t + \tau + h_{nn})\| \\
&+ \|f(t + \tau + h_{nn}) - f(r_j + h_{nn})\| \\
&+ \|f(r_j + h_{nn}) - f(r_j + h_{mm})\| \\
&+ \|f(r_j + h_{mm}) - f(t + \tau + h_{mm})\| \\
&+ \|f(t + \tau + h_{mm}) - f(t + h_{mm})\| \\
&< 5\frac{\varepsilon}{5} = \varepsilon, \qquad t \in R.
\end{aligned}
$$

The last inequality shows that the sequence $\{f(t + h_{nn}); \ n \geq 1\}$ is satisfying Cauchy's conditions for uniform convergence on R. Therefore, the initial sequence of translates $\{f(t + h_n); \ n \geq 1\}$ contains a subsequence that converges uniformly on R, which is the kind of convergence for the space $AP(R, X)$.

Theorem 3.2 is thereby proven.

Remark 3.13. See S. Zaidman [106] and A. Haraux [49] for a weaker form of the Bochner condition of almost periodicity.

A simple consequence of Theorem 3.2 is the following property of almost periodic functions with values in a Banach space.

Proposition 3.19. *The sum of two almost periodic functions with values in the Banach space X (real or complex) is almost periodic.*

Proof. Instead of using Definition 3.1, we shall rely on Theorem 3.2. Let $f_1(t)$ and $f_2(t)$ be almost periodic. We need to show that $f(t) = f_1(t) + f_2(t)$, $t \in R$, is also almost periodic (i.e., to prove that the family $\mathcal{F} = \{f(t + h); \ h \in R\}$ is relatively compact in the topology of uniform convergence on R).

Let us consider the sequence of translates $\{f(t + h_n); \ n \geq 1\} \subset \mathcal{F}$. First, $\{f_1(t + h_n); \ n \geq 1\}$ contains a subsequence, say $\{f_1(t + h_{1n}); \ n \geq 1\}$, that is uniformly convergent on R. Next, the sequence $\{f_2(t + h_{1n}); \ n \geq 1\}$ will contain a subsequence, say $\{f_2(t + h_{2n}); \ n \geq 1\}$, that converges uniformly on R. Obviously, $\{f(t + h_n); \ n \geq 1\}$ has the subsequence $\{f(t + h_{2n}); \ n \geq 1\}$ uniformly convergent on R. This is true because both sequences $\{f_1(t + h_{1n}); \ n \geq 1\}$ and $\{f_2(t + h_{2n}); \ n \geq 1\}$ are uniformly convergent on R ($\{h_{2n}\}$ is a subsequence of $\{h_{1n}\}$!), and $f(t + h_{2n}) = f_1(t + h_{2n}) + f_2(t + h_{2n})$, $n \geq 1$.

Proposition 3.19 is thus proven.

It is obvious that it implies the fact that the sum of any finite number of almost periodic functions is also almost periodic. This was necessary to conclude that AP(R, X) is a Banach space.

So far, we have seen in this section that almost periodicity in Bohr's sense and the relative compactness of the family of translates are equivalent. We did not make reference to the approximation property by trigonometric polynomials. This is what the next proposition does.

Proposition 3.20. *Let f be a map from R into the Banach space X such that it can be uniformly approximated in the norm by trigonometric polynomials of the form*

$$T(t) = a_1 e^{i\lambda_1 t} + a_2 e^{i\lambda_2 t} + \cdots + a_m e^{i\lambda_m t}, \qquad t \in R, \qquad (3.68)$$

with $\lambda_j \in R$, $a_j \in X$, $j = 1, 2, \ldots, m$. Then f is almost periodic: $f \in$ AP(R, X).

Proof. Let us notice, first, that any term in equation (3.68) is periodic of period $2\pi/\lambda_j$ when $\lambda_j \neq 0$ and constant when $\lambda_j = 0$ (also periodic). Hence, $T(t)$ given by equation (3.68) is almost periodic because it is a finite sum of almost periodic functions (Proposition 3.19 and Remark 3.10 to Definition 3.1). Since $f : R \to X$ can be uniformly approximated by trigonometric polynomials of the form (3.68), there remains to prove that AP(R, X) is closed with respect to uniform convergence on R. This fact can be easily proven.

Indeed, let $f(t) = \lim f_n(t)$, as $n \to \infty$, uniformly on R, where $f_n(t)$, $n \geq 1$, are in AP(R, X). For arbitrary $\varepsilon > 0$, there exists $N = N(\varepsilon) > 0$ such that, for $n \geq N$,

$$\|f_n(t) - f(t)\| < \varepsilon/3, \qquad t \in R. \qquad (3.69)$$

Let $\ell = \ell(\varepsilon/3)$ be the number corresponding to $\varepsilon/3$ according to Definition 3.1. Denote by τ any $(\varepsilon/3)$-translation number of $f_N(t)$. For arbitrary $t \in R$, one can write the inequalities

$$\|f(t + \tau) - f(t)\| \leq \|f(t + \tau) - f_N(t + \tau)\|$$
$$+ \|f_N(t + \tau) - f_N(t)\| + \|f_N(t) - f(t)\| < \frac{\varepsilon}{3} + \frac{\varepsilon}{3} + \frac{\varepsilon}{3} = \varepsilon, \qquad (3.70)$$

taking into account the definition of translation numbers. But formula (3.70) proves that any $(\varepsilon/3)$-translation number for $f_N(t)$ is an ε-translation number for $f(t)$.

This ends the proof of Proposition 3.20.

Remark 3.14. As in the case of the spaces $AP(R, R)$ or $AP(R, C)$, the converse proposition to Proposition 3.20 is postponed until we get into Chapter 4 and present the theory of Fourier series associated to almost periodic functions in various senses.

We have seen above that the property of relative compactness of the set $\mathcal{F} = \{f(t + h); \ h \in R\}$ of an almost periodic function is a characteristic property of almost periodicity (Bohr).

We shall now look for a characterization of relative compactness for an arbitrary set in $AP(R, X)$. This has important applications to functional equations.

Before obtaining a criterion of relative compactness in the space $AP(R, X)$, we shall establish an auxiliary result.

Proposition 3.21. *Let $f_k(t) \in AP(R, X)$, $k = 1, 2, \ldots, n$. Then, for every $\varepsilon > 0$, there exist common ε-translation numbers for these functions.*

Proof. We shall associate to these functions an almost periodic function with values in X^n by letting $f(t) = \mathrm{col}(f_1(t), \ldots, f_n(t))$. It is well known that X^n is also a Banach space, the norm being defined (for instance) by $\|x\| = \sum_{k=1}^{n} \|x_k\|$, where $x = \mathrm{col}(x_1, x_2, \ldots, x_n)$, and $x_j \in X$, $j = 1, 2, \ldots, n$. For the almost periodicity of $f(t)$ (i.e., $f \in AP(R, X^n)$) we shall prove the relative compactness of the family $\mathcal{F} = \{f(t+h); \ h \in R\}$. Indeed, let $\{h_m; \ m \geq 1\} \subset R$ be an arbitrary sequence. Then $\{f_1(t+h_m); \ m \geq 1\}$ contains a subsequence uniformly convergent on R. Let this subsequence be denoted by $\{f_1(t+h_{1m}); \ m \geq 1\}$. We now consider the sequence $\{f_2(t+h_{1m}); \ m \geq 1\}$, and based on the almost periodicity of f_2, we can find a subsequence, say $\{f_2(t + h_{2m}); \ m \geq 1\}$, that converges uniformly on R.

We proceed further in the same way as above and finally end up with a sequence $\{f_n(t+h_{nm}); \ m \geq 1\}$ that is uniformly convergent on R. But $\{h_{nm}; \ m \geq 1\}$ is a subsequence of each $\{h_{km}; \ m \geq 1\}$, $k = 1, 2, \ldots, n - 1$, which means that each sequence $\{f_k(t+h_{nm}); \ m \geq 1\}$ is uniformly convergent on R. Obviously, this means that the sequence (in $AP(R, X^n)$) $\{f(t + h_m); \ m \geq 1\}$ contains a subsequence $\{f(t + h_{nm}); \ m \geq 1\}$ that is uniformly convergent on R. Therefore, $f \in AP(R, X^n)$.

Let us now take $\varepsilon > 0$ and consider the number $\ell = \ell(\varepsilon)$ according to Definition 3.1. Then each interval of length ℓ, say $(a, a + \ell) \subset R$, contains a number τ with an ε-translation number for $f \in AP(R, X^n)$:

$$\|f(t + \tau) - f(t)\| < \varepsilon, \qquad t \in R. \tag{3.71}$$

Taking into account the definition of the norm in the product space, formula (3.71) implies

$$\|f_k(t + \tau) - f_k(t)\| < \varepsilon, \qquad t \in R, \ k = 1, 2, \ldots, n. \tag{3.72}$$

The inequality (3.72) shows that, for any finite number of functions in $AP(R, X)$, there are common ε-translation numbers for each $\varepsilon > 0$.

Proposition 3.21 is thus proven.

We are prepared to state and prove a criterion of relative compactness for subsets of AP(R, X).

Theorem 3.3. *Let $\mathcal{M} \subset$ AP(R, X) be a subset. Then $\mathcal{M}$ is relatively compact in* AP(R, X) *if and only if the following properties hold true:*

(1) *$\mathcal{M}$ is equicontinuous, which means that for any $\varepsilon > 0$ there exists $\delta(\varepsilon) > 0$ such that for any $f \in \mathcal{M}$ one has $\|f(t) - f(s)\| < \varepsilon$ if $|t - s| < \delta(\varepsilon)$.*
(2) *$\mathcal{M}$ is equi–almost-periodic, which means that for each $\varepsilon > 0$ there exists $\ell(\varepsilon) > 0$ such that any interval $(a, a + \ell) \subset R$ contains a number τ that is an ε-translation number for all $f \in \mathcal{M}$.*
(3) *For fixed $t \in R$, the set $\{f(t); \ f \in \mathcal{M}\}$ is relatively compact in the space X.*

Proof. We will prove first that these conditions are necessary. We rely on Proposition 2.5 of Chapter 2 regarding the characterization of relatively compact sets in metric spaces. If $\mathcal{M} \subset$ AP(R, X) is relatively compact, then we can find, for each $\varepsilon > 0$, a finite number of functions $f_1, f_2, \ldots, f_m$ that form an ε-net. In other words, each $f \in \mathcal{M}$ is contained in one of the balls defined by $\|f - f_k\|_{\text{AP}} < \varepsilon$, $k = 1, 2, \ldots, m$. This finite set of functions, $f_1, f_2, \ldots, f_m$, is both equicontinuous and equi–almost-periodic, according to Proposition 3.21. Now let $f \in \mathcal{M}$ and p be an integer such that $1 \le p \le m$, for which $\|f - f_p\|_{\text{AP}} < \varepsilon$. Then, for any ε-translation number common to all f_k, $1 \le k \le m$, we have

$$\|f(t + \tau) - f(t)\| \le \|f(t + \tau) - f_p(t + \tau)\| + \|f_p(t + \tau) - f_p(t)\|$$

$$+ \|f_p(t) - f(t)\| < 3\varepsilon, \qquad t \in R,$$

which shows that $\mathcal{M}$ is equi–almost-periodic. It is obvious from the inequalities above that $\mathcal{M}$ is also equicontinuous (τ need not be an ε-translation number but a "small" real number). To finish with the necessity, we have to consider now property (3) from the statement of Theorem 3.3. The relative compactness of $\mathcal{M}$ allows us to extract from any sequence $\{f_k; \ k \ge 1\} \subset \mathcal{M}$ a subsequence that converges uniformly on R.

In particular, fixing $t \in R$, from the sequence $\{f_k(t); \ k \ge 1\} \subset X$, we can extract a convergent (in X) subsequence. Therefore, the set $\{f(t); \ f \in \mathcal{M}\}$, for fixed $t \in R$, is relatively compact.

To prove the sufficiency of conditions (1), (2), (3) for the relative compactness of the set $\mathcal{M} \subset$ AP(R, X), we will consider a countable subset of R, say S, such that S is dense in R. At this point, the proof goes along the same lines as the proof of Theorem 3.2. In other words, from any sequence $\{f_k(t); \ k \ge 1\} \subseteq \mathcal{M}$, we can extract, by using the diagonal procedure, a subsequence that converges at each point of $S = \{s_n; \ n \ge 1\} \subset R$. In the proof of Theorem 3.2, we dealt with a sequence $\{f(t + h_n); \ n \ge 1\} \subset \mathcal{F}$ and proved it does contain a subsequence $\{f(t + h_{nn}); \ n \ge 1\}$ that converges at each point $t = s_j \in S$. This time we deal with $\{f_k(t); \ k \ge 1\} \subset \mathcal{M}$.

We shall prove that a sequence that is pointwise convergent on S is in fact uniformly convergent on R. Without loss of generality, we can assume that the sequence $\{f_k(t); \ k \geq 1\} \subset \mathcal{M}$ is pointwise convergent on S. For $\varepsilon > 0$, denote by $\ell(\varepsilon)$ the corresponding number according to assumption (2) of equi–almost-periodicity of $\mathcal{M}$ and by $\delta(\varepsilon)$ the corresponding (to ε) number according to assumption (1) of equicontinuity of $\mathcal{M}$. Again, as in the proof of Theorem 3.2, we choose a subset $S_0 = \{s_1, s_2, \ldots, s_p\} \subset S$ such that each point $t \in [0, \ell]$ will be at a distance smaller than $\delta(\varepsilon)$ from S_0. This fact is possible because of the density of S on R. It is obvious that $\{f_k(t); \ k \geq 1\}$ is uniformly convergent in S_0. By $N(\varepsilon)$, we denote the number corresponding to $\varepsilon > 0$ from the definition of uniform convergence. For $f \in R$ arbitrary, there exists $\tau \in [-t, -t + \ell]$ that is an ε-translation number common to all functions of $\mathcal{M}$. Because $t + \tau \in [0, \ell]$, there results the existence of an $s_j \in S_0$ for which $|t + \tau - s_j| < \delta$. Now, for $n, m \geq N(\varepsilon)$, we can write the sequence of inequalities

$$\|f_n(t) - f_m(t)\| \leq \|f_n(t) - f_n(t + \tau)\| + \|f_n(t + \tau) - f_n(s_j)\|$$

$$+ \|f_n(s_j) - f_m(s_j)\| + \|f_m(s_j) - f_m(t + \tau)\| + \|f_m(t + \tau) - f_m(t)\| < 5\varepsilon.$$

This shows that $\{f_k(t); \ k \geq 1\} \subset \mathcal{M}$ satisfies Cauchy's criterion, which implies its uniform convergence on R.

The proof of Theorem 3.3 is now complete.

We shall now prove, as we've done for each space of almost periodic functions with scalar (real or complex) values, that the *mean value* does exist for all functions in $\mathrm{AP}(R, X)$. The approach will be different from that encountered in Section 3.4, where the definition of almost periodicity was based on the approximation property by trigonometric polynomials (for which the existence of the mean value is trivial). But the result will be basically the same.

Theorem 3.4. *For any $f \in \mathrm{AP}(R, X)$, there exists the mean value*

$$M\{f\} = \lim_{T \to \infty} T^{-1} \int_0^T f(t)dt \in X. \tag{3.73}$$

Proof. For $\varepsilon > 0$, let $\ell = \ell(\varepsilon)$ be the length of the interval on R such that each $(a, a + \ell)$, $a \in R$, contains an $(\varepsilon/2)$-translation number for f. Denote $A = \sup\{\|f(t)\|; \ t \in R\}$, and for a given a let $\tau \in (a, a + \ell)$ be the $(\varepsilon/2)$-translation number for f. Then

$$\int_a^{a+T} f(t)dt = \int_a^\tau f(t)dt + \int_\tau^{\tau+T} f(t)dt + \int_{\tau+T}^{a+T} f(t)dt,$$

which implies

$$T^{-1}\left\|\int_0^T f(t)dt - \int_a^{a+T} f(t)dt\right\| \leq T^{-1}\left\|\int_0^T f(t)dt - \int_\tau^{\tau+T} f(t)dt\right\|$$

$$+T^{-1}\left\|\int_a^\tau f(t)dt\right\| + T^{-1}\left\|\int_{\tau+T}^{a+T} f(t)dt\right\| \leq T^{-1}\int_0^T \|f(t) - f(t+\tau)\|dt$$

$$+T^{-1}\int_a^\tau \|f(t)\|dt + T^{-1}\int_{\tau+T}^{a+T} \|f(t)\|dt \leq \frac{\varepsilon}{2} + \frac{A\ell}{T} + \frac{A\ell}{T} = \frac{\varepsilon}{2} + \frac{2A\ell}{T}.$$

Taking $a = (k-1)T$, $k = 1, 2, \ldots, n$, we obtain from above the inequalities

$$T^{-1}\left\|\int_0^T f(t)dt - \int_{(k-1)T}^{kT} f(t)dt\right\| \leq \frac{\varepsilon}{2} + \frac{2A\ell}{T}, \qquad k = 1, 2, \ldots, n. \quad (3.74)$$

The inequalities (3.74) lead to

$$T^{-1}\left\|\int_0^T f(t)dt - n^{-1}\int_0^{nT} f(t)dt\right\| = T^{-1}\left\|\int_0^T f(t)dt - n^{-1}\sum_{k=1}^n \int_{(k-1)T}^{kT} f(t)dt\right\|$$

$$\leq n^{-1}T^{-1}\sum_{k=1}^n\left\|\int_0^T f(t)dt - \int_{(k-1)T}^{kT} f(t)dt\right\| < n^{-1}n\left(\frac{\varepsilon}{2} + \frac{2A\ell}{T}\right) = \frac{\varepsilon}{2} + \frac{2A\ell}{T}.$$

If we now take two positive numbers T_1 and T_2 such that $m_1T_1 = m_2T_2$ for some natural numbers m_1 and m_2, then from the last inequality above we obtain

$$\left\|T_1^{-1}\int_0^{T_1} f(t)dt - T_2^{-1}\int_0^{T_2} f(t)dt\right\| < \varepsilon + 2A\ell(T_1^{-1} + T_2^{-1}). \quad (3.75)$$

Let us notice now that the set of pairs (T_1, T_2) for which the ratio T_1/T_2 is rational (as requested above), $T_1 > 0$, $T_2 > 0$, is dense in $(0, \infty) \times (0, \infty)$. This remark allows us to conclude that formula (3.75) holds true in general (i.e., regardless of the rationality or irrationality of the ratio T_1/T_2).

Now choose $T_1, T_2 > 4A\ell/\varepsilon$. One obtains from formula (3.75)

$$\left\|T_1^{-1}\int_0^{T_1} f(t)dt - T_2^{-1}\int_0^{T_2} f(t)dt\right\| < 2\varepsilon, \quad (3.76)$$

which proves the existence of the limit in equation (3.73) (i.e., of the mean value $M\{f\}$).

Remark 3.15. Actually, a somewhat stronger result,

$$\lim_{T\to\infty} T^{-1}\int_a^{a+T} f(t)dt = M\{f\}, \quad (3.77)$$

is valid uniformly with respect to $a \in R$. Indeed, since

$$\int_a^{a+T} f(t)dt = \int_0^T f(t+a)dt, \qquad a \in R,$$

we conclude that $M\{f(t)\}$ and $M\{f(t+a)\}$ exist simultaneously and are equal. In the proof of Theorem 3.4, we have obtained as a by-product the inequality

$$T^{-1} \left\| \int_0^T f(t)dt - \int_a^{a+T} f(t)dt \right\| \leq \frac{\varepsilon}{2} + \frac{2A\ell}{T},$$

where ℓ is an $(\varepsilon/2)$-translation number of f and $A = \sup\{\|f(t)\|;\ t \in R\}$. As we see, the right-hand side can be less than ε if $T > 4A\ell/\varepsilon = T_0(\varepsilon)$ with $T_0(\varepsilon)$ independent of $a \in R$.

Remark 3.16. The formula for $M\{f\}$ used in Section 3.3 was

$$M\{f\} = \lim_{\ell \to \infty} (2\ell)^{-1} \int_{-\ell}^{\ell} f(t)dt. \qquad (3.78)$$

Is this different from the one used above? We can show that the same $M\{f\}$ is defined by both equations (3.77) and (3.78). Indeed, if we let $a = -T$ in equation (3.77), we obtain

$$\lim_{T \to \infty} T^{-1} \int_{-T}^{0} f(t)dt = M\{f\},$$

which combined with

$$\lim_{T \to \infty} T^{-1} \int_0^T f(t)dt = M\{f\}$$

leads immediately to equation (3.78).

Remark 3.17. It is natural to ask whether the number $M\{f\}$ attached to an almost periodic function reduces to the usual mean value if f is a periodic function. It's easy to see that this is indeed the case.

Assume that $f \in BC(R, X)$ is periodic of period $\omega > 0$. Then the mean value of f is given by

$$M(f) = \omega^{-1} \int_0^{\omega} f(t)dt. \qquad (3.79)$$

Now let's estimate $M\{f\}$ according to the definition adopted in this section. One has

$$M\{f\} = \lim_{T \to \infty} T^{-1} \int_0^T f(t)dt = \lim_{T \to \infty} T^{-1} \left[\int_0^\omega + \int_\omega^{2\omega} + \cdots + \int_{k\omega}^T \right]$$

$$= \lim_{T \to \infty} T^{-1} \left[k \int_0^\omega f(t)dt + \int_{k\omega}^T f(t)dt \right]$$

$$= \lim_{k \to \infty} \left[\frac{k}{k\omega + \alpha} \int_0^\omega f(t)dt + \frac{1}{k\omega + \alpha} \int_{k\omega}^{k\omega + \alpha} f(t)dt \right],$$

where $0 \leq \alpha < \omega$. From the relations above, we obtain

$$M\{f\} = \omega^{-1} \int_0^\omega f(t)dt = M(f). \tag{3.80}$$

Therefore, the concept of the mean value for almost periodic functions is the natural extension of the usual concept of the mean value of a function on a given interval.

3.6 Almost Periodic Functions Depending on Parameters

Several applied problems, particularly related to the theories of oscillations and waves, require some concepts that can be described within the framework of almost periodic functions depending on parameters.

For instance, in oscillation theory, one deals with ordinary differential equations of the form

$$\ddot{x}_k(t) = f_k(t, x_1(t), \ldots, x_n(t)), \ k = 1, 2, \ldots, n,$$

engaging the functions $f_k(t, x_1, \ldots, x_n)$ with $(x_1, x_2, \ldots, x_n) \in \Omega \in R^n$. In this case, if almost periodicity is understood in regard to the variable $t \in R$, the variables $x_1, x_2, \ldots, x_n$ appear as parameters.

In the theory of waves, in which the basic mathematical tool is the nonhomogeneous equation

$$u_{tt} = \Delta u + f(t, x_1, \ldots, x_n)$$

with the Laplacian

$$\Delta u = u_{x_1 x_2} + u_{x_2 x_2} + \cdots + u_{x_n x_n},$$

a solution will be of the form $u = u(t, x_1, x_2, \ldots, x_n)$. This fact again leads to the concept of an almost periodic function (in t) depending on the parameters $x_1, x_2, \ldots, x_n$.

We can, of course, consider almost periodic functions in t of the form $u = u(t, \lambda)$, where λ is a parameter belonging to a certain topological space (in particular, to a metric space in which convergence is defined by means of the distance function, or metric). Instead of this rather abstract approach, we shall deal in this section with particular dependencies on parameters. More precisely, we shall present the cases where we have *uniform* dependency or where the dependency is "measured" in the *mean*. This last case is linked to the name of C. F. Muckenhoupt [73] and plays a central role in regard to the problem of almost periodicity of solutions of partial differential equations (Chapter 7); in particular, in connection with almost periodic waves (i.e., with solutions of the wave equations).

We shall start with the concept of almost periodic functions uniformly depending on parameters. These are nothing but almost periodic functions, say in the space $AP(R, X)$, where X is a function space of the form $C(\Omega, R)$, $\Omega \subset R^n$ being a compact subset. In other words, we are interested in a family of functions that can be represented in the form

$$f = f(t, x_1, x_2, \ldots, x_n), \tag{3.81}$$

with $t \in R$ and $(x_1, x_2, \ldots, x_n) \in \Omega \subset R^n$. The values of f may be in R, in C, or in multidimensional spaces R^m, C^m, $m \geq 1$.

Obviously, from equation (3.81) we realize that f can be regarded as a map from R into the space $C(\Omega, R)$, which is a Banach space with the supremum norm (always finite due to the continuity of f and the compactness of Ω).

Therefore, we are in the framework described in Section 3.5. Any result established for almost periodic functions with values in a Banach space can be transposed to the case of functions depending on parameters. In our case of uniform dependency, we may say that the parameters are in the space $C(\Omega, R)$.

To simplify the notation somewhat, we shall denote $x = (x_1, x_2, \ldots, x_n)$, which leads to the formula

$$f = f(t, x), \ t \in R, \ x \in \Omega, \tag{3.82}$$

as an equivalent for equation (3.1). Hence, the definition of almost periodic (Bohr) functions depending uniformly on the parameter x can be formulated as follows.

Definition 3.2. The continuous function $f(t, x)$ from $R \times \Omega$ to R is called almost periodic in t uniformly with respect to the parameter $x \in \Omega$ if for every $\varepsilon > 0$ there exists $\ell = \ell(\varepsilon)$ such that any interval $(a, a+\ell) \subset R$ contains a number τ with Bohr's property

$$\|f(t+\tau, x) - f(t, x)\| < \varepsilon, \qquad t \in R, \ x \in \Omega, \tag{3.83}$$

where $\| \cdot \|$ is the norm in $C(\Omega, R)$:

$$\|f(x)\| = \sup(|f(x)|; \ x \in \Omega). \tag{3.84}$$

Remark 3.18. The uniform character of the dependency with respect to parameters is assured by the fact that $\ell = \ell(\varepsilon)$ is independent of $x \in \Omega$.

Since an almost periodic function in the meaning of Definition 3.2 is an almost periodic function from R into $C(\Omega, R)$ (i.e., in a Banach space), the following properties are direct consequences of the properties established in Section 3.5.

Proposition 3.22. *If $f \in AP(R, C(\Omega, R))$, then f is bounded on $R \times \Omega$ and uniformly continuous on $R \times \Omega$.*

Indeed, from Proposition 3.16 there results the uniform continuity of $f(t, x)$ in t uniformly with respect to $x \in \Omega$, while the uniform continuity of $f(t, x)$ in x, $x \in \Omega$, for fixed $t \in R$, is the consequence of a classical result. (Each continuous function on a compact set is uniformly continuous.) Let $\varepsilon > 0$ be arbitrary and $\ell = \ell(\varepsilon/3)$ be the number corresponding to almost periodicity of f. The function $f(t, x)$ is obviously uniformly continuous on the compact set $[-1, 1+\ell] \times \Omega$. Consider the positive number $\delta(\varepsilon/3) < 1$ such that

$$|f(t, x) - f(s, y)| < \varepsilon/3$$

for $(t, x), (t, y) \in [-1, 1 + \ell] \times \Omega$ such that $|t - s| < \delta$, $|x - y| < \delta$. If τ is an $(\varepsilon/3)$-translation number in the interval $[-t, -t+\ell]$, which means $0 \le t+\tau \le \ell$ and $-1 \le s + \tau \le 1 + \ell$, then we have

$$|f(t, x) - f(s, y)| \le |f(t, x) - f(t + \tau, x)| + |f(t + \tau, x) - f(s + \tau, y)|$$
$$+ |f(s + \tau, y) - f(s, y)| < \frac{\varepsilon}{3} + \frac{\varepsilon}{3} + \frac{\varepsilon}{3} = \varepsilon$$

because in the middle term of the right-hand side we have $|t + \tau - (s + \tau)| = |t - s| < \delta$ and $|x - y| < \delta$.

This shows the validity of Proposition 3.22.

Proposition 3.23. *If $f \in AP(R, C(\Omega, R))$, then λf, with $\lambda \in R$, $f(t + h, x)$, $h \in R$, $f(at, x)$, $a \in R$, $a \ne 0$, and $\|f(t, x)\|$ are also in $AP(R, C(\Omega, R))$, resp. $AP(R, R)$.*

Proposition 3.24. *Definition 3.2 of almost periodicity uniformly with respect to $x \in \Omega$ is equivalent to the relative compactness of the family $\{f(t + h, x);$ $h \in R\}$ in the space $BC(R, C(\Omega, R))$.*

This is a consequence of Theorem 3.2 in the preceding section. The notation $BC(R, C(\Omega, R))$ stands for the space of continuous and bounded maps from R into $C(\Omega, R)$ with the supremum norm.

Proposition 3.25. *The sum of a finite number of functions in $AP(R, C(\Omega, R))$ belongs to this space.*

Proposition 3.26. *A finite number of functions in $AP(R, C(\Omega, R))$ has the property that for each $\varepsilon > 0$ there exist common ε-translation numbers.*

Proposition 3.27. *Any function $f(t,x)$, $t \in R$, $x \in \Omega$, that can be approximated in the supremum norm by trigonometric polynomials of the form*

$$T(t,x) = a_1(x)e^{i\lambda_1 t} + a_2(x)e^{i\lambda_2 t} + \cdots + a_p(x)e^{i\lambda_p t}$$

with $a_k(x)$, $k = 1, 2, \ldots, p$, in $C(\Omega, R)$ and $\lambda_k \in R$, $k = 1, 2, \ldots, p$ belongs to $AP(R, C(\Omega, R))$.

The converse of Proposition 3.27 is also true, but the proof is postponed (Chapter 4).

Proposition 3.28. *If $\{f_n(t,x);\ n \geq 1\} \subset AP(R, C(\Omega, R))$ and $f_n(t,x)$ converges uniformly on $R \times \Omega$ to $f(t,x)$, then $f(t,x) \in AP(R, C(\Omega, R))$.*

This proposition is actually a reformulation of the fact that $AP(R, C(\Omega, R))$ is a complete space.

Proposition 3.29. *Let $f \in AP(R, C(\Omega, R))$ be such that $f_t = \partial f / \partial t$. If $f_t(t,x)$ is uniformly continuous in t uniformly with respect to $x \in \Omega$, then $f_t \in AP(R, C(\Omega, R))$.*

The requirement in Proposition 3.29 means that for each $\varepsilon > 0$ there exists $\delta(\varepsilon) > 0$ such that $|t - s| < \delta$ implies $|f_t(t,x) - f_t(s,x)| < \varepsilon$ for all $x \in \Omega$.

Proposition 3.30. *Let $f \in AP(R, C(\Omega, R))$ and $x = x(t) \in AP(R, R^n)$ such that $x(t) \in \Omega$, $t \in R$. Then $f(t, x(t)) \in AP(R, R)$.*

Proof. We shall prove that $f(t, x(t)) \in AP(R, R)$ using Proposition 3.24 above. In other words, if $\{h_k;\ k \geq 1\} \subset R$, then there exists a subsequence $\{h_{1k};\ k \geq 1\}$ with the property that $f(t + h_{1k}, x(t + h_{1k}))$ converges uniformly on R. This will suffice to conclude that $f(t, x(t)) \in AP(R, R)$.

Indeed, we know that $\{h_k;\ k \geq 1\}$ contains a subsequence $\{h_{1k};\ k \geq 1\}$ such that the sequence $\{x(t + h_{1k});\ k \geq 1\}$ is uniformly convergent on R. Without loss of generality, we can state that $\{f(t + h_{1k}, x);\ k \geq 1\}$ is also uniformly convergent on R uniformly with respect to $x \in \Omega$. If $\varepsilon > 0$, then there exist $\delta(\varepsilon) > 0$ with the property

$$|f(t,x) - f(t,y)| < \frac{\varepsilon}{2}, \qquad t \in R,$$

provided $|x - y| < \delta$, $x, y \in \Omega$. On the other hand, we can write

$$|f(t + h_{1k}, x) - f(t + h_{1p}, x)| < \frac{\varepsilon}{2}, \qquad x \in \Omega,$$

for $k, p \geq N_1(\varepsilon)$, $t \in R$, and

$$|x(t + h_{1k}) - x(t + h_{1p})| < \delta(\varepsilon)$$

for $k, p \geq N_2(\varepsilon)$. Denote $N = \max(N_1, N_2)$, and combining the inequalities above, we obtain

$$|f(t + h_{1k}, x(t + h_{1k})) - f(t + h_{1p}, x(t + h_{1p}))|$$
$$\leq |f(t + h_{1k}, x(t + h_{1k})) - f(t + h_{1k}, x(t + h_{1p}))|$$
$$+|f(t + h_{1k}, x(t + h_{1p})) - f(t + h_{1p}, x(t + h_{1p}))|$$
$$< \frac{\varepsilon}{2} + \frac{\varepsilon}{2} = \varepsilon, \qquad \text{for } k, p \geq N(\varepsilon).$$

This ends the proof of Proposition 3.30.

Remark 3.19. Proposition 3.30 can be reformulated as follows.

Niemytskii's operator

$$x(t) \longrightarrow f(t, x(t)), \tag{3.85}$$

under conditions specified above, is acting from the space $\mathrm{AP}(R, R^n)$ *into* $\mathrm{AP}(R, R)$.

Other types of operators with this property will be considered in subsequent chapters.

We shall now investigate another type of dependency of parameters, in which the supremum norm encountered above (in parameter space) is substituted by integral norms (basically L^p-norms).

Definition 3.3. Let $\mathrm{AP}(R, L^p(\Omega, \mathcal{C}))$, $1 \leq p < \infty$, $\Omega \subset R^m$ be a measurable set, the space of almost periodic functions from R into the (Banach) space $L^p(\Omega, \mathcal{C})$. Then $f = f(t, x)$, $t \in R$, $x \in \Omega$, is called *almost periodic in the mean of order* $p \geq 1$ if $f \in \mathrm{AP}(R, L^p(\Omega, \mathcal{C}))$.

The definition above means the following properties are satisfied by $f \in \mathrm{AP}(R, L^p(\Omega, \mathcal{C}))$:

(1) For each $\varepsilon > 0$ and $t_0 \in R$, there exists $\delta = \delta(\varepsilon, t_0) > 0$ such that

$$|t - t_0| < \delta \text{ implies } \int_\Omega |f(t, x) - f(t_0, x)|^p dx < \varepsilon^p.$$

(2) For each $\varepsilon > 0$, there exists $\ell = \ell(\varepsilon) > 0$ with the property that any interval $(a, a + \ell) \in R$ contains a point τ for which

$$\int_\Omega |f(t + \tau, x) - f(t, x)|^p dx < \varepsilon^p, \qquad t \in R.$$

The integrals above are understood in Lebesgue's sense, and dx stands for Lebesgue's measure.

The properties established in Section 3.5 for the space $\mathrm{AP}(R, X)$ apply in the case $X = L^p(\Omega, \mathcal{C})$, $1 \leq p < \infty$, and the following properties of the almost periodic functions, depending on the mean of the parameters, can be stated as follows.

Proposition 3.31. *Let* $f = f(t, x) \in \text{AP}(R, L^p(\Omega, \mathcal{C}))$, $1 \le p < \infty$. *Then* f *is bounded on the mean of order* p; *i.e., there exists* $M > 0$ *such that*

$$\int_\Omega |f(t, x)|^p dx \le M, \qquad t \in R,$$

and it is uniformly continuous in the sense that, for each $\varepsilon > 0$, *there exists* $\delta = \delta(\varepsilon) > 0$ *with the property*

$$|t - s| < \delta \text{ implies } \int_\Omega |f(t, x) - f(s, x)|^p dx < \varepsilon^p.$$

Proposition 3.31 is the consequence of Proposition 3.16.

Proposition 3.32. *If* $f = f(t, x) \in \text{AP}(R, L^p(R, \mathcal{C}))$, $1 \le p < \infty$, *then any sequence* $\{f(t + h_k); \ k \ge 1\}$ *contains a subsequence* $\{f(t + h_{1k}); \ k \ge 1\}$ *that converges uniformly with respect to* $t \in R$ *in the space* $L^p(\Omega, \mathcal{C})$ *as* $k \to \infty$.

Proposition 3.32 has the following meaning: For each $\varepsilon > 0$, there exists $N = N(\varepsilon)$ such that

$$\int_\Omega |f(t + h_{1k}, x) - f(t + h_{1p}, x)|^p dx < \varepsilon^p, \qquad t \in R,$$

as soon as $k, p \ge N(\varepsilon)$.

Proposition 3.33. *The converse of Proposition 3.32 is true. This means that Definition 3.3 is equivalent to the relative compactness of the family of translates of* f *in the sense made precise in Proposition 3.32.*

Other properties similar to those stated in Propositions 3.18–3.21 above can be transposed in the case of almost periodic functions in $\text{AP}(R, L^p(\Omega, \mathcal{C}))$, $1 \le p < \infty$.

The concept of almost periodic functions depending on the mean with respect to the parameters is very significant in connection with the almost periodicity of solutions of hyperbolic (waves) or parabolic (heat transfer) partial differential equations. These problems will be investigated in subsequent chapters.

Let us mention the fact that the *mean value* does exist for the almost periodic functions depending on the mean with respect to the parameters. From Theorem 3.1, there results the existence of the limit

$$\lim_{T \to \infty} T^{-1} \int_0^T f(t, x) ds \qquad (3.86)$$

in the sense of convergence in $L^p(\Omega, \mathcal{C})$, $1 \le p < \infty$.

In concluding this section, we shall prove that Stepanov's almost periodic functions can be regarded as almost periodic functions in Bohr's sense with values in a Banach space.

We shall limit our considerations to the space $S = S^1(R, C)$, the case of the space $S^p(R, C)$, $1 < p < \infty$, being the same. Moreover, the space $S(R, C)$ is the richest among Stepanov's spaces.

Let $f = f(t)$, $t \in R$, a map from R into C defined for almost $t \in R$ and locally integrable (i.e., Lebesgue integrable on any bounded interval of R). To the function f we associate the function $\varphi(t, x) = f(t+x)$, $t \in R$, $x \in [0, 1] \subset R$. It is obvious that, for almost all $t \in R$, the map $t \to \varphi(t, x)$, $0 \le x \le 1$, is from R into $L^1 = L([0, 1], C)$ due to the local integrability of f.

What we want to prove now is the fact that the almost periodicity in the mean (of order 1) of $\varphi(t, x)$ with $\Omega = [0, 1]$ is actually the Stepanov type of almost periodicity for f. In other words, if $\varphi \in AP(R, L([0, 1], C))$, then $f \in S(R, C)$.

Let us write the fact that $\varphi \in AP(R, L([0, 1], C))$; i.e., for each $\varepsilon > 0$, one can find $\ell = \ell(\varepsilon)$ such that any interval $(a, a + \ell) \subset R$ contains a number τ that is an ε-translation number for φ. This leads to the formula

$$\int_0^1 |\varphi(t + \tau, x) - \varphi(t, x)| dx < \varepsilon, \qquad t \in R.$$

In terms of f, the inequality above means

$$\int_0^1 |f(t + \tau + x) - f(t + x)| dx < \varepsilon, \qquad t \in R.$$

By the change of variable $t + x = u$, the inequality above becomes

$$\int_t^{t+1} |f(u + \tau) - f(u)| du < \varepsilon, \qquad t \in R.$$

By taking the supremum with respect to t, the last inequality implies

$$|f(t + \tau) - f(t)|_S \le \varepsilon,$$

which proves that $f \in S(R, C)$.

Conversely, if $f \in S(R, C)$, one easily gets $\varphi \in AP(R, L([0, 1], C))$. The continuity of φ, required in the definition of almost periodicity, follows from the well-known property of $L([0, 1], C)$

$$\lim \int_0^1 |f(t + h) - f(t)| dt = 0 \quad \text{as } h \to 0.$$

3.7 Variations on the Theme of Almost Periodicity

Besides the spaces of almost periodic functions discussed so far in this book, namely AP_1, AP, S, $B^2 = AP_2$, and B, there are in the literature many more types and generalizations. It is our aim, in this last section of Chapter 3, to present a sort of inventory of such extensions of the concept of almost periodic functions, without paying full attention to the details. Most of these variations on the concept of almost periodicity are motivated by the applications of these types of functions in the investigation of applied problems in such fields as mechanics, electricity theory, electronics, stochastic processes, statistics and probability, discrete processes, and others. The full list of references would be beyond the framework of this book. Adequate references will be provided for the topics presented below.

The first class of almost periodic functions we shall briefly discuss here is that introduced by H. Weyl [100]. These functions are also present in the book by A. S. Besicovitch [10], where the definition relies on the completion of the space T of trigonometric polynomials with respect to a norm defined in the following manner. We shall limit ourselves to the case $p = 1$, the case of a general p, $1 \leq p < \infty$, being rather straightforward.

Let $f \in L_{\mathrm{loc}}(R, C)$ be such that, for $\ell > 0$,

$$\sup_{t \in R} \left\{ \ell^{-1} \int_t^{t+\ell} |f(s)| ds \right\} < \infty. \tag{3.87}$$

It can be easily seen that if formula (3.87) is true for a single $\ell > 0$, then it is true for any $\ell > 0$. Actually, formula (3.87) expresses the fact that $f \in M(R, C)$, and the supremum defines a norm that is equivalent to the norm $| \cdot |_M$ defined by formula (3.27) in Section 3.3.

Weyl had the idea to take the limit as $\ell \to \infty$ in formula (3.87):

$$|f|_W = \limsup_{\ell \to \infty} \left\{ \sup_{t \in R} \left(\ell^{-1} \int_t^{t+\ell} |f(s)| ds \right) \right\}. \tag{3.88}$$

For those $f \in L_{\mathrm{loc}}(R, C)$, for which the lim sup in equation (3.88) is *finite*, it can be shown (see A. S. Besicovitch [10]) that the limit (not only the lim sup!) does exist on the right-hand side of equation (3.88); i.e., equation (3.88) is the same as

$$|f|_W = \lim_{\ell \to \infty} \left\{ \sup_{t \in R} \left(\ell^{-1} \int_t^{t+\ell} |f(s)| ds \right) \right\}. \tag{3.89}$$

The linear space W consisting of these $f \in L_{\mathrm{loc}}(R, C)$ for which $|f|_W$ defined by equation (3.89) is finite becomes a seminormed space with the seminorm $f \longrightarrow |f|_W$.

Indeed, $f \longrightarrow |f|_W$ is a seminorm because $|f|_W = 0$ does not imply $f = 0$ a.e. For instance, if $f \in L^1(R, C)$, then $|f|_W = 0$.

As we have done in the cases B^2 or B, we should consider the factor space W/W_0, where $W_0 = W_0(R, C)$ is defined by $W_0 = \{f; \ f \in W, \ |f|_W = 0\}$.

Definition 3.4. The space $\mathrm{APW}(R, C)$ of *almost periodic functions in Weyl's sense* is the closure in W/W_0 of the set $T = T(R, C)$ of trigonometric polynomials.

One may rephrase the definition above as follows: $f \in \text{APW}(R,\mathcal{C})$ if for each $\varepsilon > 0$ there exists $T_\varepsilon \in T$ such that $f = T_\varepsilon + r_\varepsilon$ with $|r_\varepsilon|_W < \varepsilon$.

The following inclusions are valid:

$$S(R,\mathcal{C}) \subset \text{APW}(R,\mathcal{C}) \subset B(R,\mathcal{C}). \qquad (3.90)$$

Formula (3.90) shows that the almost periodic functions in Weyl's sense are more general than Stepanov's almost periodic functions, but they constitute special cases of the Besicovitch type of almost periodic functions.

In order to prove the first inclusion (3.90), one proceeds in the following manner. Let $\ell = k + \theta_k$ with k positive integer and $0 \le \theta_k < 1$. Then $\ell \to \infty$ if and only if $k \to \infty$. From equation (3.89), we see that for $f \in M$ one has

$$\ell^{-1} \int_t^{t+\ell} |f(s)|ds = (k + \theta_k)^{-1} \left\{ \int_t^{t+1} |f(s)|ds + \int_{t+1}^{t+2} |f(s)|ds \right.$$
$$\left. + \cdots + \int_{t+k-1}^{t+k} |f(s)|ds + \int_{t+k}^{t+k+\theta_k} |f(s)|ds \right\} \le (k+\theta_k)^{-1}(k+1)|f|_M,$$

which implies

$$\sup_{t \in R} \ell^{-1} \int_t^{t+\ell} |f(s)|ds < k^{-1}(k+1)|f|_M.$$

Letting $\ell \to \infty$ in the inequality above, which implies $k \to \infty$, one obtains

$$|f|_W \le |f|_M, \qquad (3.91)$$

and since, for $f \in S(R,\mathcal{C})$, $|f|_S = |f|_M$, there results, from inequality (3.91), $S(R,\mathcal{C}) \subset \text{APW}(R,\mathcal{C})$.

The second inclusion in formula (3.90) follows immediately from the inequality

$$\ell^{-1} \int_0^\ell |f(s)|ds \le \sup_{t \in R} \left\{ \ell^{-1} \int_t^{t+\ell} |f(s)|ds \right\},$$

which leads, as $\ell \to \infty$, to

$$|f|_B \le |f|_W. \qquad (3.92)$$

More properties of functions in $\text{APW}(R,\mathcal{C})$ can be obtained by using the same kinds of arguments as in Sections 3.3 and 3.4.

Another class of functions that we shall mention here was introduced relatively recently by C. Zhang [108] under the name of *pseudo almost periodic functions*.

Definition 3.5. A function $f \in \text{BC}(R,\mathcal{C})$ is called pseudo almost periodic if it can be represented as

$$f(t) = g(t) + r(t), \qquad t \in R, \qquad (3.93)$$

with $g \in \mathrm{AP}(R,\mathcal{C})$ and r such that

$$\lim_{\ell \to \infty} (2\ell)^{-1} \int_{-\ell}^{\ell} |r(t)| dt = 0. \tag{3.94}$$

Remark 3.20. We notice that $r = f - g \in \mathrm{BC}(R,\mathcal{C})$. One can also reformulate the definition above by saying that $f \in \mathrm{PAP}(R,\mathcal{C}) = $ the class of pseudo almost periodic functions if f can be represented in the form (3.93), with $g \in \mathrm{AP}(R,\mathcal{C})$ and $r \in \mathrm{BC}(R,\mathcal{C}) \cap B_0(R,\mathcal{C})$, where $B_0(R,\mathcal{C})$ denotes the zero linear manifold of $B(R,\mathcal{C})$ as defined by relation (3.64) in Section 3.4.

Remark 3.21. It is obvious from Definition 3.5 that $\mathrm{PAP}(R,\mathcal{C}) \subset B(R,\mathcal{C})$. In other words, any pseudo almost periodic function is also a Besicovitch almost periodic function. Since $\mathrm{PAP}(R,\mathcal{C})$ contains only continuous functions, this space is considerably "smaller" than the space B/B_0 but richer than the space $\mathrm{AP}(R,\mathcal{C})$. The mean value does exist for any $f \in \mathrm{PAP}(R,\mathcal{C})$, and if we take into account equation (3.93), then obviously $M\{f\} = M\{g\}$.

The pseudo almost periodic functions were thoroughly investigated in a series of papers by C. Zhang [108], [109]. The last paper quoted contains further references, as well as indications concerning applications to the theory of differential equations.

It is obvious how the concept of pseudo almost periodicity can be extended to deal with spaces like $\mathrm{PAP}(R, R^n)$ or even $\mathrm{PAP}(R, X)$, where X stands for a Banach space. A characterization of functions in $\mathrm{PAP}(R,\mathcal{C})$ will be provided in Chapter 4 in terms of Fourier series.

In the category of almost periodic functions depending on parameters, we will briefly deal with *random functions almost periodic in probability*. This concept was introduced by O. Onicescu and V. Istrățescu [75] and investigated by many authors. Our book *Almost Periodic Functions* [23] contains a concise presentation of the basic facts related to this concept, as well as adequate references (see also A. Precupanu [83]).

If we denote by (Ω, K, P) a probability space, then a map $f : \Omega \to \mathcal{C}$ is called a *random variable* if the inverse image of any Borel set in $\mathcal{C}$ belongs to K. In other words, the map f is measurable with respect to the probability P.

A *random function* is a map $x : S \times \Omega \longrightarrow \mathcal{C}$ such that for every $s \in S = $ a topological space, $x(s;\omega)$ is a random variable. We shall have in mind the case $S = R$. In this case, one speaks about a *random process*.

The *continuity in probability* is defined as follows. Let $x(t;\omega)$ be a random function defined for all t in a neighborhood of $t_0 \in R$. Then $x(t;\omega)$ is called continuous in probability at t_0 if, for any $\varepsilon > 0$, $\eta > 0$, there corresponds $\delta(\varepsilon, \eta) > 0$ such that $P(\omega; |x(t;\omega) - x(t_0;\omega)| \geq \varepsilon) < \eta$, provided $|t - t_0| < \delta(\varepsilon, \eta)$.

Definition 3.6. A random function $x(t;\omega)$ continuous on R in probability is called almost periodic in probability if to any $\varepsilon > 0$, $\eta > 0$, there corresponds a number $\ell = \ell(\varepsilon, \eta) > 0$ such that any interval of length ℓ on the real axis contains a number τ with the property

$$P(\omega; |x(t + \tau; \omega) - x(t; \omega)| \geq \varepsilon) < \eta, \qquad t \in R.$$

Definition 3.6 is similar to Bohr's definition. Other classes of almost periodicity for random functions can be tailored in accordance with the definitions of other types of almost periodicity.

Most of the basic properties encountered in the spaces $AP(R, \mathcal{C})$ or $AP(R, R)$ can be formulated accordingly for random functions almost periodic in probability. For instance, a necessary and sufficient condition for a random function $x(t; \omega)$ to be almost periodic in probability is the relative compactness of its set of translations $\{x(t + h; \omega); h \in R\}$ in the sense of uniform convergence in probability on the real axis. This means that from any sequence $\{x(t + h_n; \omega); n \geq 1\}$ one can extract a subsequence $\{x(t + h_{1n}; \omega); n \geq 1\}$ such that for any $\varepsilon > 0$, $\eta > 0$, one can find $N = N(\varepsilon, \eta)$ with the property

$$P(\omega; |x(t + h_{1n}; \omega) - x(t + h_{1m}; \omega)| \geq \varepsilon) < \eta, \qquad t \in R,$$

provided $n, m \geq N(\varepsilon, \eta)$.

A *random trigonometric polynomial* is a random function or process of the form

$$T(t; \omega) = \sum_{k=1}^{m} c_k(\omega) e^{i\lambda_k t}, \qquad t \in R,$$

with $c_k(\omega)$, $h = 1, 2, \ldots, m$, random variables.

The usual approximation theorem for almost periodic functions is valid in the following formulation: Any random function $x(t; \omega)$ almost periodic in probability can be approximated by random trigonometric polynomials with any degree of accuracy. This means that, for each $\varepsilon > 0$, $\eta > 0$, there exists a random trigonometric polynomial $T(t; \omega)$ such that

$$P(\omega; |x(t; \omega) - T(t; \omega)| \geq \varepsilon) < \eta, \qquad t \in R.$$

Let us conclude this brief consideration of random functions almost periodic in probability with the remark that they can be viewed as almost periodic functions (Bohr) with values in a space of random variables.

The next generalization of almost periodicity we shall consider relies on the idea of perturbation. The *asymptotic almost periodic functions* were introduced by M. Fréchet [41], [42]. They have been investigated by many authors and have found interesting applications in the theory of differential equations. The following simple example will illustrate the natural occurrence of these functions. If we consider the ordinary differential equation $\dot{x}(t) = -x(t) + f(t)$ with $f \in AP(R, R)$, then the general solution is given by the formula

$$x(t) = e^{-t} \left[c + \int_0^t e^s f(s) ds \right]. \qquad (3.95)$$

If we now choose the constant c as

$$c = \int_{-\infty}^{0} e^{s} f(s) ds, \tag{3.96}$$

we find from equation (3.95) the solution

$$x_0(t) = \int_{-\infty}^{t} e^{-(t-s)} f(s) ds, \tag{3.97}$$

which is in $AP(R, R)$. Indeed, equation (3.97) can be rewritten as the convolution product $x_0(t) = (K * f)(t)$, where $K(t) = 0$ for $t < 0$ and $K(t) = e^{-t}$ for $t \geq 0$. Since $K \subset L^1(R, R)$, there results from Proposition 3.8 that $x_0 \in AP(R, R)$. Returning to equation (3.95), we can write the general solution in the form

$$x(t) = Ae^{-t} + x_0(t), \qquad t \in R. \tag{3.98}$$

This formula is naturally conducive to the following definition (Fréchet).

Definition 3.7. A continuous function $x(t)$ on R is called *asymptotically almost periodic* if it can be represented in the form

$$x(t) = f(t) + r(t), \qquad t \in R, \tag{3.99}$$

with $f \in AP(R, R)$ and $r(t)$ continuous on R, and

$$\lim_{t \to \infty} r(t) = 0. \tag{3.100}$$

We denote by $AAP(R, R)$ the space of asymptotically almost periodic functions.

Remark 3.22. Similar definitions can be formulated for spaces like $AAP(R, \mathcal{C})$, $AAP(R, R^n)$, or $AAP(R, X)$ with X a Banach space.

Remark 3.23. The representation (3.99) is unique. If we admit that $f_1(t) + r_1(t) = f_2(t) + r_2(t)$, $t \in R$, with $f_1, f_2 \in AP(R, R)$ and $r_1(t), r_2(t)$ continuous and vanishing at ∞, then from $f_1 - f_2 = r_2 - r_1$ we find that $|f_1(t) - f_2(t)| \longrightarrow 0$ as $t \longrightarrow \infty$. This implies $M\{|f_1(t) - f_2(t)|\} = 0$, which means $f_1(t) \equiv f_2(t)$ on R.

Remark 3.24. The space $AAP(R, R)$ is obviously a linear space, with the usual addition and scalar multiplication, and can be organized as a Fréchet space using the topology given by the seminorms

$$|x|_n = \sup\{|x(t)|;\ t \in [-n, \infty)\}, \qquad n \geq 0. \tag{3.101}$$

As we know, the family of seminorms $\{|x|_n;\ n \geq 0\}$ generates an invariant metric (see Proposition 2.12 of Chapter 2) on $AAP(R, R)$. It is worth mentioning the fact that $AAP(R, R)$ can be organized as an algebra with respect to the usual operations of pointwise addition, multiplication, and scalar multiplication. It can be seen that these operations are continuous in the topology defined by the seminorms $\{|x|_n;\ n \geq 0\}$.

Other important properties of the asymptotically almost periodic functions are the existence of the mean value

$$\lim_{\ell \to \infty} \left(\ell^{-1} \int_a^{a+\ell} f(t)dt \right),$$

the relative compactness of the family of translates $\{f(t+h);\ h > 0\}$, and that the product of a function in $\mathrm{AAP}(R, R)$ by a function in $\mathrm{AP}(R, R)$ is in $\mathrm{AAP}(R, R)$.

We note that similar generalizations of almost periodicity could be formulated for other types of almost periodic functions. For instance, if $x \in S(R, R)$, one may consider functions that can be represented in the form

$$y(t) = x(t) + r(t), \qquad t \in R, \tag{3.102}$$

with $r \in M_0(R, R)$; i.e., such that

$$\lim_{t \to \infty} \int_t^{t+1} |r(s)|ds = 0. \tag{3.103}$$

For this class or space of functions, which seems to be naturally denoted by $\mathrm{AS}(R, R)$, we do not find too much information in the existing literature. Their investigation appears to be an open subject.

The last concept related to almost periodicity we shall discuss now is that of an *almost periodic sequence* or equivalently, the *almost periodic functions of an integer variable*: $x_n = f(n)$, $n \in Z$, with $x_n \in R$, $n \in Z$. This concept appeared early in the theory of almost periodic functions (see A. Walther [98]). It has gained significance mainly due to the increased use of discrete models in various applied areas.

Definition 3.8. The map $f : Z \to R$ is called *almost periodic* if to any $\varepsilon > 0$ there corresponds a positive integer $N(\varepsilon)$ with the property that in any set of N consecutive integers there exists an integer p such that

$$|f(n+p) - f(n)| < \varepsilon, \qquad n \in Z. \tag{3.104}$$

The number p is called an ε-translation number of f.

Proposition 3.34. *A necessary and sufficient condition for a function $f : Z \to R$ to be almost periodic in the sense of Definition 3.8 is that any sequence of translates $\{f(n + m_k);\ k \geq 1\}$ contain a subsequence $\{f(n + m_{1k});\ k \geq 1\}$ that is uniformly (with respect to $n \in Z$) convergent.*

The proof is given in our book [23]. If we define a norm in the set of almost periodic sequences, say $\|f\| = \sup\{|f(n)|;\ n \in Z\}$, then we get the space $\mathrm{AP}(Z, R)$. The condition in Proposition 3.34 is then the relative compactness of the family of translates, a feature we have encountered in other types of almost periodicity.

Another interesting property, characteristic for almost periodic elements in $\mathrm{AP}(Z, R)$, can be stated as follows.

Proposition 3.35. *A necessary and sufficient condition for $f \in \mathrm{AP}(Z, R)$ is the existence of a function $\varphi \in \mathrm{AP}(R, R)$ such that $f(n) = \varphi(n)$, $n \in Z$.*

Proof. If $f \in \mathrm{AP}(Z, R)$ is given, then $\varphi(t) \in \mathrm{AP}(R, R)$ can be constructed by letting $\varphi(t) = f(n) + (t - n)[f(n + 1) - f(n)]$, $t \in [n, n + 1)$, $n \in Z$. The remaining part of the proof can be carried out without difficulty.

Obviously, the graph of the function $\varphi(t)$ is obtained by successively joining the points $(n, f(n)) \subset R^2$ by straight line segments.

The mean value for functions in $\mathrm{AP}(Z, R)$ is defined by the formula

$$M(f) = \lim_{k \to \infty} \frac{f(n + 1) + f(n + 2) + \cdots + f(n + k)}{k}. \tag{3.105}$$

Proposition 3.36. *For each $f \in \mathrm{AP}(Z, R)$, the mean value defined by equation (3.105) exists uniformly with respect to $n \in Z$.*

Proof. Consider the function $\varphi \in \mathrm{AP}(R, R)$ associated with f as shown above. Then

$$M\{\varphi\} = \lim_{k \to \infty} \left(k^{-1} \int_n^{n+k} \varphi(t) dt \right) \tag{3.106}$$

exists uniformly for $n \in Z$. But

$$k^{-1} \int_n^{n+k} \varphi(t) dt = \frac{f(n + 1) + f(n + 2) + \cdots + f(n + k)}{k} + \frac{f(n) - f(n + k)}{2k}.$$

The relationship above tells us that the limit in equation (3.105) does exist because $[f(n) - f(n + k)](2k)^{-1} \longrightarrow 0$ as $k \longrightarrow \infty$ due to the fact that $|f(n)|$, $n \in Z$, is bounded. Moreover, $M\{\varphi\} = M(f)$.

Proposition 3.37. *If $f \in \mathrm{AP}(Z, R)$ and $F(n)$, $n \in Z$ is defined by $F(0)$ arbitrary in R, $F(n + 1) - F(n) = f(n)$, $n \in Z$, then $F \in \mathrm{AP}(Z, R)$ if and only if it is bounded in Z.*

Proof. Let $\varphi \in \mathrm{AP}(R, R)$ be defined as above, and note that

$$\int_0^t \varphi(s) ds = F([t] + 1) - \int_{[t]}^t \varphi(s) ds - \frac{1}{2} [f(0) + f([t])].$$

The last two terms on the right-hand side are bounded, and this means that $F(n)$ and $\int_0^t \varphi(s) ds$ are simultaneously bounded or unbounded on Z, resp. R. If $F(n)$, $n \in Z$, are bounded, then $\int_0^t \varphi(s) ds$ is bounded, which means it is in $\mathrm{AP}(R, R)$. This implies $F \in \mathrm{AP}(Z, R)$. The converse is obviously true. We rely here on Theorem 5.1, which will be proved in Chapter 5. Besides the space $\mathrm{AP}(Z, \mathcal{C})$, there are other spaces of almost periodic functions on Z similar to those encountered in the case of a continuous variable $t \in R$.

We shall briefly deal with the space $B^2(Z,\mathcal{C})$, which corresponds to the Besicovitch space $B^2(R,\mathcal{C})$. The simplest way to introduce the space $B^2(Z,\mathcal{C})$ is by the completion of the Banach space $AP(Z,\mathcal{C})$ with respect to the norm

$$\|f\| = \{M\{|f|^2\}\}^{1/2} \tag{3.107}$$

with $M\{f\}$ defined by equation (3.105).

In order to prove that equation (3.107) makes sense for any $f \in AP(Z,\mathcal{C})$, it suffices to restrict our consideration to the case of the space $AP(Z,R)$.

If $f \in AP(Z,R)$, one has

$$\begin{aligned}
\left|f^2(n+p) - f^2(n)\right| &= \left|[f(n+p) - f(n)][f(n+p) + f(n)]\right| \\
&\leq 2A|f(n+p) - f(n)|
\end{aligned} \tag{3.108}$$

with $A = \sup\{|f(n)|;\ n \in Z\}$. A is finite, a fact resulting, for instance, from Proposition 3.35 and the boundedness of almost periodic functions in $AP(R,R)$. The inequality (3.108) shows that $f \in AP(Z,R)$ implies $f^2 \in AP(Z,R)$, from which the existence of the mean value follows (see Proposition 3.36). Therefore, the map

$$f \longrightarrow \|f\| = \{M\{|f|^2\}\}^{1/2} \tag{3.109}$$

of $AP(Z,\mathcal{C})$ into $\mathbb{R}_+$ is defined, and it's an elementary fact to check the conditions for a seminorm. Moreover, we can easily check that it is a norm. What's to be shown is that $M\{f\} = 0$, $f \in AP(Z,R)$, $f \geq 0$ on Z imply $f(n) = 0$ for $n \in Z$. Indeed, if $\varphi \in AP(R,R)$ and $f \in AP(Z,R)$ are related as shown in Proposition 3.35, then $M\{f\} = M\{\varphi\}$ according to Proposition 3.36. If we assume $M\{f\} = 0$, $f \geq 0$, then we obtain $M\{\varphi\} = 0$. Obviously, $\varphi \geq 0$ on R, and from Proposition 3.15 we obtain $\varphi \equiv 0$. This implies $f(n) = 0$, $n \in Z$. The fact that the map (3.109) is a norm on $AP(Z,\mathcal{C})$ enables us to conclude that the completion of $AP(Z,\mathcal{C})$ is also a Banach space, which we denote by $B^2(Z,\mathcal{C})$. And this space is the *discrete Besicovitch space*.

Remark 3.25. We can prove that the space $B^2(Z,\mathcal{C})$ is a Hilbert space. We invite the reader to show that the inner product can be defined by $\langle f, g \rangle = M\{f\bar{g}\}$, $f, g \in B^2(Z,\mathcal{C})$. First, note that for $f, g \in AP(Z,\mathcal{C})$ one has $f\bar{g} \in AP(Z,\mathcal{C})$. The inner product is extended to $B^2(Z,\mathcal{C})$ by continuity.

Remark 3.26. If we do not assume $f \in AP(Z,\mathcal{C})$, then $M\{f\} = 0$ and $f \geq 0$ do not imply $f(n) = 0$, $n \in Z$.

Indeed, if $\{f(n);\ n \in Z\}$ is in $\ell^1(Z,\mathcal{C})$ (i.e., $\sum_{n \in Z}|f(n)| < \infty$), then obviously $M\{f\} = 0$, while f may not be zero at any $n \in Z$. A space of almost periodic functions richer than $B^2(Z,\mathcal{C})$ is the Besicovitch space $B^1(Z,\mathcal{C})$, or simply $B(Z,\mathcal{C})$. The norm with respect to which we need to complete $AP(Z,\mathcal{C})$, in order to obtain $B(Z,\mathcal{C})$ is $f \longrightarrow M\{|f|\}$, $f \in B(Z,\mathcal{C})$.

Another type of almost periodicity, known as *weak almost periodicity*, will be defined and discussed in Chapter 5 when we investigate the conditions assuring the almost periodicity of the primitive of an almost periodic function.

4

Fourier Analysis of Almost Periodic Functions

4.1 General Remarks

The Fourier analysis implies the study of such concepts as Fourier series (which constitutes a special case of the general concept of trigonometric series), Fourier transforms, Fourier integrals, and other related topics. It is well known that these concepts and related methods play a central role in investigating such phenomena as oscillations and waves. They are also relevant in systems theory and other branches of applied science. Fourier analysis also has been for almost two centuries one of the most powerful engines in advancing real analysis, approximation theory, and other fields.

The original concept of a Fourier series preoccupied not only Fourier and his contemporaries but many other mathematicians and scientists, the name of Euler being one of the first to be mentioned.

Fourier series can be briefly described by means of the formula

$$f(t) \sim \frac{a_0}{2} + \sum_{k=1}^{\infty}(a_k \cos kt + b_k \sin kt), \tag{4.1}$$

where $a_k, b_k, \ k \geq 0$, are given by the formulas

$$a_k = \frac{1}{\pi}\int_{-\pi}^{\pi} f(t) \cos kt \, dt, \qquad b_k = \frac{1}{\pi}\int_{-\pi}^{\pi} f(t) \sin kt \, dt. \tag{4.2}$$

The series in formula (4.1) is called the *Fourier series* of the function $f(t)$. Generally, $f(t)$ is assumed to be periodic of period 2π and integrable on $[-\pi, \pi]$, which means that formulas (4.2) make sense. Formulas (4.2) are known as Euler's formulas.

The $\sim$ sign marks the association between the function $f(t)$ and the series on the right-hand side of formula (4.1). In some circumstances, the $\sim$ sign can be replaced by the $=$ sign. Criteria of this nature abound in the literature, and we shall not dwell here on the classical theory of Fourier series. Excellent

C. Corduneanu, *Almost Periodic Oscillations and Waves*,
DOI 10.1007/978-0-387-09819-7_4, © Springer Science+Business Media, LLC 2009

presentations of this theory can be found in the books by R. E. Edwards [36] and N. K. Bary [6].

It is useful to mention here the complex version of formula (4.1),

$$f(t) \sim \sum_{k=0}^{\infty} a_k e^{ikt}, \qquad t \in R, \tag{4.3}$$

with $a_k \in \mathcal{C}$.

A more general case than formula (4.3) was considered in Section 3.1, where instead of the integers k at the exponents, we have *arbitrary real numbers* λ_k; i.e.,

$$f(t) \sim \sum_{k=0}^{\infty} a_k e^{i\lambda_k t}, \qquad t \in R. \tag{4.4}$$

If the series on the right-hand side of formula (4.4) is uniformly convergent (in particular, if it is absolutely convergent), then the $\sim$ sign in formula (4.4) can be replaced by the $=$ sign, and the connection between $f(t)$ and the coefficient a_k is given by

$$a_k = \lim_{\ell \to \infty} (2\ell)^{-1} \int_{-\ell}^{\ell} f(t) e^{-i\lambda_k t} dt = M\{f(t) e^{-i\lambda_k t}\}, \qquad k \geq 0. \tag{4.5}$$

The case of absolute convergence of formula (4.4) leads to the space $AP_1(R, \mathcal{C})$ we investigated in Section 3.1.

What we shall attempt to do in this chapter is to show how a "Fourier series" can be attached to any almost periodic functionregardless of the particular type and what kind of connection exists between the almost periodic function and its Fourier series. A very deep connection does exist between the function and its Fourier series, which will help us to reconstruct the function starting from its Fourier series. We have in mind, of course, the general type of series given by formula (4.4) and not only the classical type represented by formula (4.1). Nevertheless, the "model" offered by the classical theory of Fourier series plays an important role in developing the concept of Fourier series for almost periodic functions. Particular attention will be paid to the space $AP(R, \mathcal{C})$, which is more often encountered in the literature and likely plays a greater role in applications. Similar considerations can be made for the space $AP(R, X)$ with X a Banach space.

Among the basic facts related to the Fourier series that can be generalized from the classical case of periodic functions that generate the form (4.1) to the case of almost periodic functions that generate series of the form (4.4), we can mention here the following:

(1) If the Fourier series is uniformly convergent, then it converges to the function that generated it.

(2) The Parseval formula relates the Fourier coefficients and the function.

For instance, in the case described by formula (4.1), Parseval's formula looks like

$$\frac{a_0^2}{4} + \sum_{k=1}^{\infty}(a_k^2 + b_k^2) = \frac{1}{\pi}\int_{-\pi}^{\pi}|f(t)|^2 dt,$$

while in the case of almost periodic functions it looks like

$$\sum_{k=0}^{\infty}|a_k|^2 = M\{|f|^2\}.$$

The last formula corresponds to the situation related to formula (4.4).

(3) Generally, the Fourier series of a periodic or almost periodic function does not converge to the generating function. The series (Cesarò–Fejer–Bochner) is always summable to the generating function.

This result will complete the proof of the equivalence between Bohr's definition, the relative compactness of the family of translates in adequate norms, and the approximation property (using trigonometric polynomials).

Some results of convergence for the Fourier series are directly applicable to the periodic case, being suggested by known criteria for the latter. There are some criteria that apply specifically only to certain classes of almost periodic functions that do not encompass the periodic ones.

4.2 The Fourier Series Associated to an Almost Periodic Function

As we saw in Section 3.4, for any function $f \in B(R, \mathcal{C})$, there exists the mean value

$$M\{f\} = \lim_{\ell \to \infty}(2\ell)^{-1}\int_{-\ell}^{\ell}f(t)dt. \tag{4.6}$$

Since $B(R, \mathcal{C})$ is the richest space among the spaces of almost periodic functions we considered in Chapter 3, any result based on the existence of the mean value in $B(R, \mathcal{C})$ is automatically valid in all these spaces of almost periodic functions.

Let us note the fact that $f \in B(R, \mathcal{C})$ implies $f(t)e^{-i\lambda t} \in B(R, \mathcal{C})$ for any $\lambda \in R$. Therefore, the quantity

$$a(f, \lambda) = M\{f(t)e^{-i\lambda t}\} \tag{4.7}$$

is well defined for any $f \in B(R, \mathcal{C})$ and $\lambda \in R$.

The following result is of utmost importance in regard to the concept of Fourier series corresponding to almost periodic functions in various senses.

Proposition 4.1. *For each $f \in B(R, \mathcal{C})$, there exists at most a countable set of values of $\lambda \in R$ such that*

$$a(f, \lambda) \neq 0. \tag{4.8}$$

Proof. We know that there exists a sequence of trigonometric polynomials $\{T_m(t); \ m \geq 1\} \subset \mathcal{T}$ such that $|f - T_m|_B \longrightarrow 0$ as $m \to \infty$. In Section 3.4, we have seen that the map $f \longrightarrow M\{f\}$ is continuous from B into $\mathcal{C}$. This implies

$$\lim_{m \to \infty} M\{T_m(t)e^{-i\lambda t}\} = a(f, \lambda). \tag{4.9}$$

But for each $m \geq 1$, $M\{T_m(t)e^{-i\lambda t}\} \neq 0$ only for a finite number of values of λ. Namely, if $T_m(t) = \sum_{k=1}^{n} a_k e^{i\lambda_k t}$, $a_k \in \mathcal{C}$, $\lambda_k \in R$, $k = 1, 2, \ldots, n$, then only for $\lambda = \lambda_k$, $k = 1, 2, \ldots, n$, can we have $M\{T_m(t)e^{-i\lambda t}\} \neq 0$. Obviously, there exists only a countable set of values of λ such that $\{T_m(t)e^{-i\lambda t}\} \neq 0$ for at least one m, $m \geq 1$, and only for such values can we obtain $a(f, \lambda) \neq 0$.

Proposition 4.1 is thereby proven.

Definition 4.1. Let $f \in B(R, \mathcal{C})$, and denote by λ_k, $k \geq 1$, the real values for which

$$a_k = a(f, \lambda_k) \neq 0, \qquad k \geq 1. \tag{4.10}$$

Then the series

$$\sum_{k=1}^{\infty} a_k e^{i\lambda_k t} \tag{4.11}$$

is called the *Fourier series* of the function f. This fact is represented by

$$f(t) \sim \sum_{k=1}^{\infty} a_k e^{i\lambda_k t}. \tag{4.12}$$

Remark 4.1. The numbers $\lambda_k \in R$, $k = 1, 2, \ldots$, are called the *Fourier exponents* of f, while $a_k \in \mathcal{C}$, $k = 1, 2, ..$, are called *Fourier coefficients* of the function f.

Remark 4.2. It is obvious that the number of Fourier exponents could be finite. In such a case, we deal with a function f that reduces to a trigonometric polynomial.

Remark 4.3. Proposition 4.1 and Definition 4.1 are valid for the space $B(R, \mathcal{C})$. But is is obvious that everything that has been said in connection with the space $B(R, \mathcal{C})$ can be extended to the space $B(R, \mathcal{C}^n)$. In this case, $a_k \in \mathcal{C}^n$, $k \geq 1$.

Remark 4.4. If $f \in B(R, \mathcal{C})$ and formula (4.12) takes place, then the following relations are valid:

$$\bar{f}(t) \sim \sum_{k=1}^{\infty} \bar{a}_k e^{-i\lambda_k t},$$

$$f(t + a) \sim \sum_{k=1}^{\infty} a_k e^{i\lambda_k a} e^{i\lambda_k t},$$

$$e^{i\lambda t}f(t) \sim \sum_{k=1}^{\infty} a_k e^{i(\lambda + \lambda_k)t},$$

$$f'(t) \sim \sum_{k=1}^{\infty} i\lambda_k a_k e^{i\lambda_k t}.$$

The last relation is true only if $f' \in B(R, \mathcal{C})$, and therefore we can speak about its Fourier series. First, the formula $M\{f'(t)e^{-i\lambda t}\} = i\lambda M\{f(t)e^{-i\lambda t}\}$ should be established (elementary). It implies that the Fourier exponents of f' are the same as those of f, with the possible exception of the zero exponent (i.e., there is a free term in the series, say a_0).

There is another approach with regard to the way we can attach a Fourier series to an almost periodic function, but this will be applicable only to the space $B^2(R, \mathcal{C})$. Of course, this approach will lead to results that are valid for such spaces as $S^2(R, \mathcal{C})$, $W^2(R, \mathcal{C})$, or $AP(R, \mathcal{C})$. At the same time, this approach will lead to the solution of a problem of best approximation by trigonometric polynomials. In other words, we will have a *variational method* of attaching the Fourier series to a given almost periodic function. See A. S. Besicovitch [10] for details.

We shall consider, for simplicity, only the case of the space $AP(R, \mathcal{C})$. Let $f \in AP(R, \mathcal{C})$, and consider the set of trigonometric polynomials of the form

$$T_m(t) = \sum_{k=1}^{m} c_k e^{i\mu_k t}, \qquad t \in R, \tag{4.13}$$

with m fixed and $c_k \in \mathcal{C}$, $\mu_k \in R$, $k = 1, 2, \ldots, m$. Denote

$$F(c_1, c_2, \ldots, c_m) = M\{|f(t) - T_m(t)|^2\}, \tag{4.14}$$

and try to minimize the function F. The real numbers μ_k, $k = 1, 2, \ldots, m$, are assumed fixed. The fact that the right-hand side in equation (4.14) is finite for each f and T_m is the consequence of $f, T_m \in AP(R, \mathcal{C})$.

Since we dealt only with the case of almost periodic functions with values in $\mathcal{C}$ (or $\mathcal{C}^n$), it is natural to ask ourselves if similar concepts can be extended to the spaces of almost periodic functions with values in a Banach space. The answer to this question is positive, but the discussion will be postponed. Taking into account the fact that in Section 3.5 we have proven the existence of the mean value for functions in $AP(R, X)$ with X a Banach space, we can say we have the main tool in providing the construction of the Fourier series for functions in $AP(R, X)$. Some auxiliary results are necessary, and until we establish these results, we are not able to proceed any further.

Based on the properties of the mean value (see Section 3.5) and the fact that $|\alpha|^2 = \alpha\bar{\alpha}$ for each $\alpha \in \mathcal{C}$, we obtain from equation (4.14)

$$M\left\{\left|f(t) - \sum_{k=1}^{m} c_k e^{i\mu_k t}\right|^2\right\} = M\{|f(t)|^2\} - \sum_{k=1}^{m} \bar{c}_k M\{f(t)e^{-i\mu_k t}\}$$

$$-\sum_{k=1}^{m} c_k M\{\bar{f}(t)e^{i\mu_k t}\} + \sum_{k=1}^{m}\sum_{j=1}^{m} c_k \bar{c}_j M\{e^{i\mu_k t - i\mu_j t}\}.$$

Elementary calculations lead to the formula

$$F(c_1, c_2, \ldots, c_m) = M\{|f(t)|^2\} + \sum_{k=1}^{m} |c_k - a(f, \mu_k)|^2 - \sum_{k=1}^{m} |a(f, \mu_k)|^2,$$

from which we see that $\min F$ is attained for $c_k = a(f, \mu_k)$, $k = 1, 2, \ldots, m$, and

$$\min F = M\{|f(t)|^2\} - \sum_{k=1}^{m} |a(f, \mu_k)|^2. \tag{4.15}$$

Because $\min F \geq 0$, as follows from equation (4.14), we obtain Bessel's inequality

$$\sum_{k=1}^{m} |a(f, \mu_k)|^2 \leq M\{|f(t)|^2\}. \tag{4.16}$$

This inequality allows us to prove the following important result (see Proposition 4.1).

Proposition 4.2. *For each $f \in \mathrm{AP}(R, \mathcal{C})$, there exists at most a countable set of values λ such that formula (4.10) takes place.*

Proof. For any $\gamma > 0$, there exists only a finite number of λ such that $|a(t, \lambda)| \geq \gamma$. Indeed, from formula (4.16) we obtain $m\gamma \leq M\{|f(t)|^2\}$, which provides an upper bound for m. This implies that there exists an integer $k > 0$ such that $(k+1)^{-1} \leq |a(f, \lambda)|^2 < k$. Hence, the set of λ's for which $|a(f, \lambda)| > 0$ is a countable union of finite sets, and therefore it is countable itself.

Since $\mathrm{AP}(R, \mathcal{C}) \subset B(R, \mathcal{C})$ and we know about the existence of Fourier exponents for $f \in B(R, \mathcal{C})$, there results once again that for each $f \in \mathrm{AP}(R, \mathcal{C})$ we have a sequence of Fourier exponents $\{\lambda_k; \ k \geq 1\}$. Of course, the Fourier coefficients are $a(f, \lambda_k) = a_k$, $k \geq 1$.

Remark 4.5. Proposition 4.2 offers an alternate proof of the existence of the Fourier series for an almost periodic function in the Bohr space $\mathrm{AP}(R, \mathcal{C})$. While it covers a space immersed in $B(R, \mathcal{C})$, it has some advantages that will be emphasized in subsequent sections.

Proposition 4.3. *Let $f \in \mathrm{AP}(R, \mathcal{C})$. If a_k, $k \geq 1$, are its Fourier coefficients, then the following (Bessel) inequality holds:*

$$\sum_{k=1}^{\infty} |a_k|^2 \leq M\{|f(t)|^2\}. \tag{4.17}$$

Proof. From inequality (4.16), we derive

$$\sum_{k=1}^{m} |a_k|^2 \leq M\{|f(t)|^2\} \tag{4.18}$$

due to the fact that $a(f, \lambda_k) = a_k$, $k \geq 1$. But the right-hand side in formula (4.18) is a constant, while the left-hand side grows with m. We can let $m \to \infty$ in the inequality (4.18), which leads to formula (4.17).

The proof of Proposition 4.3 is complete.

The next result is an improvement of Proposition 4.3 with important implications in the Fourier analysis of almost periodic functions.

Theorem 4.1. *Let $f \in \mathrm{AP}(R, C)$ with Fourier series*

$$f(t) \sim \sum_{k=1}^{\infty} a_k e^{i\lambda_k t}. \tag{4.19}$$

Then, Parseval's equality is valid:

$$\sum_{k=1}^{\infty} |a_k|^2 = M\{|f(t)|^2\}. \tag{4.20}$$

Proof. If we take into account formula (4.17), there remains to prove the inequality

$$M\{|f(t)|^2\} \leq \sum_{k=1}^{\infty} |a_k|^2. \tag{4.21}$$

Since $f \in \mathrm{AP}(R, C)$, there exists for every $n \geq 0$ a trigonometric polynomial $T_n(t)$ such that

$$M\{|f(t) - T_n(t)|^2\} < \frac{1}{n}. \tag{4.22}$$

Denote by $T_n^*(t)$, $n \geq 1$, the polynomial defined as follows:

$T_n^*(t) \equiv 0$ if none of the Fourier exponents of $f(t)$ appears in $T_n(t)$.

$T_n^*(t) = \sum a_k e^{i\lambda_k t}$, the summation being extended to those k's

for which λ_k is a common Fourier exponent to f and T_n.

From formula (4.22), we derive from formula (4.16)

$$M\{|f(t) - T_n^*(t)|^2\} \leq M\{|f(t) - T_n(t)|^2\} \leq \frac{1}{n}. \tag{4.23}$$

But

$$M\{|f(t) - T_n^*(t)|^2\} = M\{|f(t)|^2\} - \sum |a_k|^2, \tag{4.24}$$

the summation being extended to those k's for which λ_k is a Fourier exponent of $T_n(t)$. From formula (4.23) and (4.24), one obtains

$$M\{|f(t)|^2\} \leq \sum |a_k|^2 + \frac{1}{n}, \tag{4.25}$$

with the summation extended to those k's for which λ_k is the exponent in $T_n(t)$. Hence, from formula (4.25) we can write the inequality

$$M\{|f(t)|^2\} \leq \sum_{k=1}^{\infty} |a_k|^2 + \frac{1}{n}.$$

The inequality (4.21) follows from above, letting $n \to \infty$.

The proof of Theorem 4.1 is complete. As mentioned above, its conclusion remains valid in the space $B^2(R, \mathcal{C})$.

Remark 4.6. As we know (Section 3.4), when we write $f \in B^2(R, \mathcal{C})$, we do not have in mind just the function f itself. It represents a whole class of equivalent functions according to the relation $f \simeq g$ iff

$$\lim_{\ell \to \infty} (2\ell)^{-1} \int_{-\ell}^{\ell} |f(t) - g(t)|^2 dt = 0. \tag{4.26}$$

If we rely on inequality (3.52) from Section 3.4, then equation (4.26) implies

$$\lim_{\ell \to \infty} (2\ell)^{-1} \int_{-\ell}^{\ell} |f(t) - g(t)| dt = 0. \tag{4.27}$$

From equation (4.27), we derive

$$\lim_{\ell \to \infty} (2\ell)^{-1} \int_{-\ell}^{\ell} f(t) e^{-i\lambda t} dt = \lim_{\ell \to \infty} (2\ell)^{-1} \int_{-\ell}^{\ell} g(t) e^{-i\lambda t} dt,$$

which means that $a(f, \lambda) = a(g, \lambda)$. Hence, f and g have the same Fourier series.

The discussion above leads us to the following conclusion.

Corollary 4.1. Let $f, g \in B^2(R, \mathcal{C})$. Then $f \sim g$, and consequently they have the same Fourier series, iff

$$M\{|f(t) - g(t)|^2\} = 0. \tag{4.28}$$

Remark 4.7. Corollary 4.1 represents a uniqueness theorem because two functions satisfying the condition (4.28) belong to the same class of equivalence and therefore are identifiable in $B^2(R, \mathcal{C})$.

Remark 4.8. Since $AP(R, \mathcal{C}) \subset B^2(R, \mathcal{C})$, there results from Corollary 4.1 the uniqueness of Fourier series for Bohr's almost periodic functions. But equation (4.28) implies in this case $f(t) \equiv g(t)$, $t \in R$.

An application of Corollary 4.1 is related to the convergence of Fourier series attached to almost periodic functions.

Proposition 4.4. *Let $f \in B^2(R, \mathcal{C})$ be such that*

$$f \sim \sum_{k=1}^{\infty} a_k e^{i\lambda_k t}. \tag{4.29}$$

If the Fourier series on the right-hand side of formula (4.29) is uniformly convergent, then $f \in AP(R, \mathcal{C})$ and the $\sim$ sign can be replaced by the $=$ sign.

Proof. Indeed, let us denote

$$g(t) = \sum_{k=1}^{\infty} a_k e^{i\lambda_k t}. \tag{4.30}$$

The function $g(t)$ is in $AP(R, \mathcal{C})$ due to the fact that it is the uniform limit of a sequence of polynomials. We obtain

$$M\{g(t)e^{-i\lambda t}\} = \lim_{k \to \infty} \sum_{j=1}^{k} a_j M\{e^{-i(\lambda - \lambda_k)t}\} = \begin{cases} a_j & \text{for } \lambda = \lambda_j, \\ 0 & \text{for } \lambda \neq \lambda_j. \end{cases}$$

This means that $g(t)$ has the same Fourier series as $f(t)$. Consequently, $f(t) \equiv g(t)$, which ends the proof of Proposition 4.4.

Remark 4.9. The way we defined the space $AP_1(R, \mathcal{C})$ in Section 3.1 implies that the series involved in the definition is the Fourier series of the function.

In particular, if $\{\lambda_k; \ k \geq 1\} \subset R$ is an arbitrary countable set, then any absolutely convergent series of the form

$$\sum_{k=1}^{\infty} a_k e^{i\lambda_k t}$$

represents a function in $AP_1(R, \mathcal{C})$. This shows the richness of the spaces of almost periodic functions if we keep in mind that $AP_1(R, \mathcal{C})$ is contained in other spaces we have investigated. It is obvious that this manner of constructing almost periodic functions has applications to spaces such as $AP_1(R, X)$, with X a Banach space.

4.3 Convergence of Fourier Series

As in Section 4.2, for each $f \in B(R, \mathcal{C})$, we can define the Fourier series

$$f(t) \sim \sum_{k=1}^{\infty} a_k e^{i\lambda_k t}, \tag{4.31}$$

with $\lambda_k \in R$ and $a_k \in C$, $k \geq 1$. Proposition 4.4 above shows the importance of the uniform convergence of a Fourier series associated with a function $f \in B^2(R, C)$.

In particular, if the Fourier series is absolutely convergent, which means $f \in AP_1(R, C)$, it is also uniformly convergent. This is the kind of convergence characteristic of the space $AP(R, C)$, and it is worth mentioning that this kind of convergence is rather strong and it is natural to look for criteria that ensure weaker kinds of convergence.

We shall establish a criterion of pointwise convergence of Fourier series, generalizing a classical result known for Fourier series of periodic functions.

Before we can prove this criterion, we need to establish some auxiliary results.

Let $\varphi_{a,b}(\lambda)$ be the function defined by $\varphi_{a,b}(\lambda) = 1$ for $|\lambda| \leq a$, $\varphi_{a,b}(\lambda) = (b-a)^{-1}(b - |\lambda|)$ for $a < |\lambda| < b$, and $\varphi_{a,b}(\lambda) = 0$ for $|\lambda| \geq b$, where $a < b$ are real positive numbers, while $\lambda \in R$.

Obviously, $\varphi_{a,b}(\lambda) \in L^1(R, R)$, which means its Fourier transform

$$\psi_{a,b}(t) = (2\pi)^{-1} \int_R \varphi_{a,b}(\lambda)e^{-i\lambda t}d\lambda$$

is well defined. Actually, one finds after elementary calculations

$$\psi_{a,b}(t) = \frac{2}{\pi(b-a)t^2} \sin \frac{b-a}{2} t \sin \frac{b+a}{2} t, \tag{4.32}$$

a formula that obviously also makes sense for $t = 0$.

One can establish the following inequality:

$$\int_R |\psi_{a,b}(t)|dt \leq \frac{2}{\pi} \left(2 + \ln \frac{b+a}{b-a} \right). \tag{4.33}$$

Indeed, letting $(b-a)x = 2v$ and $(b+a) = \eta(b-a)$, one finds

$$\int_R |\psi_{a,b}(t)|dt = \frac{2}{\pi} \int_0^\infty \frac{|\sin v \sin \eta v|}{v^2} dv$$

$$\leq \frac{2}{\pi} \int_0^1 \frac{|\sin \eta v|}{v} dv + \frac{2}{\pi} \int_1^\infty \frac{dv}{v^2} = \frac{2}{\pi} \int_0^\eta \frac{|\sin v|}{v} dv + \frac{2}{\pi}$$

$$= \frac{2}{\pi} \int_0^1 \frac{|\sin v|}{v} dv + \frac{2}{\pi} \int_1^\eta \frac{|\sin v|}{v} dv + \frac{2}{\pi} \leq \frac{2}{\pi} + \frac{2}{\pi} \ln \eta + \frac{2}{\pi}$$

$$= \frac{2}{\pi} (2 + \ln \eta) = \frac{2}{\pi} \left(2 + \ln \frac{b+a}{b-a} \right).$$

The inequality (4.33) is thus proven.

Consider now the convolution

$$f_{a,b}(t) = \int_R f(t+x)\psi_{a,b}(x)dx \tag{4.34}$$

with $f \in AP(R,C)$, as indicated by formula (4.31). The convolution (4.34) obviously exists because $\psi_{a,b} \in L^1(R,R)$. Moreover, $f_{a,b}(t) \in AP(R,C)$, as we saw in Section 3.2.

Let us now calculate the Fourier exponents and Fourier coefficients of the function $f_{a,b}(t)$ given by equation (4.34). It is an elementary matter to show that

$$M\left\{f_{a,b}(t)e^{-i\lambda t}\right\} = \varphi_{a,b}(\lambda)M\left\{f(t)e^{-i\lambda t}\right\}. \tag{4.35}$$

The relation (4.35) shows that any Fourier exponent of f is also an exponent for $f_{a,b}(t)$, provided $\varphi_{a,b}(\lambda) \neq 0$. On the other hand, if $\varphi_{\alpha,b}(\lambda_k) = 0$, then $M\left\{f_{a,b}(t)e^{-i\lambda_k t}\right\} = 0$. This means that we can write

$$f_{a,b}(t) \sim \sum_{k=1}^{\infty} a_k \varphi_{a,b}(\lambda_k)e^{i\lambda_k t}. \tag{4.36}$$

Proposition 4.5. *Let $f \in AP(R,C)$, and assume there is a $\theta \in (0,1)$ such that $0 < a < \theta b$, while $b \to \infty$. Then*

$$\lim_{b \to \infty} f_{a,b}(t) = f(t) \tag{4.37}$$

uniformly on R (i.e., in $AP(R,C)$).

Proof. The inversion formula for the Fourier transform of functions in $L^1(R,C)$ gives

$$\varphi_{a,b}(\lambda) = \int_R \psi_{a,b}(x)e^{i\lambda x}dx, \tag{4.38}$$

and for $\lambda = 0$ we obtain

$$\int_R \psi_{a,b}(x)dx = 1. \tag{4.39}$$

From equations (4.34) and (4.39), by elementary transformations, one obtains the formula

$$f_{a,b}(t) - f(t) = \int_R \{f(t+x) - f(t)\}\psi_{a,b}(x)dx.$$

Taking equation (4.32) into account and making the substitution $(b-a)x = 2v$, one obtains

$$f_{a,b}(t) - f(t) = \frac{1}{\pi}\int_R \left\{f\left(t + \frac{2v}{b-a}\right) - f(t)\right\}\frac{\sin v}{v^2}\sin\frac{b+a}{b-a}\,v\,dv.$$

Consider now $N > 0$, and split the last integral:

$$|f_{a,b}(t) - f(t)| \leq \frac{2M}{\pi} \int_{-\infty}^{-N} \frac{dv}{v^2} + \frac{2M}{\pi} \int_{N}^{\infty} \frac{dv}{v^2}$$

$$+ \frac{1}{\pi} \int_{-N}^{N} \left| f\left(t + \frac{2v}{b-a}\right) - f(t) \right| \frac{1}{v^2} \left| \sin v \sin \frac{b+a}{b-a} v \right| dv,$$

with $M = \sup |f(t)|$, $t \in R$. But each of the first two integrals on the right-hand side equals N^{-1}. Therefore, given $\varepsilon > 0$, one can write

$$\frac{2M}{\pi} \int_{-\infty}^{-N} \frac{dv}{v^2} + \frac{2M}{\pi} \int_{N}^{\infty} \frac{dv}{v^2} < \frac{\varepsilon}{2}, \tag{4.40}$$

provided $N > 8M(\pi\varepsilon)^{-1}$. We shall now fix N such that formula (4.40) takes place. Because $a < \theta b$, $0 < \theta < 1$, we have $b - a > b - \theta b = (1 - \theta)b$. But $(1-\theta)b \longrightarrow \infty$ as $b \to \infty$. If we now consider $|v| \leq N$, we obtain for sufficiently large b the inequality

$$\left| f\left(t + \frac{2v}{b-a}\right) - f(t) \right| < \frac{\varepsilon}{4\{2 + \ln[2/(1-\theta)]\}}$$

due to the uniform continuity of f on $[-N, N]$. Hence,

$$\frac{1}{\pi} \int_{-N}^{N} \left| f\left(t + \frac{2v}{b-a}\right) - f(t) \right| \frac{1}{v^2} \left| \sin v \sin \frac{b+a}{b-a} v \right| dv$$

$$\leq \frac{\pi\varepsilon}{4\pi\{2 + \ln[2/(1-\theta)]\}} \int_{R} \frac{1}{v^2} \left| \sin v \sin \frac{b+a}{b-a} v \right| dv$$

$$\leq \frac{\varepsilon}{2} \frac{2 + \ln[(b+a)/(b-a)]}{2 + \ln[2/(1-\theta)]},$$

if one also takes into account formula (4.33).

Let us now observe that $(b + a)/(b - a) < (b + b)/(b - \theta b) = 2/(1 - \theta)$, which allows us to write for sufficiently large b

$$\frac{1}{\pi} \int_{-N}^{N} \left| f\left(t + \frac{2v}{b-a}\right) - f(t) \right| \frac{1}{v^2} \left| \sin v \sin \frac{b+a}{b-a} v \right| dv < \frac{\varepsilon}{2}.$$

From the inequalities above, starting with the one for $|f_{a,b}(t) - f(t)|$, we derive

$$|f_{a,b}(t) - f(t)| < \frac{\varepsilon}{2} + \frac{\varepsilon}{2} = \varepsilon$$

for sufficiently large b.

This ends the proof of Proposition 4.5.

We shall now consider a class of almost periodic functions in $AP(R, \mathcal{C})$ such that the Fourier exponents satisfy the condition

$$\lim_{k \to \infty} |\lambda_k| = +\infty. \tag{4.41}$$

We may say that the set of Fourier exponents has only a limit point, and this limit is at infinity.

It is useful to observe that the periodic functions, regardless of the period, belong to the class characterized by the condition (4.41).

In order to provide some symmetry to the formulas, we agree to add to the Fourier exponents the numbers $-\lambda_k$ for those k's that do not appear among the Fourier exponents. We number the exponents, including the added ones, in such a way that $\lambda_k = -\lambda_{-k}$. Under this assumption, it is natural to write the Fourier series of $f(t)$ in the form

$$f(t) \sim \sum_{k=-\infty}^{\infty} a_k e^{i\lambda_k t}, \tag{4.42}$$

noting that, for the added exponents, the corresponding coefficients $a_k = 0$. So, on the right-hand side of formula (4.42), we really have the same terms as in formula (4.31).

The following result is in accordance with the similar situation encountered in the case of periodic functions.

Theorem 4.2. *Let* $f \in AP(R, C)$ *be such that*

$$\lambda_{k+1} - \lambda_k \geq \alpha > 0, \qquad k \in Z. \tag{4.43}$$

If $t_0 \in R$ *is a point in the neighborhood of which* $f(t)$ *has bounded variation, then*

$$\lim_{n \to \infty} \sum_{k=-n}^{n} a_k e^{i\lambda_k t_0} = f(t_0). \tag{4.44}$$

Proof. Let us take $a = \lambda_n$, $b = \lambda_n + \alpha$ in formula (4.43), which leads to the following relations:

$$S_n(t) = \sum_{k=-n}^{n} a_k e^{i\lambda_k t} = f_{a,b}(t)$$

$$= \frac{2}{\pi\alpha} \int_R f(t+x) \, \frac{\sin \frac{\alpha x}{2} \sin\left(\lambda_n + \frac{\alpha}{2}\right) x}{x^2} \, dx. \tag{4.45}$$

Let $h \in R$ be a positive number, and note that

$$S_n(t_0) = \frac{2}{\pi\alpha} \left(\int_{-\infty}^{-h} + \int_{-h}^{h} + \int_{h}^{\infty} \right) = I_1 + I_2 + I_3, \tag{4.46}$$

with obvious notations. The behavior of the second term I_2, for $h \longrightarrow 0$, is the same as that of the integral

$$\bar{I}_2 = \frac{1}{\pi} \int_{-h}^{h} f(t_0 + x) \, \frac{\sin\left(\lambda_n + \frac{\alpha}{2}\right) x}{x} \, dx \tag{4.47}$$

due to the fact that $\sin \frac{\alpha x}{2} / \left(\frac{\alpha x}{2}\right) \longrightarrow 1$ as $x \longrightarrow 0$.

Indeed, we have

$$I_2 - \overline{I}_2 = \frac{1}{\pi} \int_{-h}^{h} f(t_0 + x) \sin\left(\lambda_n + \frac{\alpha}{2}\right) x \frac{2 \sin \frac{\alpha x}{2} - \alpha x}{\alpha x^2} \, dx,$$

and since the fraction under the integral tends to 0 with x (or h), we can write

$$|I_2 - \overline{I}_2| < \frac{\varepsilon}{3} \qquad \text{for } h < \delta(\varepsilon). \tag{4.48}$$

From now on, we shall fix such an h.

On the other hand, for $x \in (-\infty, -h) \cup (h, \infty)$, the function

$$f(t_0 + x) \sin \frac{\alpha u}{2} / u^2$$

is absolutely integrable. Therefore, for each I_1 and I_3 that appear as sines of Fourier transforms of an absolutely integrable function (i.e., in the space $L^1(R, R)$), we can write

$$|I_1|, |I_3| < \frac{\varepsilon}{3}, \qquad \text{for } n \geq N(\varepsilon). \tag{4.49}$$

Based on formulas (4.46)–(4.48), we can write

$$|S_n(t_0) - \overline{I}_2| < 3\,\frac{\varepsilon}{3} = \varepsilon \qquad \text{for } n \geq N(\varepsilon). \tag{4.50}$$

But

$$\lim_{n \to \infty} \overline{I}_2 = \lim_{n \to \infty} \frac{1}{\pi} \int_{-h}^{h} f(t_0 + x) \frac{\sin\left(\lambda_n + \frac{\alpha}{2}\right) x}{x} \, dx$$
$$= \frac{1}{2} \left[f(t_0 + 0) + f(t_0 - 0) \right] = f(t_0). \tag{4.51}$$

Therefore

$$|S_n(t_0) - f(t_0)| < \varepsilon \qquad \text{for } n \geq N_1(\varepsilon),$$

which ends the proof of Theorem 4.2.

Remark 4.10. For formulas (4.49) and (4.51), see R. E. Edwards [36] or N. K. Bary [6]. In equation (4.51), we took into account the hypothesis that stipulates the property of local bounded variation of $f(t)$ in a neighborhood of t_0.

We shall now consider a case of uniform convergence of the Fourier series under another condition concerning the distribution of the Fourier exponents.

Theorem 4.3. *Consider $f \in AP(R, C)$ with Fourier series (4.42). Assume there exists $\theta \in (0, 1)$ with the property $\lambda_k \leq \theta \lambda_{k+1}$, $k \geq 0$, and $\lambda_k \longrightarrow \infty$ as $k \longrightarrow \infty$. Then the Fourier series (4.42) is uniformly convergent on R.*

Proof. The proof of Theorem 4.3 follows easily from Proposition 4.5. Indeed, we take $a = \lambda_k$, $b = \lambda_{k+1}$, $k \geq 0$, and note that

$$f_{a,b}(t) = \sum_{|\lambda_j| < \lambda_k} a_j \, e^{i\lambda_j t}.$$

The last relation is the consequence of the fact that for $a = \lambda_k$ and $b = \lambda_{k+1}$ we have $\varphi_{a,b}(\lambda_j) = 1$ when $|\lambda_j| < \lambda_k$. Then Proposition 4.5 applies if we take into account the hypothesis $\lambda_k \leq \theta \lambda_{k+1}$, $k \geq 0$, and $\lambda_k \longrightarrow \infty$ as $k \longrightarrow \infty$.

Remark 4.11. In the proof of Proposition 4.5, one assumes that $f \in AP(R, \mathcal{C})$. But, as we know, the Fourier series can be attached to almost periodic functions that belong to much larger classes than $AP(R, \mathcal{C})$. Does Theorem 4.3 remain valid if we assume only $f \in B(R, \mathcal{C})$? Or $f \in S(R, \mathcal{C})$?

4.4 Summability of Fourier Series

It has been shown by Cesaro and Fejér that, in the case of periodic functions, the Fourier series can be used to reconstruct the generating function. Instead of looking for the sum of the Fourier series, which does not exist in many cases, a summability method can be used in order to obtain the original function. Bochner has shown that the same procedure of summability can be applied to almost periodic functions. The case of functions in $AP(R, \mathcal{C})$ is the most commonly encountered, even though summability methods can be used for other classes of almost periodic functions. The book by A. S. Besicovitch [10] is a source in this regard.

We shall rely on Theorem 3.3 of Chapter 3, which has been proven for the space $AP(R, X)$ with X a Banach space, which can be applied to the space $AP(R, \mathcal{C})$. The result we need can be formulated as follows.

Proposition 4.6. *A set $\mathcal{M} \subset AP(R, \mathcal{C})$ is relatively compact if and only if the following conditions are satisfied:*

(1) *The family $\mathcal{M}$ is equicontinuous on R.*
(2) *The family $\mathcal{M}$ is equi–almost-periodic.*
(3) *For each $t \in R$, the set $\{f(t); \; f \in \mathcal{M}\}$ is bounded.*

Remark 4.12. Condition (3) is the condition of relative compactness in $\mathcal{C}$. Of course, it would be valid in any space R^n or $\mathcal{C}^n$. The definitions of equicontinuity and equi–almost-periodicity are provided in Theorem 3.3 of Chapter 3.

The following result is concerned with the convergence in $AP(R, \mathcal{C})$ and relies on the concept of convergence in $B^2(R, \mathcal{C})$. This last type of convergence is certainly weaker than the uniform convergence.

Theorem 4.4. *Let $\{f_k(t); \; k \geq 1\} \subset AP(R, \mathcal{C})$ be a sequence convergent in $B^2(R, \mathcal{C})$. If this sequence is equicontinuous and equi–almost-periodic, then it is uniformly convergent on R.*

Proof. The assumption of convergence in $B^2(R, C)$ implies that for any $\varepsilon > 0$ there exists a natural number $N(\varepsilon)$ such that

$$M\{|f_n(t) - f_m(t)|^2\} < \varepsilon, \tag{4.52}$$

provided $m, n \geq N(\varepsilon)$.

We shall now prove that formula (4.52), which represents Cauchy's condition in $B^2(R, C)$, implies the following inequality representing Cauchy's condition in $AP(R, C)$: For each $\eta > 0$, there exists $N_1(\eta) > 0$ such that

$$|f_n(t) - f_m(t)| < \eta, \qquad t \in R, \tag{4.53}$$

as soon as $m, n \geq N(\eta)$.

The following property we shall establish implies Cauchy's condition (4.53):

If

$$|f_n(t_0) - f_m(t_0)| > c, \qquad t_0 \in R, \tag{4.54}$$

with fixed t_0, then we can find a constant $c_1 = c_1(c) > 0$ such that

$$M\{|f_n(t) - f_m(t)|^2\} > c_1. \tag{4.55}$$

Indeed, based on the property of equicontinuity of the sequence $\{f_k(t); \ k \geq 1\}$, we shall have

$$|f_n(t_1) - f_n(t_2)| < \frac{c}{8} \qquad \text{for } |t_1 - t_2| < \delta(c). \tag{4.56}$$

If τ is a point where

$$|f_n(\tau) - f_m(\tau)| > \frac{c}{2}, \tag{4.57}$$

then there exists a segment of length 2δ, centered at τ, such that for $|t - \tau| < \delta$ we have

$$|f_n(t) - f_m(t)| > \frac{c}{4}. \tag{4.58}$$

This is due to the fact that if there is a point $\bar{t}$, $|\bar{t} - \tau| < \delta$, where formula (4.58) is violated, then

$$|f_n(\tau) - f_m(\tau)| \leq |f_n(\tau) - f_n(\bar{t})| + |f_n(\bar{t}) - f_m(\bar{t})|$$

$$+ |f_m(\bar{t}) - f_m(\tau)| < \frac{c}{8} + \frac{c}{4} + \frac{c}{8} = \frac{c}{2},$$

which is in contradiction with formula (4.57).

Further, there exists a number $\ell_1 = \ell(c/4) > 0$ such that each interval $(a, a + \ell_1) \subset R$ contains a $(c/4)$-translation number common to all terms of the sequence $\{f_k(t); \ k \geq 1\}$. Therefore, any interval $(a, a + \ell_1)$ will contain a point t_0' for which

$$|f_k(t_0') - f_k(t_0)| \leq \frac{c}{4}, \qquad k \geq 1. \tag{4.59}$$

Consequently, taking formulas (4.54) and (4.59) into account,

$$|f_n(t_0') - f_m(t_0')| \geq |f_n(t_0) - f_m(t_0)| - |f_m(t_0) - f_m(t_0')|$$

$$-|f_n(t_0') - f_n(t_0)| \geq |f_n(t_0) - f_m(t_0)| - 2\,\frac{c}{4} > \frac{c}{2}.$$

This implies that formula (4.57) holds for $\tau = t_0'$, with $t_0' \in (a, a+\ell_1)$, $a \in R$, and in each interval $(a, a+\ell) \in R$, $\ell = \ell_1 + 2\delta$, the inequality (4.58) holds on a subinterval of length 2δ. Hence,

$$\frac{1}{n\ell} \int_0^{n\ell} |f_n(t) - f_m(t)|^2 dt = \frac{1}{n\ell} \sum_{k=0}^{n-1} \int_{k\ell}^{(k+1)\ell} |f_n(t) - f_m(t)|^2 dt$$

$$\geq \frac{1}{n\ell} \left(\frac{c}{4}\right)^2 \cdot n \cdot 2\delta = \frac{2\delta}{\ell} \left(\frac{c}{4}\right)^2 = c_1(c).$$

Letting $\ell \longrightarrow \infty$, one obtains from above that

$$M\{|f_n(t) - f_m(t)|^2\} \geq c_1 > 0. \tag{4.60}$$

But formula (4.60) contradicts Cauchy's condition for $\{f_k(t);\ k \geq 1\}$ in $B^2(R, C)$.

One of the most remarkable applications of Theorem 4.4 is the proof of the so-called *approximation theorem* for almost periodic functions using trigonometric polynomials, starting with Bohr's definition. Once this theorem is proven, we shall be able to state that we have three equivalent properties (or definitions) of the functions $f \in AP(R, C)$: Bohr's property, the approximation in $BC(R, C)$ with any degree of accuracy using trigonometric polynomials, and the relative compactness in $BC(R, C)$ of the family of its translates $\mathcal{F} = \{f(t+h) : h \in R\}$.

Since the equivalence of Bohr's property and the relative compactness of the family of translates have been established (see Section 3.5 for the case of Banach spaces; this also covers the spaces $AP(R, C)$ or $AP(R, C^n)$), as well as the fact that the approximation property implies Bohr's property, there remains to show only the fact that Bohr's property implies the approximation property.

It is useful to point out the fact that from Bohr's property we derived the existence of the mean value (Section 3.5), which allows us to introduce the Fourier series.

Now, relying on the concept of Fourier series, we shall show that the approximation property for functions in $AP(R, C)$ can be derived from Bohr's property without the help of other properties that have been emphasized. Usually we have started with the approximation property using trigonometric polynomials as the definition for various spaces of almost periodic functions.

Theorem 4.5. *Let $f \in AP(R, \mathcal{C})$ with Fourier series*

$$f(t) \sim \sum_{k=1}^{\infty} a_k e^{i\lambda_k t}. \tag{4.61}$$

Then, there exists a sequence of trigonometric polynomials

$$\sigma_m(t) = \sum_{k=1}^{n} r_{k,m} a_k e^{i\lambda_k t}, \qquad n = n(m),$$

such that $\sigma_m(t) \longrightarrow f(t)$ in $AP(R, \mathcal{C})$. The rational numbers $r_{k,m}$ depend on λ_k and m but not on a_k's. Moreover, $\lim_{m \to \infty} r_{k,m} = 1$.

Proof. Before we get into the convergence part of the proof, we shall dwell first on an auxiliary topic necessary in constructing the polynomials $\sigma_m(t)$, $m \geq 1$.

Let $E \subset R$ be a set of reals. We shall say that the linearly independent numbers β_k, $k \geq 1$, form a *basis* for E if any $\alpha \in E$ can be represented in the form

$$\alpha = r_1 \beta_1 + r_2 \beta_2 + \cdots + r_n \beta_n, \tag{4.62}$$

with $r_k \in Q$ the set of rational numbers. The independence is meant here with respect to rational coefficients.

The following statement is proven easily: Any countable set of real numbers admits a basis with elements of its own. Indeed, let $\{\lambda_k; \ k \geq 1\} \subset R$ be a countable set. Denote $\beta_1 = \lambda_1$, and eliminate from the set all λ_k's that are of the form $r_1 \beta_1$ with $r_1 \in Q$. Then let β_2 be the first λ_k that is not of the form $r_1 \beta_1, r_1 \in Q$. Then we eliminate from the remaining λ_k's those of the form $r_1 \beta_1 + r_2 \beta_2$, if any. We proceed in the same way, letting β_3 be the first λ_k not yet eliminated, and so on. We construct in this way the sequence $\{\beta_j; \ j \geq 1\}$, which constitutes the basis for $\{\lambda_k; \ k \geq 1\}$. It implies that any element $\alpha \in E$ can be represented in the form (4.62) with convenient numbers $r_1, r_2, \ldots, r_n$. The representation is unique because, from equation (4.62) and $\alpha = r'_1 \beta_1 + r'_2 \beta_2 + \cdots + r'_n \beta_n$, we derive

$$(r_1 - r'_1)\beta_1 + (r_2 - r'_2)\beta + \cdots + (r_n - r'_n)\beta_n = 0,$$

and from the independence of the β''_i's, we get $r_j = r'_j$, $j = 1, 2, \ldots, n$.

In order to construct the polynomials $\sigma_m(t)$, $m \geq 1$, we shall introduce the Fejér kernels

$$K_n(t) = \frac{1}{n} \frac{\sin^2(nt/2)}{\sin^2(t/2)}. \tag{4.63}$$

For $t = 2k\pi$, $k \in Z$, the values of $K_n(t)$ are defined in such a way that $K_n(t)$ is continuous at such points (and hence everywhere on R).

It can be easily proven that

$$K_n(t) = \sum_{j=-n}^{n} \left(1 - \frac{|j|}{n}\right) e^{-ijt}. \qquad (4.64)$$

The following properties of $K_n(t)$ are easily derived from equation (4.64) or (4.63):

$$K_n(t) \geq 0; \quad M\{K_n(t)\} = 1, \qquad n \geq 1. \qquad (4.65)$$

Now, consider a basis $\{\beta_j; \; j \geq 1\}$ for the set of Fourier exponents of f, $\{\lambda_k; \; k \geq 1\}$, and define the polynomials $\sigma_m(t)$, $m \geq 1$, by

$$\sigma_m(t) = M_s\left\{f(t+s)K_{(m!)^2}\left(\frac{\beta_1 s}{m!}\right) \cdots K_{(m!)^2}\left(\frac{\beta_m s}{m!}\right)\right\}, \qquad (4.66)$$

where M_s means that the mean value of the function inside $\{\cdot\}$ must be taken with respect to s.

In order to obtain a more adequate formula for $\sigma_m(t)$, we shall denote for simplicity $a(\lambda) = M\{f(t)e^{-i\lambda t}\}$, and taking into account equations (4.64) and (4.66), we get

$$\sigma_m(t) = \sum_{j_1=-(m!)^2}^{(m!)^2} \cdots \sum_{j_m=-(m!)^2}^{(m!)^2} \left(1 - \frac{|j_1|}{(m!)^2}\right) \cdots \left(1 - \frac{|j_m|}{(m!)^2}\right)$$

$$\cdot a\left(\frac{j_1\beta_1 + \cdots + j_m\beta_m}{m!}\right) \exp\left[i\left(\frac{j_1\beta_1 + \cdots + j_m\beta_m}{m!}\right)t\right].$$

We shall now define the rationals $r_{k,m}$ by

$$r_{k,m} = \left(1 - \frac{|j_1|}{(m!)^2}\right) \cdots \left(1 - \frac{|j_m|}{(m!)^2}\right),$$

where the numbers j_k, $k = 1, 2, \ldots, m$, are determined from the representation

$$\lambda_k = \frac{j_1\beta_1 + j_2\beta_2 + \cdots + j_m\beta_m}{m!}. \qquad (4.67)$$

Let us note that such a representation of any λ_k is possible if m is sufficiently large. One must have

$$\frac{|j_k|}{(m!)^2} = \frac{|r_k|}{m!}, \qquad k = 1, 2, \ldots, m,$$

if $\lambda_k = j_1\beta_1 + j_2\beta_2 + \cdots + j_m\beta_m$. Then, for $r_{k,m}$ above, one obtains

$$\lim_{m\to\infty} r_{k,m} = 1, \qquad (4.68)$$

as required in the statement of the theorem.

We shall now apply Parseval's formula to the function

$$f(t) - \sigma_m(t) \in \text{AP}(R, \mathcal{C}).$$

There results

$$M\{|f(t) - \sigma_m(t)|^2\} = \sum_{k=1}^{\infty}(r_{k,m} - 1)^2|a_k|^2. \tag{4.69}$$

Now choose N large enough that

$$\sum_{k=N+1}^{\infty}|a_k|^2 < \frac{\varepsilon}{2} \tag{4.70}$$

with $\varepsilon > 0$ given, and then take m large enough that for fixed N with equation (4.69) valid

$$\sum_{k=1}^{N}(r_{k,m} - 1)^2|a_k|^2 < \frac{\varepsilon}{2}, \qquad m \geq N_1(\varepsilon). \tag{4.71}$$

Formula (4.71) follows from property (4.68). Combining formulas (4.69), (4.70), and (4.71), there results

$$M\{|f(t) - \sigma_m(t)|^2\} < \frac{\varepsilon}{2} + \frac{\varepsilon}{2} = \varepsilon, \tag{4.72}$$

provided $m \geq N_1(\varepsilon)$.

The inequality (4.71) shows that the sequence $\{\sigma_m(t); \ m \geq 1\}$ is convergent in $B^2(R, \mathcal{C})$ to $f(t)$. We need to show that the sequence $\{\sigma_m(t); \ m \geq 1\}$ is equicontinuous and equi–almost-periodic on R. Then the proof will be complete based on Theorem 4.4.

First, we have

$$M\left\{K_{(m!)^2}\left(\frac{\beta_1 t}{m!}\right)\cdots K_{(m!)^2}\left(\frac{\beta_m t}{m!}\right)\right\} = 1 \tag{4.73}$$

because the mean value is taken for a trigonometric polynomial whose term without the exponential is obviously 1.

Second, relying on the first relation in formula (4.65), we observe that

$$\sup_{t \in R}|\sigma_m(t+s) - \sigma_m(t)| \leq \sup_{t \in R}|f(t+s) - f(t)|. \tag{4.74}$$

From formula (4.74), we derive that $\{\sigma(t); \ m \geq 1\}$ is equicontinuous on R (because f is uniformly continuous) and any ε-translation number for f is an ε-translation number common to all $\sigma_m(t)$, $m \geq 1$.

Theorem 4.5 is thereby proven.

In concluding this section, we shall show how a summability criterion like the one given by Theorem 4.5 can help us to obtain a criterion of *convergence* in $\text{AP}(R, \mathcal{C})$.

Theorem 4.6. *Let* $f \in AP(R, \mathcal{C})$ *be a function whose Fourier exponents are linearly independent (over rationals). Then the Fourier series of* f *is convergent in* $AP_1(R, \mathcal{C})$ *to* f.

Proof. Assume f is as shown in formula (4.61). Denote $\theta_k = \operatorname{Arg} a_k$, $k \geq 1$, and consider the functions

$$\varphi_k(t) = 1 + \frac{1}{2}\left[e^{i(\lambda_k t + \theta_k)} + e^{-i(\lambda_k t + \theta_k)}\right],$$

$$\Phi_n(t) = \prod_{k=1}^{n} \varphi_k(t) = 1 + \frac{1}{2}\sum_{k=1}^{n} e^{-i(\lambda_k t + \theta_k)} + \sum_{j} c_j e^{i\mu_j t}.$$

Obviously, the numbers μ_j are linear combinations of λ_k's with integer coefficients, at least one of them being nonzero. Hence, none of the μ_j can vanish and none can be equal to a λ_k due to their linear independence. We obtain

$$M\{\Phi_n(t)\} = 1, \qquad M\{f(t)\Phi_n(t)\} = \frac{1}{2}\sum_{k=1}^{n}|a_k|.$$

But $\Phi_n(t) \geq 0$, $t \in R$, because $\varphi_n(t) \geq 0$, $k \geq 1$. Hence

$$|M\{f(t)\Phi_n(t)\}| \leq \sup_{t \in R}|f(t)| \cdot M\{\Phi_n(t)\} = \sup_{t \in R}|f(t)|.$$

Since

$$\sum_{k=1}^{n}|a_k| \leq 2\sup_{t \in R}|f(t)|$$

for any integer n, there results

$$\sum_{k=1}^{\infty}|a_k| \leq 2|f|_{\mathrm{AP}},$$

which means that the Fourier series of f is absolutely convergent. Therefore, $f \in AP_1(R, \mathcal{C})$, which proves the theorem (absolute convergence always implies uniform convergence for trigonometric series).

4.5 The Fourier Series of Almost Periodic Functions in Banach Spaces

The case of the space $AP(R, X)$ with X a Banach space needs special consideration while introducing the Fourier series. There are some features in common with the finite-dimensional cases $AP(R, \mathcal{C})$ or $AP(R, \mathcal{C}^n)$, but the case of $AP(R, X)$ has to be dealt with separately, which is exactly what we are going to do in this section. We shall rely on some auxiliary facts that are of interest in themselves. The case of $AP(R, X)$ could be reduced to the scalar case of the space $AP(R, \mathcal{C})$, but we shall pursue another venue.

Proposition 4.7. *Let* $\varphi \in \text{AP}(R,C)$ *be such that*

$$\varphi(t) \sim \sum_{k=1}^{\infty} a_k e^{i\lambda_k t}. \tag{4.75}$$

Then, for each $\varepsilon > 0$, *one can determine a natural number* n *and a number* $\delta > 0$ $(\delta < \pi)$ *such that any solution* τ *of the system of Diophantine inequalities*

$$|\lambda_k \tau| < \delta \pmod{2\pi}, \qquad k = 1, 2, \ldots, n, \tag{4.76}$$

is an ε-*translation number of* φ.

Proof. For arbitrary $\varepsilon > 0$, there exists a trigonometric polynomial $S(t) \in \mathcal{T}$ such that

$$|\varphi(t) - S(t)| < \frac{\varepsilon}{3}, \qquad t \in R. \tag{4.77}$$

Without loss of generality, from Theorem 4.5, we can assume that all the exponents of $S(t)$ are chosen among the λ_k's. From formula (4.77), we easily derive that any $(\varepsilon/3)$-translation number of $S(t)$ is an ε-translation number for $\varphi(t)$. Indeed, we can write for any $t, \tau \in R$

$$|\varphi(t + \tau) - \varphi(t)| \leq |\varphi(t + \tau) - S(t + \tau)|$$

$$+ |S(t + \tau) - S(t)| + |S(t) - \varphi(t)|,$$

which immediately leads to our assertion.

Now, if τ satisfies formula (4.76), one has

$$\left| e^{i\lambda_k \tau} - 1 \right| = \sqrt{(1 - \cos \lambda_k \tau)^2 + \sin^2 \lambda_k \tau} = 2 \left| \sin \frac{\lambda_k \tau}{2} \right| < \delta.$$

Consequently, if $S(t) = \sum_{k=1}^{n} b_k e^{i\lambda_k t}$, one obtains

$$|S(t + \tau) - S(t)| \leq \sum_{k=1}^{n} |b_k| \left| e^{i\lambda_k \tau} - 1 \right| < \delta \sum_{k=1}^{n} |b_k| < \frac{\varepsilon}{3},$$

provided we have $3\delta \sum_{k=1}^{n} |b_k| < \varepsilon$. This inequality shows how large the number δ can be in terms of ε Because the b_k's are determined by φ. It suffices to choose the polynomial $S(t)$ as indicated in Theorem 4.5. This means that the b_k's are known because the a_k's are known as Fourier coefficients of φ.

We shall now deal with another auxiliary proposition, which is entirely concerned with Diophantine inequalities and is due to Kronecker. It can be stated as follows.

Proposition 4.8. *Let* λ_k, μ_k, $k = 1, 2, \ldots, n$, *be arbitrary real numbers. The system of Diophantine inequalities*

$$|\lambda_k \tau - \mu_k| < \delta \ (\text{mod } 2\pi), \qquad k = 1, 2, \ldots, n,$$

admits solutions τ, for any $\delta > 0$, if and only if any relationship of the form

$$\sum_{k=1}^{n} m_k \lambda_k = 0,$$

with $m_k \in Z$, $k = 1, 2, \ldots, n$, implies the relationship

$$\sum_{k=1}^{n} m_k \mu_k = 0 \ (\text{mod } 2\pi).$$

Proof. We shall first prove the necessity of the condition. If we assume that for any $\delta > 0$ the system of inequalities in the statement admits solutions, then we can write for such a solution τ the relationships $\lambda_k \tau - \mu_k = \delta_k - 2p_k \pi$ with $|\delta_k| < \delta$, $k = 1, 2, \ldots, n$, and $p_k \in Z$, $k = 1, 2, \ldots, n$. If we multiply both sides of the relationships above by m_k and add with respect to k, $k = 1, 2, \ldots, n$, we obtain

$$\sum_{k=1}^{n} m_k \mu_k - 2\pi \sum_{k=1}^{n} m_k p_k = -\sum_{k=1}^{n} m_k \delta_k,$$

taking into account $\sum_{k=1}^{n} m_k \lambda_k = 0$. We further derive

$$\left| \sum_{k=1}^{n} m_k \mu_k - 2\pi \sum_{k=1}^{n} m_k p_k \right| < \delta \sum_{k=1}^{n} |m_k|$$

because $|\delta_k| < \delta$, $k = 1, 2, \ldots, n$. But δ is arbitrary, which means that

$$\sum_{k=1}^{n} m_k \mu_k - 2\pi \sum_{k=1}^{n} m_k p_k = 0,$$

a relationship equivalent to $\sum_{k=1}^{n} m_k \mu_k = 0 \ (\text{mod } 2\pi)$. This ends the proof of the necessity.

In order to prove the sufficiency of the condition, we will show that the minimum of the modulus of the function

$$f(t) = 1 + \sum_{k=1}^{n} e^{i(\lambda_k t - \mu_k)}$$

is equal to $n + 1$, which is actually the value of the function

$$F(u_1, u_2, \ldots, u_n) = 1 + u_1 + u_2 + \cdots + u_n$$

at the point $u_k = 1$, $k = 1, 2, \ldots, n$. For any natural number p, we set

$$[f(t)]^p = \sum_j d_j e^{i\beta_j t}, \tag{4.78}$$

$$[F(u_1, \ldots, u_n)]^p = \sum a_{j_1, j_2, \ldots, j_n} u_1^{j_1} u_2^{j_2} \ldots u_n^{j_n}. \tag{4.79}$$

It is obvious that the expansion in equation (4.78) can be obtained from the expansion in equation (4.79), with $u_k = \exp\{i(\lambda_k \tau - \mu_k)\}$, $k = 1, 2, \ldots, n$, by adding terms with the same β_j. According to our assumption, if

$$\sum_{k=1}^{n} m'_k \lambda_k = \sum_{k=1}^{n} m''_k \lambda_k,$$

then from the definition of $f(t)$ we have

$$\sum_{k=1}^{n} m'_k \mu_k = \sum_{k=1}^{n} m''_k \mu_k \pmod{2\pi}.$$

This means that similar terms in the expansion have equal arguments $\pmod{2\pi}$. Since the modulus of a sum of complex numbers with the same argument equals the sum of moduli of these numbers (a fact that is obvious if we write each term in trigonometric form, $\rho \exp(i\alpha)$), we are led to the relation

$$\sum_j |d_j| = \sum a_{j_1, j_1, \ldots, j_n} = [F]^p = (n+1)^p. \tag{4.80}$$

But from equation (4.78) we derive

$$|d_j| = \left| M\{|f(t)|^p e^{-i\beta_j t}\} \right| \leq A^p \tag{4.81}$$

with $A = \sup\{|f(t)|;\ t \in R\} = |f|_{\mathrm{AP}}$. If we now admit that $|f|_{\mathrm{AP}} = h < n+1$, then we obtain

$$\sum_j |d_j| \leq (p+1)^n h^p \tag{4.82}$$

if we also take into account the fact that the number of terms in equation (4.79) is at most $(p+1)^n$. Indeed, this is obvious for $n = 1$ and any p natural when the number of terms in the expansions (4.79) is $p + 1$. Now we assume that the statement is valid for any p and for a fixed n (i.e., the number of terms in the expansion (4.79) is not greater than $(p+1)^n$). Then we can write

$$(1 + u_1 + \cdots + u_n + v)^p = \sum_{k=0}^{p} \binom{p}{k} (1 + u_1 + \cdots + u_n)^{p-k} v^k,$$

the expansion having $(p + 1)$ terms, products of power in parentheses, and powers of v. The number of terms in each power in parentheses is at most $(p + 1)^n$ according to our hypothesis of induction and taking into account the fact that this number grows with the exponent. Therefore, the number of distinct terms in the last expansion above is at most $(p + 1)(p + 1)^n = (p + 1)^{n+1}$. This proves the statement according to the induction principle.

We shall now compare formula (4.80) with formula (4.82). There results the inequality

$$(n + 1)^p \leq (p + 1)^n h^p \tag{4.83}$$

for fixed natural n and any $p \in N$. But formula (4.83) is not possible if we choose p sufficiently large. The left-hand side has exponential growth with respect to p, while the right-hand side has only polynomial growth.

The contradiction above proves the sufficiency of the condition in Proposition 4.8, which ends the proof.

Remark 4.13. Proposition 4.8 is a useful tool to better understand the relationship between the Fourier exponents of an almost periodic function and its almost period. We shall dwell on this topic before we define the Fourier series of functions in $AP(R, X)$.

Proposition 4.9. Let $\varphi \in AP(R, \mathcal{C})$ be such that

$$\varphi(t) \sim \sum_{k=1}^{\infty} a_k e^{i\lambda_k t}$$

and $\lambda \in R$ a number that is not a linear combination of λ_k's with rational coefficients (i.e., λ does not belong to the module generated by the λ_k's over the rationals). Then, for each $\varepsilon > 0$, there exist δ, $0 < \delta < (\pi/2)$, and an integer n such that φ admits ε-translation numbers τ, verifying the system of inequalities

$$|\lambda_k \tau| < \delta, \ k = 1, 2, \ldots, n, \ |\lambda\tau - \pi| < \delta \ (\text{mod} \, 2\pi). \tag{4.84}$$

Proof. According to Proposition 4.7, there exist δ and n, which depend on ε only, such that any solution of the system consisting of the first n inequalities in formula (4.84) is an ε-translation number for $\varphi(t)$. We must prove that among them there is at least one satisfying the last inequality in formula (4.84). This fact is a direct consequence of Proposition 4.7. Indeed, a relationship of the form

$$\sum_{k=1}^{n} m_k \lambda_k + m\lambda = 0, \qquad m_k \text{ and } m \text{ integers}, \tag{4.85}$$

is not possible. In Proposition 4.7, we can take $\mu_k = 0$, $k = 1, 2, \ldots, n$, $\mu_{k+1} = \pi$, which would constitute a contradiction to equation (4.85).

This ends the proof of Proposition 4.9.

Proposition 4.10. Let $f \in AP(R, X)$, where X stands for a Banach space (complex or real). Then

$$a(\lambda, f) = M\{f(t)e^{-i\lambda t}\}, \qquad \lambda \in R, \tag{4.86}$$

is different from the null element in X only for a set of λ's at most countable.

Proof. Let us define the function

$$\varphi(t) = \sup_{u \in R} |f(t+u) - f(u)|_X, \qquad t \in R.$$

Since

$$
\begin{aligned}
|\varphi(t+\tau) - \varphi(t)| &= \left| \sup_{u \in R} |f(t+\tau+u) - f(u)|_X - \sup_{u \in R} |f(t+u) - f(u)|_X \right| \\
&\leq \sup_{u \in R} \left| |f(t+\tau+u) - f(u)|_X - |f(t+u) - f(u)|_X \right| \\
&\leq \sup_{u \in R} |f(t+\tau+u) - f(t+u)|_X = \sup_{u \in R} |f(\tau+u) - f(u)|_X,
\end{aligned}
$$

we see that $\varphi(t)$ is uniformly continuous on R and almost periodic (Bohr). An ε-translation number for f is also an ε-translation number for φ. If λ_k, $k = 1, 2, \ldots$ are the Fourier exponents of φ, we denote by $\mathcal{M}(\varphi)$ the module generated by the set $\{\lambda_k\} \subset R$ over the rationals. We shall now prove that

$$\lambda \notin \mathcal{M}(\varphi) \implies a(\lambda, f) = \theta \in X. \tag{4.87}$$

Since the module $\mathcal{M}(\varphi)$ consists of all real numbers of the form

$$r_1 \lambda_1 + r_2 \lambda_2 + \cdots + r_m \lambda_m, \qquad m \geq 1,$$

and r_k rationals, there results that the cardinal number of $\mathcal{M}(\varphi)$ is at most $\aleph_0$. Hence, $\mathcal{M}(\varphi) \subset R$ is at most countable.

Now let $\lambda \notin \mathcal{M}(\varphi)$. According to Proposition 4.8, for each $\varepsilon > 0$, we can find ε-translation numbers τ of φ such that

$$|\lambda \tau - \pi| < \frac{\pi}{2} \ (\mathrm{mod}\, 2\pi). \tag{4.88}$$

But formula (4.88) implies

$$\left| 1 - e^{-i\lambda t} \right| > 1. \tag{4.89}$$

The relations

$$
\begin{aligned}
a(\lambda, f) &= \lim_{T \to \infty} T^{-1} \int_{\tau}^{\tau+T} f(t) e^{-i\lambda t} dt = e^{-i\lambda \tau} M\{f(t+\tau) e^{-i\lambda t}\} \\
&= e^{-i\lambda \tau} a(\lambda, f) + e^{-i\lambda \tau} M\{[f(t+\tau) - f(t)] e^{-i\lambda t}\}
\end{aligned}
$$

are obvious, and keeping in mind that τ is an ε-translation number for f, one obtains, based also on formula (4.88),

$$|a(\lambda, f)|_X |1 - e^{-i\lambda \tau}| \leq \varepsilon, \qquad |a(\lambda, f)|_X < \varepsilon.$$

The arbitrariness of $\varepsilon > 0$ proves Proposition 4.10.

Definition 4.2. Let $f \in AP(R, X)$, with X a Banach space. Denote by λ_k, $k \geq 1$, those real λ's for which $a_k = a(\lambda_k, f) \neq \theta$. Then we write

$$f(t) \sim \sum_{k=1}^{\infty} a_k e^{i\lambda_k t}, \tag{4.90}$$

and we shall call the series on the right-hand side of formula (4.90) the *Fourier series associated to* f.

Remark 4.14. The way we have defined the Fourier series for $f \in AP(R, X)$ coincides with the case $f \in AP(R, \mathcal{C})$, despite the fact that we have proceeded in a different way to prove the existence of this series. The cases of the spaces $S(R, X)$ or $B(R, X)$ will not be discussed here.

An important feature related to the concept of Fourier series consists in the fact that this series completely characterizes the generating functions. This result is known as the *uniqueness theorem*. It will be established by reducing the problem to the case of scalar functions. In other words, the result concerning $AP(R, X)$ will be derived from the one corresponding to the space $AP(R, \mathcal{C})$. The scalar case has been discussed in Corollary 4.1.

Proposition 4.11. *Let* $f \in AP(R, X)$ *and* $\varphi \in X^* =$ *the dual of* X. *Then* $\varphi(f(t)) \in AP(R, \mathcal{C})$, *and* $\varphi(M\{f\}) = M\{\varphi(f)\}$.

Proof. The proof follows directly from the obvious inequalities

$$|\varphi(f(t + \tau)) - \varphi(t)| \leq \|\varphi\| |f(t + \tau) - f(t)|_X,$$

where $\|\varphi\|$ is the norm in X^* and

$$\varphi\left(\frac{1}{T} \int_0^T f(t) dt\right) = \frac{1}{T} \int_0^T \varphi(f(t)) dt.$$

Theorem 4.7 (Uniqueness). *Let* $f, g \in AP(R, X)$, *with* X *a Banach space. If they have identical Fourier series, then* $f(t) \equiv g(t)$, $t \in R$.

Proof. Our hypothesis leads to the relationship $M\{f(t)e^{-i\lambda t}\} = M\{g(t)e^{-i\lambda t}\}$, which tells us that the Fourier exponents are the same and so are the Fourier coefficients. Both sides are zero, for λ real, except the values λ_k, $k \geq 1$, which denote the Fourier exponents of both functions. For $\lambda = \lambda_k$, one must have $a(\lambda_k; f) = a(\lambda_k, g)$, which expresses the equality of Fourier coefficients of f and g. For any $\varphi \in X^*$, one has $M\{\varphi(f)e^{-i\lambda t}\} = M\{\varphi(g)e^{-i\lambda t}\}$. This means that the scalar functions $\varphi(f(t))$ and $\varphi(g(t))$ have the same Fourier series. According to Theorem 4.5, we get $\varphi(f(t)) = \varphi(g(t))$, and since $\varphi \in X^*$ is arbitrary, one obtains $f(t) = g(t)$, $t \in R$.

We shall now prove the *approximation theorem* for functions in $AP(R, X)$ by using the Bochner–Fejér polynomials defined in Section 4.4 for almost periodic functions in $AP(R, C)$. Before we state the result, let us point out the fact that the trigonometric polynomials $\sigma_m(t)$, $m \geq 1$, defined by formula (4.66), make sense when $f \in AP(R, X)$. Of course, their coefficients now belong to X.

Theorem 4.8. *Let* $f \in AP(R, X)$ *with*

$$f(t) \sim \sum_{k=1}^{\infty} a_k e^{i\lambda_k t},$$

where X denotes a Banach space (over the complex field). Then, the sequence of polynomials

$$\sigma_m(t) = \sum_{k=1}^{n} r_{k,m} a_k e^{i\lambda_k t}, \qquad n = n(m), \qquad (4.91)$$

converges in $AP(R, X)$, *as* $m \to \infty$, *to the generating function f. The rational numbers $r_{k,m}$ depend on k and m but not on a_k. Moreover,*

$$r_{k,m} \longrightarrow 1 \qquad as \; m \longrightarrow \infty.$$

Proof. The proof is rather similar to that of Theorem 4.5 in the case of scalar-valued functions. There is a significant difference at the moment we apply Theorem 3.3 of relative compactness of sets in $AP(R, X)$. In the case of $AP(R, C)$, it has been much simpler to deal with because the boundedness is equivalent to relative compactness in finite-dimensional spaces. We shall use the function $K_n(t)$, $n \geq 1$, defined above by equation (4.63). We can write, for each $m \geq 1$,

$$\sigma_m(t) = M_s \left\{ f(t+s) K_{(m!)^2} \left(\frac{\beta_1 s}{m!} \right) \cdots K_{(m!)^2} \left(\frac{\beta_m s}{m!} \right) \right\}, \qquad (4.92)$$

where $\{\beta_j; \, j \geq 1\}$ is a basis for the Fourier exponents of the function $f \in AP(R, X)$. If we take into account properties (4.64) and (4.65) of the functions $K_n(t)$, $n \geq 1$, and proceed as in the proof of Theorem 4.5, one obtains the expression (4.91) for the polynomials $\sigma_m(t)$, $m \geq 1$.

First, we shall prove that the sequence $\{\sigma_m(t); \, m \geq 1\}$ is weakly convergent to f. Indeed, for each $\varphi \in X^*$, one obtains for $\varphi(\sigma_m(t))$, $m \geq 1$, the formula

$$\varphi(\sigma_m(t)) = \lim_{T \to \infty} T^{-1} \int_t^{t+T} K_m(s-t) \varphi(f(s)) ds \qquad (4.93)$$

with

$$K_m(u) = K_{(m!)^2} \left(\frac{\beta_1 u}{m!} \right) \cdots K_{(m!)^2} \left(\frac{\beta_m u}{m!} \right), \qquad m \geq 1. \qquad (4.94)$$

Taking into account the properties of $K_m(u)$, one sees from equation (4.93) that $\varphi(\sigma_m(t))$ converges uniformly on R to $\varphi(f(t))$ for any $\varphi \in X^*$. We also have to take into account Proposition 4.11 above as well as the fact that the Fourier coefficients of $\varphi(f(u))$ are, according to Proposition 4.11, equal to $\varphi(a_k)$, $k \geq 1$. This concludes the first part of the proof (i.e., the fact that $\{\sigma_m(t), m \geq 1\}$ converges weakly to $f(t)$ for each $t \in R$).

Second, in order to obtain the fact that the convergence is uniform on R in the norm of X, we shall apply the criterion of relative compactness given in Proposition 4.6. Let us note now that equation (4.92) implies the following estimates for the polynomials $\sigma_m(t)$, $m \geq 1$:

$$|\sigma_m(t)|_X \leq \sup_{t \in R} |f|_X = |f|_{AP}, \qquad t \in R, \qquad (4.95)$$

$$|\sigma_m(t + \tau) - \sigma_m(t)| \leq \sup_{t \in R} |f(t + \tau) - f(t)|, \qquad (4.96)$$

where $t, \tau \in R$ are arbitrary. While equation (4.95) shows the equiboundedness of the set $\{\sigma_m(t); m \geq 1\} \subset X$ for $t \in R$, the inequality (4.96) proves the equicontinuity and equi–almost-periodicity of that set.

In order to draw a conclusion about the relative compactness of the set $\{\sigma_m(t); m \geq 1\}$ for $t \in R$, we will apply a result of Phillips (see our book [23], p. 170).

According to this result, we must prove that, from any bounded sequence $\{\varphi_n; n \geq 1\} \subset X^*$, one can extract a subsequence $\{\varphi_{n_k}; k \geq 1\}$ such that $\{\varphi_{n_k}(\sigma_m(t)); k \geq 1\}$ is uniformly convergent on R also uniformly with respect to $m \geq 1$. Indeed, from equality (4.94), we derive the following inequalities for any bounded sequence $\{\varphi_n; n \geq 1\} \subset X^*$, where say $\|\varphi_n\| \leq M < +\infty$, $n \geq 1$, and $m \geq 1$ hold true due to formulas (4.95) and (4.96):

$$|\varphi_n(\sigma_m(t))| \leq M|f|_{AP}, \qquad (4.97)$$

$$|\varphi_n(\sigma_m(t + \tau)) - \varphi_n(\sigma_m(t))| \leq M|f(t + \tau) - f(t)|_{AP}. \qquad (4.98)$$

These inequalities imply the equiboundedness, equicontinuity, and equi–almost-periodicity of the sequence $\{\varphi_n(\sigma_m(t)); n \geq 1\} \subset AP(R, C)$. They hold uniformly when $m \geq 1$. From Proposition 4.6, there exists a subsequence $\{\varphi_{n_k}(\sigma_m(t)); k \geq 1\}$ such that it converges in $AP(R, C)$ uniformly when $m \geq 1$. This means that for any sequence $\{\varphi_n; n \geq 1\} \in X^*$, there exists a subsequence $\{\varphi_{n_k}; k \geq 1\}$ such that it is uniformly convergent on R, as $k \longrightarrow \infty$, uniformly on the set $\{x; x \in X, x = \sigma_m(t), t \in R, m \geq 1\} = A$. In other words, the map $\varphi \longrightarrow \varphi(x)$ from X^* into $B(A, D)$ is compact. Hence, the set $A \subset X$ is relatively compact in X.

In order to conclude the proof, we need to show that the sequence $\{\sigma_m(t); m \geq 1\} \subset AP(R, X)$ converges in the same topology uniformly on R (i.e., in $AP(R, X)$). We have seen above that the sequence $\{\sigma_m(t); m \geq 1\}$ converges weakly in X to the function $f(t)$. On the other hand, due to the fact that $\{\sigma_m(t); m \geq 1\} \subset AP(R, X)$ is relatively compact, there results that any convergent subsequence, say $\{\sigma_{m_p}(t); \ p \geq 1\}$, must have $f(t)$ as its limit. Indeed, the convergence in $AP(R, X)$ implies the weak convergence, and since the limit is unique (even for pointwise weak convergence!), we must have $\lim \sigma_{m_p}(t) = f(t)$ as $p \longrightarrow \infty$, $t \in R$. Therefore, any convergent subsequence of $\{\sigma_m(t); \ m \geq 1\}$ in $AP(R, X)$ must converge to $f(t)$. This means that $\lim \sigma_m(t) = f(t)$ as $m \longrightarrow \infty$ in $AP(R, X)$.

Theorem 4.8 is thereby proven.

Remark 4.15. Theorem 4.8 allows us to conclude that the approximation property using trigonometric polynomials is characteristic for functions in $AP(R, X)$. In other words, we have established the equivalence of the Bohr definition, the relative compactness of the family of translates $\{f(t + h); \ h \in R\}$, and the approximation property using trigonometric polynomials. Hence, one can define $AP(R, X)$ as the closure of the set of trigonometric polynomials of the form $\sum_{k=1}^{n} a_k \exp(i\lambda_k t)$ with $a_k \in X$ and $\lambda_k \in R$, $k \geq 1$.

Remark 4.16. There are several proofs for the remarkable Theorem 4.8. In L. Amerio [2], one can find a proof of another constructive type. Also, in M. A. Krasnoselskii et al. [55], a more direct proof is suggested based on the use of Minkowski's functionals.

Remark 4.17. Hilbert spaces occur in applications and, as mentioned already, for the space $AP(R, H)$, where is a H is a Hilbert space, the theory of Fourier series is the same as the case of the space $AP(R, \mathcal{C}) \subset B^2(R, \mathcal{C})$, as developed in Section 4.2. In particular, Parseval's formula $\sum_{k=1}^{\infty} \|a_k\|^2 = M\{\|f\|^2\}$ is valid for any $f \in AP(R, H)$.

4.6 Further Topics Related to Fourier Series

In this section, we will discuss several facts and properties related to the concept of Fourier series for various classes of almost periodic functions. We shall also attempt a classification of almost periodic functions based on the nature of the module generated by the sequence of Fourier coefficients. This approach will lead us to the concept of *quasi-periodic functions*. They constitute a subspace of $AP(R, \mathcal{C})$ and were investigated much earlier than the space $AP(R, \mathcal{C})$ by P. Bohl [13], [14]. Some of the procedures used by Bohl, and also by E. Esclangon [38], have been extended by Bohr to the case of functions in $AP(R, \mathcal{C})$. Part of the results will be given without proof but with adequate references. Such results will not be used in forthcoming chapters.

First, we shall deal with the space $AP_1(R, \mathcal{C})$. The following result is easily proven: If $\{\lambda_k; \ k \geq 1\} \subset R$ is an arbitrary sequence and $\{a_k; \ k \geq 1\} \subset \ell^1(Z_+, \mathcal{C})$, then

$$f(t) = \sum_{k=1}^{\infty} a_k e^{i\lambda_k t} \in AP_1(R, \mathcal{C}). \tag{4.99}$$

The inclusion (4.99) is obvious, and we only note the fact that, choosing λ_k's and a_k's as indicated, we will obtain the whole space $AP_1(R,C)$. It is worth mentioning that the distribution of λ_k's on R can be totally arbitrary.

Second, we shall mention a result due to A. S. Besicovitch that is of the Riesz–Fischer type: If $\{\lambda_k; \ k \geq 1\} \subset R$ is an arbitrary sequence and $\{a_k; \ k \geq 1\} \subset \ell^2(Z_+,C)$, then there exists f such that

$$f(t) \sim \sum_{k=1}^{\infty} a_k e^{i\lambda_k t} \in AP_2(R,C). \tag{4.100}$$

The proof of the existence of $f(t)$ can be found in A. S. Besicovitch [10].

In connection with the results mentioned above, it appears natural to check the validity of a similar result for the case of the space $AP_r(R,C)$, $1 < r < 2$, introduced in Section 2.5.

As seen above, the distribution of Fourier exponents of an almost periodic function can be completely arbitrary (on the real axis R). In Sections 4.3 and 4.4, we have seen that the distribution of the Fourier exponents plays a certain role in regard to the convergence of the Fourier series to the generating function. When talking about the distribution of Fourier exponents, we have in mind the *geometric distribution*. But instead of looking at the geometric aspect, we shall now consider some *algebraic* properties of the set of Fourier exponents. In Section 4.5, we have dealt with the module of the Fourier exponents of an almost periodic function over the field Q of rationals.

Let $f \in AP(R,C)$ be such that

$$f(t) \sim \sum_{k=1}^{\infty} a_k e^{i\lambda_k t}. \tag{4.101}$$

The smallest additive group in R that contains all exponents λ_k, $k \geq 1$, is called the *module* of f, and we shall denote it by $\mathrm{mod}(f)$.

It is a simple exercise to see that the elements of $\mathrm{mod}(f)$ are of the form

$$m_1\lambda_1 + m_2\lambda_2 + \cdots + m_p\lambda_p \tag{4.102}$$

with $m_j \in Z$, $j=1,2,\ldots,p$, $p \geq 1$. The representation (4.102) is unique only if the λ_k's are linearly independent (i.e., from $m_1\lambda_1+m_2\lambda_2 + \cdots + m_p\lambda_p = 0$ there results $m_1 = m_1 = \cdots = m_p = 0$). Generally speaking, this is not the case for arbitrary sequences $\{\lambda_k; \ k \geq 1\} \subset R$. When only a finite number of λ_k's are linearly independent, say $p \geq 1$, then each element of $\mathrm{mod}(f)$ can be uniquely represented in the form (4.102).

Definition 4.3. Let $f \in AP(R,C)$ be such that the maximum number of linearly independent Fourier exponents is finite, say $p \geq 1$. Then f is called *quasi-periodic*. The set of all quasi-periodic functions in $AP(R,C)$ will be denoted by $AP_Q(R,C)$.

It is obvious that the Fourier series of a function $g \in AP_Q(R, C)$ has the form

$$g \sim \Sigma a_{m_1, m_2, \ldots, m_p} e^{i(m_1 \lambda_1 + \cdots + m_p \lambda_p)t}, \tag{4.103}$$

the summation being extended from 1 to ∞ for each m_j, $j = 1, 2, \ldots, p$. The fact that we have p indices for summation does not raise any problems since the number of terms on the right-hand side of formula (4.102) is countable, and terms can be rearranged in such a way as to obtain a series with a single summation index. One can easily see that $AP_Q(R, C)$ is a linear manifold in $AP(R, C)$. Instead of the space $AP(R, C)$, one could start with another space of almost periodic functions like $AP_1(R, C)$, $AP_2(R, C) = B^2(R, C)$, or even $B(R, C)$.

If there are infinitely many linearly independent λ_k's, one obtains almost periodic functions of the most general type. Let us point out the fact that if the whole sequence of λ_k's consists of linearly independent elements, we have proven Theorem 4.6, stating that such a function necessarily belongs to the space $AP_1(R, C)$.

A remarkable case of quasi-periodicity corresponds to the situation $p = 1$. This means that a single Fourier exponent generates the whole set of exponents. In this case, $\text{mod}(f)$ consists of the set of real numbers that are multiples of a given (positive!) number, say $\omega > 0$; i.e., $\text{mod}(f) = \{k\omega; k \in Z\}$. The Fourier series has the form

$$f \sim \sum_{k \in Z} a_k e^{ik\omega t}, \tag{4.104}$$

which represents the Fourier series of a *periodic function* with period $2\pi/\omega$. Again, the terms can be rearranged in order to present the series on the right-hand side of formula (4.104) in the form (4.101).

It is interesting to present a method to generate the quasi-periodic functions starting from periodic functions depending on several variables. The periodicity is assumed with respect to each argument, and the period is the same, equal to 2π.

Let $F(t_1, t_2, \ldots, t_p)$ be such a continuous map from R^p into C. The Fourier series of F is given by

$$F(t_1, t_2, \ldots, t_p) \sim \Sigma a_{m_1, m_2, \ldots, m_p} e^{i(m_1 t_1 + m_2 t_2 + \cdots + m_p t_p)}, \tag{4.105}$$

where the summation is extended to all $m_k \in Z_+$, $k = 1, 2, \ldots, p$. The coefficients can be expressed by the formula

$$a_{m_1, m_2, \ldots, m_p} = (2\pi)^{-p} \underbrace{\int_{-\pi}^{\pi} \cdots \int_{-\pi}^{\pi}}_{p \text{ times}} e^{-(m_1 t_1 + \cdots + m_p t_p)} F(t_1, t_2, \ldots, t_p) dt_1 dt_2 \ldots dt_p,$$

which constitutes a generalization of the well-known Euler formulas.

It turns out that *any quasi-periodic function can be obtained from a periodic function of the type mentioned above, taking its values on a straight line.* Indeed, with the notation used above, let us consider in R^p the line whose parametric equations are $t_k = \omega_k t$, $k = 1, 2, \ldots, p$, $\omega_k \neq 0$, and $t \in R$. Then

$$f(t) = F(\omega_1 t, \omega_2 t, \ldots, \omega_p t), \qquad t \in R, \qquad (4.106)$$

is a continuous function in $CB(R, \mathcal{C})$, periodic in each argument. Moreover, from formula (4.105), we see that $f(t)$ has a Fourier series, namely

$$f(t) \sim \Sigma a_{m_1, m_2, \ldots, m_p} e^{i(m_1\omega_1 + m_2\omega_2 + \cdots + m_p a_p)t}. \qquad (4.107)$$

From formula (4.107), we notice that we are in the circumstances described by formula (4.103), which means that $f(t)$ is a quasi-periodic function. Let us point out the fact that our discussion has a formal character. A proper proof of the fact mentioned above can be found in Besicovitch [10] or Levitan [62].

Another class of almost periodic functions is the class of the so-called *limit periodic functions*. The definition is very simple. The function $f \in AP(R, \mathcal{C})$ is called limit periodic iff $f(t) = \lim f_k(t)$, $k \to \infty$, uniformly on R, where each $f_k(t)$, $k \geq 1$, is periodic and hence in $AP(R, \mathcal{C})$. Of course, the periods need not be the same for each $f_k(t)$. According to property III from Section 3.2, the limit function must be in $AP(R, \mathcal{C})$.

The connection of the property of limit periodicity with the Fourier series is expressed by the formula

$$f(t) \sim \sum_{k=1} a_k e^{ir_k \omega t}$$

with $r_k \in Q$ the field of rationals and $\omega \neq 0$. Obviously, this case includes the periodic one, corresponding to the situation where all r_k are integers. For details, see Besicovitch [10].

Another important concept related to Fourier series is *spectrum*, also attached to an almost periodic function. Actually, the spectrum can be defined for functions in $BC(R, \mathcal{C})$ or even $BC(R, X)$, with X a Banach space. The spectrum of almost periodic functions enjoys some special properties. Roughly speaking, the spectrum of almost periodic functions is determined by the set of its Fourier exponents. There are several (equivalent) definitions for the concept of a spectrum of an almost periodic function, and we shall follow the presentation of Levitan and Zhikov [63].

Let $f \in BC(R, X)$ with X a Banach space. A point $\lambda \in R$ is called *regular* if there is a neighborhood $V_\lambda \subset R$ such that

$$(f * \varphi)(t) = \int_R f(t - s)\varphi(s)ds = 0 \qquad \text{for every } \varphi \in L^1(R, \mathcal{C})$$

for which its Fourier transform $\widehat{\varphi} \in C_0^\infty$ with support in V_λ.

The *spectrum* of f, denoted by $\sigma(f)$, is the complement (in R) of the set of all regular points of f. Since the set of regular points is obviously open, there results that $\sigma(f)$ is a closed set in R. We note that the Fourier transform $\widehat{\varphi}(\lambda)$ is defined for $\varphi \in L^1(R,\mathcal{C})$ by $\widehat{\varphi}(\lambda) = \int_R \varphi(t)e^{-i\lambda t}dt$, while the convolution product has been discussed in Section 2.5. Other definitions for the spectrum are formulated in Y. Katznelson [54] and L. H. Loomis [65]. The last reference covers the case of measurable functions, and the line R is replaced by a locally compact topological group G.

Again let $f \in \mathrm{BC}(R,X)$, and define a *point of almost periodicity* $\lambda \in R$ by the following property: There exists a neighborhood W_λ of $\lambda \in R$ such that $f * \varphi \in \mathrm{AP}(R,X)$ for each $\varphi \in L^1(R,\mathcal{C})$ with $\widehat{\varphi} \in C_0^\infty$ and support in W_λ. The set of points of almost periodicity of a function $f \in \mathrm{BC}(R,X)$ is obviously open, and its complement in R is denoted by $\Delta(f)$. Obviously, $\Delta(f) \subset \sigma(f)$.

The following result (see Levitan and Zhikov [63]) provides conditions of almost periodicity in terms of the spectral properties. Let $f \in \mathrm{BC}(R,X)$ be uniformly continuous, and assume $\Delta(f)$ is a rarified set (i.e., it is closed in R and does not contain any nonempty perfect subset). Then $f \in \mathrm{AP}(R,X)$ if at least one of the following properties is verified:

(a) The space X does not contain $c_0(Z_+, R)$.

(b) The set $\{x; \; x = f(t), \; t \in R\} \subset X$ is weakly compact.

Remark 4.18. If instead of a Banach space X we consider one of the spaces R^n or $\mathcal{C}^n$, then condition (b) is automatically verified.

Remark 4.19. By $c_0(Z_+, R)$, one denotes the space of sequences $\{x_n; \; n \geq 1\}$ such that $|x_n| \longrightarrow 0$ as $n \longrightarrow \infty$. The norm is the usual supremum norm, and $c_0(Z_+, R)$ is a Banach space.

Other topics on almost periodicity will be discussed later as they are needed.

5

Linear Oscillations

5.1 The Equation $\dot{x}(t) = f(t)$

Before one can investigate linear oscillations described by linear differential
equations or related linear equations (such as integro-differential ones), we
need to clarify the matter of almost periodicity in regard to the simplest
differential equation,

$$\dot{x}(t) = f(t), \qquad t \in R, \tag{5.1}$$

with f chosen from various spaces of almost periodic functions. In other words,
we want to answer the question: *When is the primitive of an almost periodic
function almost periodic?*

The answer is negative in general. For instance, if we assume $f(t)$ to be
periodic (and continuous) of period 2π, then it is easy to see that

$$x(t) = C + \int_0^t f(s)ds \tag{5.2}$$

is periodic of the same period iff the mean value of $f(t)$ is equal to zero:

$$\int_0^{2\pi} f(t)dt = 0. \tag{5.3}$$

A simple example illustrating the fact that the primitive of a periodic function may not be almost periodic is given by $f(t) = 1 + \cos t$, whose primitive
$F(t) = C + t + \sin t$ is unbounded. Hence, it does not belong to $AP(R, R)$ or
even to $S(R, R)$, $B(R, R)$.

There are important categories of almost periodic functions for which the
answer to the question above is positive. In Section 3.1, we noted that for $f \in
AP_1(R, C)$ such that the Fourier exponents satisfy $|\lambda_k| \geq \delta > 0$, $k = 1, 2, \ldots$,
the primitive is also in $AP_1(R, C)$.

The following classical result provides an answer to the question of almost
periodicity of the primitive of an almost periodic function in the case of the
space $AP(R, C)$.

C. Corduneanu, *Almost Periodic Oscillations and Waves*,
DOI 10.1007/978-0-387-09819-7_5, © Springer Science+Business Media, LLC 2009

Theorem 5.1. *Let* $f \in AP(R, \mathcal{C})$. *Then its primitive* $F \in AP(R, \mathcal{C})$ *iff it belongs to* $BC(R, \mathcal{C})$.

Proof. Since $f \in AP(R, \mathcal{C})$, it means that $f = f_1 + if_2$ with $g_1, g_2 \in AP(R, R)$, it suffices to prove the theorem for $f \in AP(R, R)$ only because

$$\int_0^t f(s)ds = \int_0^t f_1(s)ds + i \int_0^t f_2(s)ds,$$

and the first term is bounded iff both integrals on the right-hand side are bounded in R.

Our hypothesis states that $F(t) \in BC(R, R)$, which implies the existence of two real numbers $m < M$ such that

$$m = \inf_{t \in R} F(t) \le \sup_{t \in R} F(t) = M. \tag{5.4}$$

Denote by $\mathcal{S}(f)$ the set of real sequences $s = \{s_n; \, n \ge 1\}$ with the property that $f(t + s_n) \longrightarrow f_s(t)$ in $AP(R, R)$ as $n \longrightarrow \infty$. Such sequences do exist according to the Bochner property of functions in $AP(R, R)$. For such a sequence, we can write

$$F(t + s_n) = F(s_n) + \int_0^t f(s + s_n)ds, \qquad n \ge 1, \, t \in R, \tag{5.5}$$

and observe that the right-hand side is uniformly convergent to $(f_s(t))$ on each finite interval of R. Without loss of generality, we can assume $F(s_n) \longrightarrow C_s = $ constant as $n \longrightarrow \infty$, substituting if necessary by a subsequence and taking into account equation (5.4).

We derive from above, for each $t \in R$,

$$\lim_{n \to \infty} F(t + s_n) = C_s + \int_0^t f_s(u)du = F_s(t). \tag{5.6}$$

From equation (5.6), we see that $F_s(t)$ is a primitive for $f_s(t)$ since the convergence in equation (5.6) is uniform on each finite interval of R. Also, from equation (5.6) we obtain

$$m \le m_s = \inf_{t \in R} F_s(t) \le \sup_{t \in R} F_s(t) = M_s \le M. \tag{5.7}$$

We shall prove now, as a subsidiary result, that $m_s = m$ and $M_s = M$. Indeed, assume $M_s < M$. Then, from $f(t + s_n) \longrightarrow f_s(t)$ in $AP(R, R)$ as $n \longrightarrow \infty$, we obtain $\lim f_s(t - s_n) = f(t)$ in $AP(R, R)$. Indeed, from $\sup |f(t + s_n) - f_s(t)| \longrightarrow 0$ as $n \longrightarrow \infty$, one derives $\sup |f_s(t - s_n) - f(t)| = 0$ as $n \longrightarrow \infty$. Without loss of generality (maybe using a subsequence of s if necessary), we can write

$$\lim_{n \to \infty} F_s(t - s_n) = C + \int_0^t f(u)du = C + F(t) \tag{5.8}$$

for some constant $C \in R$. Now, taking equations (5.7) and (5.8) into account, we can write

$$\sup_{t\in R}[C + F(t)] = C + M \le M_s < M,$$

and similarly

$$\inf_{t\in R}[C + F(t)] = C + m \ge m_s \ge m.$$

From these inequalities, we obtain first $C < 0$ and then $C \ge 0$. This is absurd, which means that equation (5.7) is valid with equal signs: $m = m_s$, $M = M_s$.

We shall now prove that $F(t) \in AP(R, R)$. Let $s = \{s_n; \ n \ge 1\} \subset R$ be arbitrary such that $s \in S(f)$; i.e., $f(t + s_n) \longrightarrow f_s(t)$ in $AP(R, R)$. We will show that the sequence of translates $\{F(t + s_n); \ n \ge 1\}$ is uniformly convergent on R. If this property is false, then we can find a constant $\rho > 0$, as well as three sequences $\{\alpha_n; \ n \ge 1\}, \{s_{n1}; \ n \ge 1\} \subset \{s_n; \ n \ge 1\}$, and $\{s_{n2}; \ n \ge 1\} \subset \{s_n; \ n \ge 1\}$, with the property

$$|F(\alpha_n + s_{n2}) - F(\alpha_n + s_{n1})| \ge \rho. \tag{5.9}$$

We can assume that $\{\alpha_n + s_{n1}; \ n \ge 1\} \in S(f)$ and $\{\alpha_n + s_{n2}; \ n \ge 1\} \in S(f)$ by going to subsequences if necessary, which translates into

$$\lim_{n\to\infty} f(t + \alpha_n + s_{n1}) = f_1(t),$$

$$\lim_{n\to\infty} f(t + \alpha_n + s_{n2}) = f_2(t),$$

both in $AP(R, R)$. From these limits, we can obtain the similar ones

$$\begin{cases} \lim_{n\to\infty} F(t + \alpha_n + s_{n1}) = F_1(t), \\ \lim_{n\to\infty} F(t + \alpha_n + s_{n2}) = F_2(t), \end{cases} \tag{5.10}$$

uniformly on each finite interval of R. Again, if necessary, one can go to subsequences, which will be denoted the same way as the original sequences. Of course,

$$F_1(t) = \int_0^t f_1(s)ds, \qquad F_2(t) = \int_0^t f_2(s)ds.$$

Let us show that

$$f_1(t) = f_2(t), \qquad t \in R. \tag{5.11}$$

Indeed, from the fact that $f(t + s_n) \longrightarrow f_s(t)$ in $AP(R, R)$, we can write

$$\sup_{t\in R}|f(t + \alpha_n + s_{n1}) - f(t + \alpha_n + s_{n2})|$$

$$= \sup_{t\in R}|f(t + s_{n1}) - f(t + s_{n2})| \longrightarrow 0.$$

This means that $F_2(t) - F_1(t) = C = $ constant, and according to formulas (5.9) and (5.10), we have $|C| \ge \rho > 0$. But this is in contradiction with the fact established above, $M_s = M$, or $\sup F_2(t) = \sup F_1(t)$, $t \in R$.

The discussion above shows that our assumption concerning the nonuniform convergence of $\{F(t + s_n); \ n \geq 1\}$ is not acceptable.

Hence, $F(t)$ satisfies the Bochner definition for almost periodicity:

$$F(t) \in \mathrm{AP}(R, R).$$

This ends the proof of Theorem 5.1.

While the case of the space $\mathrm{AP}(R, \mathcal{C})$ did not pose very difficult problems, the case of the space $\mathrm{AP}(R, X)$ with X a Banach space will require some preparation and some new concepts. We shall discuss this case in the next section.

In closing the discussion in this section, we shall provide a result that guarantees the almost periodicity of the primitive of an almost periodic function from $\mathrm{AP}(R, \mathcal{C})$ in terms of the Fourier series.

First, we shall prove an auxiliary result related to trigonometric polynomials.

Proposition 5.1. *Let us consider the trigonometric polynomial*

$$T(t) = \sum_{k=1}^{n} a_k e^{i\lambda_k t}, \qquad t \in R,$$

whose exponents satisfy the condition $|\lambda_k| \geq M > 0$, $k = 1, 2, \ldots, n$. *Denote*

$$\widetilde{T}(t) = \sum_{k=1}^{n} \frac{a_k}{i\lambda_k} e^{i\lambda_k t}, \qquad t \in R.$$

Then, there exists a constant $K > 0$, *independent of* T *and* M, *such that*

$$\sup_{t \in R} |\widetilde{T}| \leq K M^{-1} \sup_{t \in R} |T|. \tag{5.12}$$

Proof. The proof relies on the use of the Fourier transform of complex-valued functions from $L^2(R, \mathcal{C})$, namely

$$\widehat{\varphi}(u) = (2\pi)^{-1} \int_R e^{-iut} \varphi(t) dt. \tag{5.13}$$

We define $\varphi(t)$ as $\varphi(t) = it$, $|t| \leq 1$, $\varphi(t) = -(it)^{-1}$, $|t| \geq 1$, and assume $M = 1$. By elementary estimates, the following inequality can be obtained:

$$|\widehat{\varphi}(u)| \leq \frac{C}{1 + u^2}, \qquad C > 0. \tag{5.14}$$

One proceeds by integrating by parts in equation (5.13) and taking into account that $2|t^3| \geq 1 + t^2$ for $|t| \geq 1$. The constant C can be determined, but its value is not material.

Since formula (5.14) implies $\widehat{\varphi} \in L^1(R, \mathcal{C})$, there results that the inversion formula holds (in the sense of principal value).

Taking into account that $|\lambda_k| \geq 1$, $k = 1, \ldots, n$, we have

$$\widetilde{T}(t) = -\sum_{k=1}^{n} a_k \varphi(\lambda_k) e^{i\lambda_k t} = -\sum_{k=1}^{n} a_k \left(\int_R \widehat{\varphi}(u) e^{i\lambda_k u} du \right) e^{i\lambda_k t}$$

$$= -\int_R \left(\sum_{k=1}^{n} a_n e^{i(u+t)\lambda_k} \right) \widehat{\varphi}(u) du = -\int_R T(u+t) \widehat{\varphi}(u) du,$$

which implies an inequality of the type (5.12):

$$\sup_{t \in R} |\widetilde{T}(t)| \leq |\widehat{\varphi}|_{L^1} \sup_{t \in R} |T(t)|. \tag{5.15}$$

In order to eliminate the restriction $M = 1$, we have to consider the trigonometric polynomial $T_1(t) = T\left(\frac{t}{M}\right)$ instead of $T(t)$. In this case, applying formula (5.15), one obviously obtains formula (5.12) with $K = |\widehat{\varphi}|_{L^1}$.

We can now state the result that assures the existence of an almost periodic primitive for a function in $AP(R, \mathcal{C})$. Let us observe that a similar result in the case of the space $AP_1(R, \mathcal{C})$ has been discussed in Section 4.1.

Theorem 5.2. *If $f \in AP(R, \mathcal{C})$ such that*

$$f(t) \sim \sum_{k=1}^{\infty} a_k e^{i\lambda_k t}, \tag{5.16}$$

then $|\lambda_k| \geq M > 0$, $k \geq 1$, implies $\int_0^t f(s)ds \in AP(R, \mathcal{C})$. Accordingly,

$$\int_0^t f(s)ds \sim C + \sum_{k=1}^{\infty} a_k (i\lambda_k)^{-1} e^{i\lambda_k t}. \tag{5.17}$$

Proof. We have shown that there exists a sequence of trigonometric polynomials $\{\sigma_n(t); n \geq 1\}$ such that $\sigma_n(t) \longrightarrow f(t)$ as $n \longrightarrow \infty$ in $AP(R, \mathcal{C})$, with $\sigma_n(t)$ of the form

$$\sigma_n(t) = \sum_{k=1}^{m(n)} a_k^{(n)} e^{i\lambda_k t}, \qquad t \in R. \tag{5.18}$$

In other words, the exponents of $\sigma_n(t)$ are chosen from those of f. This has been proven in Theorem 4.5 (Bochner's approximation theorem).

Let us now define

$$\widetilde{\sigma}_n(t) = \sum_{k=1}^{m(n)} a_k^{(n)} (i\lambda_k)^{-1} e^{i\lambda_k t}, \qquad t \in R. \tag{5.19}$$

Taking into account Proposition 5.1, we can write from equations (5.18) and (5.19) the inequality

$$\sup_{t\in R} |\tilde{\sigma}_n(t) - \tilde{\sigma}_m(t)| \leq KM^{-1} \sup_{t\in R} |\sigma_n(t) - \sigma_m(t)|, \tag{5.20}$$

where K is a constant independent of $\{\sigma_n(t);\ n \geq 1\}$ and M. The inequality (5.20) shows that the sequence $\{\tilde{\sigma}_n(t);\ n \geq 1\}$ is a Cauchy sequence in $AP(R,\mathcal{C})$. Therefore, we can find $g \in AP(R,\mathcal{C})$ such that $\tilde{\sigma}_n(t) \longrightarrow g(t) \in AP(R,\mathcal{C})$. In the same way we derived formula (5.20), one can get the inequality

$$\sup_{t\in R} |\tilde{\sigma}_n(t)| \leq KM^{-1} \sup_{t\in R} |\sigma_n(t)| \tag{5.21}$$

for each $n \geq 1$. This implies, for $n \longrightarrow \infty$,

$$|g|_{AP} \leq KM^{-1}|f|_{AP}. \tag{5.22}$$

Since $a_k^{(n)} = r_{k,n}a_k$, $k, n \geq 1$, and $r_{k,n} \longrightarrow 1$ as $n \longrightarrow \infty$, there results $a_k^{(n)} \longrightarrow a_k$ as $n \longrightarrow \infty$.

Taking equation (5.19) into account, we see that

$$a_k^{(n)}(i\lambda_k)^{-1} \longrightarrow a_k(i\lambda_k)^{-1}$$

as $n \longrightarrow \infty$ for each $k \geq 1$. This means that

$$g \sim C + \sum_{k=1}^{\infty} a_k(i\lambda_k)^{-1}e^{i\lambda_k t},$$

which motivates formula (5.17) due to the fact that $\tilde{\sigma}'_n(t) = \sigma_n(t)$, $n \geq 1$, while the convergence of both sequences $\{\sigma_n(t);\ n \geq 1\}$ and $\{\tilde{\sigma}_n(t);\ n \geq 1\}$ is uniform on R.

Remark 5.1. Formula (5.17) tells us that the indefinite integral has the same exponents as the function itself, plus the exponent 0 (zero).

Remark 5.2. The inequality (5.22) plays an important role in searching almost periodic solutions to linear differential and related equations.

The simplest mechanical interpretation of the equation $\dot{x}(t) = f(t)$ is the following: The velocity of a point moving on a straight line being almost periodic, can we say that the motion itself is almost periodic?

As we already know from Section 3.1, the answer is positive when $f \in AP_1(R, R)$ and the Fourier exponents of f stay away from zero. The same is true in the case $f \in AP(R, R)$. Generally speaking, the motion may not be almost periodic, even though the velocity is almost periodic. Under certain restrictions, the motion itself is almost periodic, as we shall see later.

5.2 Weakly Almost Periodic Functions

The concept of *weak almost periodicity* is important in itself, but it turns out to be very helpful when investigating the almost periodicity of the primitive of an almost periodic function from $AP(R, X)$ with X a Banach space.

This concept of almost periodicity also represents an illustration of the more general situation when the Banach space X is substituted by a locally convex space. Namely, we shall deal with the dual of X, denoted by X^*, and the local convex space X_w with the same elements as X but a different type of convergence than the norm convergence in X.

Let us briefly discuss these concepts before we formulate the definition of weak almost periodicity.

First, the dual X^* of the Banach space X is defined as the set of linear and continuous maps from X into $\mathcal{C}$ endowed with the operations of pointwise addition of maps and scalar multiplication; the norm is defined by

$$\|\varphi\| = \sup\{|\varphi(x)|;\ x \in X,\ \|x\| \le 1\}.$$

It is an elementary exercise to check that X^* is also a Banach space (even if X is only a normed space).

Second, the weak convergence in X, which makes X a locally convex space, denoted by X_w, is defined by means of the seminorms

$$X \ni x \longrightarrow |\varphi(x)|, \qquad \varphi \in X^*. \tag{5.23}$$

These seminorms constitute a sufficient family because $\varphi(x) = 0$, $\forall \varphi \in X^*$, implies $x = \theta =$ the null element of X. This is a consequence of the Hahn–Banach theorem (see, for instance, E. Zeidler [110]).

Definition 5.1. A map $x : R \longrightarrow X$ is called *weakly almost periodic* if, for each $\varphi \in X^*$, the map (function) $\varphi(x(t)) \in AP(R, \mathcal{C})$.

Let us denote by $WAP(R, X)$ the set of weakly almost periodic functions with values in the Banach space X.

We have seen in Section 4.5 that for any $x \in AP(R, X)$ and $\varphi \in X^*$ one has $\varphi(x(t)) \in AP(R, \mathcal{C})$. This means that the elements of $AP(R, X)$ are also weakly almost periodic.

It is worth mentioning that some basic properties of almost periodic functions in $AP(R, X)$ are also valid for weakly almost periodic functions.

Proposition 5.2. *For each $x \in WAP(R, X)$, the set $\mathcal{R}_x = \{y \in X,\ y = x(t),\ t \in R\}$ is bounded and separable.*

Proof. The boundedness of $\mathcal{R}_x$ follows from the well-known result (see, for instance, V. Trénoguine [95]) that states that boundedness and weak boundedness are equivalent concepts.

Or, for each $\varphi \in X^*$, $\varphi(x(t)) \in AP(R, \mathcal{C})$, which implies its boundedness on R. In other words, the set $\mathcal{R}_x$ is weakly bounded and hence bounded.

To prove the fact that $\mathcal{R}_x$ is separable, we need to show that $\mathcal{R}_x$ belongs to the closure of a countable set of elements of the space X. Indeed, let us consider the countable set $\{x(r); \; r \in Q = \text{the set of rationals}\}$. We know that each $t_0 \in R$ is the limit of a sequence $\{r_m; \; m \geq 1\} \subset Q$. This implies $\lim \varphi(x(r_m)) = \varphi(x(t_0))$ for each $\varphi \in X^*$. According to a theorem of S. Mazur (see again V. Trénoguine [95]), there exists a sequence of linear combinations of the form

$$y_k = \sum_{j=1}^{m_k} \lambda_{jk} x(r_j) \tag{5.24}$$

with $\lambda_{jk} \in \mathcal{C}$ such that $y_k \longrightarrow x(t_0)$ as $k \longrightarrow \infty$ in the norm topology (i.e., in the space X). Without any loss of generality, one can assume that all λ_{jk} are rational numbers, still maintaining the validity of $y_k \longrightarrow x(t_0)$ as $k \longrightarrow \infty$. But the set of all y_k of the form (5.24) is at most countable. Hence, the separability of $\mathcal{R}_x$ is proven.

There are several features encountered in the case of the space $AP(R, X)$ that remain valid for the space $WAP(R, X)$. We shall state here a few of them, without proof, sending the interested reader to the book by L. Amerio and G. Prouse [3]. Let us mention that L. Amerio first considered the space $WAP(R, X)$, establishing its basic properties.

Proposition 5.3. *The space $WAP(R, X)$ is closed with respect to the weak uniform (on R) convergence.*

This means that from $\{x_k(t); \; k \geq 1\} \subset WAP(R, X)$ with $x_k(t) \longrightarrow x(t)$ as $k \longrightarrow \infty$ uniformly with respect to $t \in R$, one derives $x(t) \in WAP(R, X)$.

In other words, $WAP(R, X)$ is complete for its kind of convergence (the weak one).

Proposition 5.4. *If $x \in WAP(R, X)$, then there exists the mean value*

$$a(\lambda, x) = \lim{}^*(2T)^{-1} \int_{-T+a}^{T+a} x(t) e^{-i\lambda t} dt \tag{5.25}$$

uniformly with respect to $a \in R$.

The $\lim^*$ stands for the weak limit, and equation (5.25) is the consequence of

$$\lim(2T)^{-1} \int_{-T+a}^{T+a} \varphi(x(t) e^{-i\lambda t}) dt = \varphi \left(\lim(2T)^{-1} \int_{-T+a}^{T+a} x(t) e^{-i\lambda t} dt \right) \tag{5.26}$$

for each $\varphi \in X^*$ (because φ is continuous).

Proposition 5.5. *For each* $x \in WAP(R, X)$, $a(\lambda, x)$ *defined by equation* (5.25) *is different from the null element of* X *only for a set of values of* λ *that is at most countable.*

This proposition shows that we can associate to any $x \in WAP(R, X)$ a Fourier series. Those λ's for which $a(\lambda, x) \neq \theta$ are the Fourier exponents of x.

As in the case of $AP(R, X)$, the Fourier series characterizes the generating function.

Proposition 5.6. *If* $x \in WAP(R, X)$, *then the (Bochner–Féjer) polynomials* $\sigma_m(t)$, *defined by equation* (4.92) *for* $x(t)$ *in Section 4.5 with the mean value substituted by the weak mean value, converge weakly to* $x(t)$ *uniformly on* R.

Proof. First, it is worth mentioning that the Fourier exponents of each $\varphi(x(t))$, $\varphi \in X^*$, are the same as for $x(t)$.

Indeed, we obtain from equation (5.26)

$$a(\lambda, \varphi(x(t))) = M\{\varphi(x(t))e^{-i\lambda t}\} = \varphi(a(\lambda, x(t))),$$

showing that $a(\lambda, x(t)) = 0$ implies $a(\lambda, \varphi(x(t))) = 0$.

Moreover, one has

$$\varphi(\sigma_m(t)) = \sum_{r=1}^{n} r_{k,m}\varphi(a_k)e^{i\lambda_k t}, \qquad n = n(m). \tag{5.27}$$

Formula (5.27) tells us that $\varphi(\sigma_m(t))$ is the Fejér–Bochner polynomial for $\varphi(x(t))$. Since $\varphi(\sigma_m(t))$, $\varphi(x(t)) \in AP(R, X)$, one has

$$\lim_{m \to \infty} \varphi(\sigma_m(t)) = \varphi(x(t)) \tag{5.28}$$

uniformly on R.

The next proposition is the analogue of the Bochner characteristic property of almost periodicity.

Proposition 5.7. *Let* $x : R \longrightarrow X_w$ *be a continuous map. The necessary and sufficient condition for having* $x \in WAP(R, X)$ *is the relative compactness of the set* $\{x(t + h); h \in R\} \subset X$ *in the following sense: From each sequence* $\{x(t + h_k); k \geq 1\} \subset X$, *one can extract a subsequence* $\{x(t + h_{1k}); k \geq 1\}$ *that is uniformly convergent on* R *in the topology of* X_w *(i.e., the sequence* $\{\varphi(x(t + h_{1k})); k \geq 1\}$ *is uniformly convergent on* R *for each* $\varphi \in X^*$).

Proof. The sufficiency of the condition is obvious. Based on the Bochner criterion in $AP(R, \mathcal{C})$, we obtain the fact that each $\varphi(x(t)) \in AP(R, \mathcal{C})$, $\varphi \in X^*$. This is the definition of functions in $WAP(R, X)$.

The necessity can be proven as follows. Assume now that one has $x(t) \in WAP(R, X)$, and consider a sequence $\{x(t + h_j); j \geq 1\} \subset X$. We shall now extract a subsequence $\{x(t + h_{1j}); j \geq 1\}$ with the property that

$$\lim_{n\to\infty} e^{i\lambda_n h_{1j}} = e^{i\alpha_j}, \qquad j \geq 1. \tag{5.29}$$

This is possible because $|e^{i\alpha}| = 1$, $\alpha \in R$, and Bolzano's criterion is applicable. There follows

$$\lim_{j\to\infty} \sigma_m(x(t + h_{1j})) = \sum_{k=1}^{n} r_{k,m} a_k e^{i\alpha_k} e^{i\lambda_k t}, \qquad r = n(m), \tag{5.30}$$

uniformly on R. We have further, for any $\varphi \in X^*$,

$$\begin{aligned}
&|\varphi(x(t + h_{1j})) - \varphi(x(t + h_{1k}))| \\
&\leq |\varphi(x(t + h_{1j})) - \varphi(\sigma_m(x(t + h_{1j})))| \\
&+ |\varphi(\sigma_m(x(t + h_{1j}))) - \varphi(\sigma_m(t + h_{1k}))| \\
&+ |\varphi(\sigma_m(x + h_{1k})) - \varphi(x(t + h_{1k}))|.
\end{aligned}$$

If we now take into account equations (5.27) and (5.28), we find that $\{\varphi(x(t + h_{1j})); \ j \geq 1\}$ is a Cauchy sequence in $\mathrm{AP}(R, \mathcal{C})$. There remains to apply Proposition 5.3 to conclude that $x(t) \in W\mathrm{AP}(R, X)$.

The following result establishes the condition for a function in $W\mathrm{AP}(R, X)$ to belong also to $\mathrm{AP}(R, X)$.

Theorem 5.3. *A necessary and sufficient condition for a function $x \in W\mathrm{AP}(R, X)$ to belong also to $\mathrm{AP}(R, X)$ is the relative compactness of its set of values (range) in X.*

Proof. The necessity of the condition is a direct consequence of Proposition 3.6.

The sufficiency will be proven by using Proposition 3.6 of Chapter 3. Let $x \in W\mathrm{AP}(R, X)$. Based on our hypothesis, the set $\mathcal{R}_x = \{x(t); \ t \in R\}$ is relatively compact in X, and the aforementioned proposition assures that from any bounded sequence $\{\varphi_n : n \geq 1\} \subset X^*$, say $\|\varphi_n\| \leq M < \infty$, $n \geq 1$, we can extract a subsequence uniformly convergent on $\mathcal{R}_x$. In other words, there exists a subsequence of $\{\varphi_n; \ n \geq 1\}$ that we can continue to denote in the same way such that $\{\varphi_n(x(t)); \ n \geq 1\}$ converges uniformly on R. Therefore, the set of functions $\{\varphi(x(t)); \ t \in R, \ \varphi \in X^*\}$ is relatively compact in the topology of uniform convergence on R. These are functions in $\mathrm{AP}(R, \mathcal{C})$, and consequently the set $\{\varphi(x(t)), \ \varphi \in X^*, \ \|\varphi\| \leq 1\}$ is equicontinuous and equi–almost-periodic. This fact is a consequence of Proposition 4.6 of Chapter 4. But we can write

$$|x(t + \tau) - x(t)|_X = \sup\{|\varphi(x(t + \tau)) - \varphi(x(t))|; \ \|\varphi\| \leq 1\},$$

which proves the sufficiency.

This ends the proof of Theorem 5.3, which characterizes Bohr's almost periodic functions in $\mathrm{AP}(R, X)$ among the weakly almost periodic functions of $W\mathrm{AP}(R, X)$.

Remark 5.3. If X is a finite-dimensional Banach space, then $AP(R,X) = WAP(R,X)$. This is a consequence of the fact that any bounded set in a finite-dimensional space is relatively compact. Proposition 5.2 above implies the equality of the two spaces.

Remark 5.4. In the case of Banach spaces of infinite dimension, only the inclusion $AP(R,X) \subset WAP(R,X)$ is valid. There are weakly almost periodic functions that are not almost periodic in Bohr's sense. In order to clarify this relationship, we need to establish some properties of the functions in $WAP(R,X)$ if X is a separable Hilbert space.

Let us first note that the separability hypothesis on the Hilbert space is not restrictive due to the fact that the range of any weakly almost periodic function is separable.

We shall assume that X is a separable Hilbert space and $\{e_n;\ n \geq 1\}$ is an orthonormal basis for X. This means that $\langle e_i, e_j \rangle = 0$ for $i \neq j$ and $\langle e_i, e_i \rangle = 1$, $i \geq 1$, while each element of X can be represented as a series, namely

$$x = \sum_{i=1}^{\infty} x_i e_i, \qquad x_i \in \mathcal{C},\ i \geq 1.$$

The series must be convergent in the norm of X, which means

$$\left| x - \sum_{i=1}^{N} x_i e_i \right|_X \longrightarrow 0 \qquad \text{as } N \longrightarrow \infty.$$

Moreover, we always have

$$\|x\|^2 = \sum_{i=1}^{\infty} |x_i|^2.$$

When $x = x(t)$, $t \in R$, there results $x_i = x_i(t)$, $t \in R$, $i \geq 1$.

We can now formulate a result that characterizes the functions in $AP(R,X)$.

Proposition 5.8. *Let* $x : R \longrightarrow X$ *be a bounded function. In order to have* $x \in AP(R,X)$ *with* X *a separable Hilbert space, it is necessary and sufficient that the following two conditions be satisfied:*

(1) $x_i(t) \in AP(R,\mathcal{C})$, $i \geq 1$.

(2) *The series*

$$|x(t)|_X^2 = \sum_{i=1}^{\infty} |x_i(t)|^2$$

converges uniformly on R.

Proof. Conditions (1) and (2) are sufficient for having $x \in AP(R, X)$. Indeed, since

$$x(t) = \sum_{i=1}^{\infty} x_i(t)e_i, \qquad t \in R, \tag{5.31}$$

the convergence being generally pointwise, we need only prove that the convergence is uniform on R in the norm of X due to the fact that each partial sum in equation (5.31) belongs to $AP(R, X)$. Or, condition (2) implies the uniform convergence on R of the series in equation (5.31). We obtain

$$\left| \sum_{i=n}^{N} x_i(t)e_i \right|^2 = \sum_{i=n}^{N} |x_i(t)|^2 < \varepsilon \qquad \text{for } N \geq n > N_0(\varepsilon), \ t \in R, \tag{5.32}$$

which represents Cauchy's criterion for the uniform convergence of series (5.31).

We shall now prove that conditions (1) and (2) are necessary for any $x \in AP(R, X)$.

First, since $x_i(t) = \langle x(t), e_i \rangle$, $i \geq 1$, we have

$$|x_i(t + \tau) - x_i(t)| \leq |x(t + \tau) - x(t)|_X, \qquad i \geq 1,$$

which shows that each $x_i(t)$, $i \geq 1$, is in $AP(R, C)$; i.e., condition (1) is necessary. Furthermore, since the equality in condition (2) is valid pointwise, we need to prove that the series is uniformly convergent on R. The equality in equation (5.32) tells us that the series

$$\sum_{i=1}^{\infty} x_i(t)e_i \quad \text{and} \quad \sum_{i=1}^{\infty} |x_i(t)|^2$$

are simultaneously uniformly convergent. Therefore, it suffices to prove that the series (5.31) converges uniformly on R.

Indeed, if we denote by $s_n(t)$, $n \geq 1$, the partial sums of the series (5.31), then we notice the fact that $s_n(t) = P_n x(t)$, $n \geq 1$, $t \in R$, with P_n the projector of the space X on the finite-dimensional subspace generated by the unit vectors e_i, $i = 1, 2, \ldots, n$. Since $\|P_n\| \leq 1$, $n \geq 1$, we can write

$$|s_n(t + \tau) - s_n(t)|_X \leq |x(t + \tau) - x(t)|_X, \tag{5.33}$$

which shows that the set of almost periodic functions $\{x(t), s_n(t), n \geq 1\} \subset AP(R, X)$ is equicontinuous and equi–almost-periodic. This will help us to prove the uniform convergence of the series (5.31). From Dini's theorem on uniform convergence, we know that the series in condition (2) is uniformly convergent on any compact interval of R. Now take $\varepsilon > 0$ arbitrary, and let $\ell = \ell(\varepsilon/3)$ be the corresponding length according to Bohr's definition. The series (5.31) converges uniformly for $t \in [0, \ell]$. If $t \in R$ is arbitrary, then we can write

$$|x(t) - s_n(t)|_X$$
$$\leq |x(t) - x(t+\tau)|_X + |x(t+\tau) - s_n(t+\tau)|_X + |s_n(t+\tau) - s_n(t)|_X. \tag{5.34}$$

Based on the equi–almost-periodicity of the set of functions $\{x(t), s_n(t); n \geq 1\}$, when τ is an $(\varepsilon/3)$-almost period common to all these functions, we can write, for any $n \geq 1$,

$$|x(t) - x(t+\tau)|_X < \frac{\varepsilon}{3}, \ |s_n(t) - s_n(t+\tau)| < \frac{\varepsilon}{3}, \qquad t \in R. \tag{5.35}$$

Concerning the middle term on the right-hand side of formula (5.34), we note that choosing the $(\varepsilon/3)$-almost period $\tau \in (-t, -t+\ell)$, we can write, for any $n \geq N(\varepsilon)$,

$$|x(t+\tau) - s_n(t+\tau)|_X < \frac{\varepsilon}{3}, \tag{5.36}$$

due to the fact that $t + \tau \in (0, \ell)$. Summing up our discussion, there results from formulas (5.34)–(5.36)

$$|x(t) - s_n(t)|_X < \varepsilon, \qquad t \in R, \ n \geq N(\varepsilon). \tag{5.37}$$

Remark 5.5. An alternate proof of Proposition 5.8 is actually contained in the proof of Theorem 5.3, where the main feature to be taken into consideration is the relative compactness of the set $\mathcal{R}_x$.

We can now construct an example of a function in $WAP(R, X)$ that does not belong to $AP(R, X)$.

Proposition 5.9. *Assume X is a separable Hilbert space of infinite dimension, and consider a sequence $\{x_n(t); \ n \geq 1\} \subset AP(R, C)$. If the following conditions hold, then the function*

$$x(t) = \sum_{i=1}^{\infty} x_i(t)e_i$$

is in $WAP(R, X)$ but not in $AP(R, X)$:

(1) *$x_i(t)$, $i \geq 1$, are uniformly bounded, and we can assume, without loss of generality, $\sup\{|x_i(t)|; \ t \in R\} = 1$, $i \geq 1$.*
(2) *The supports of $x_i(t)$ and $x_j(t)$, $i \neq j$, are disjoint sets.*

Proof. Let us remind the reader that the support of a function defined on R is the set of those values $t \in R$ such that the function does not vanish at t. It is not difficult to produce such functions (e.g., $x_i(t)$ is periodic of period 1, and its support in $(0, 1)$ is the interval $((i+1)^{-1}, i^{-1})$).

The function $x(t)$ is in $WAP(R, X)$, but cannot be in $AP(R, X)$, due to the fact that condition (2) in Proposition 5.8 is obviously violated:

$$\sup \sum_{i=1}^{\infty} |x_i(t)|^2 = 1, \qquad t \in R.$$

The series cannot converge uniformly.

5.3 The Equation $\dot{x}(t) = f(t)$ in Banach Spaces

The preceding section, besides illustrating another concept of almost periodicity, is providing us with some tools for investigating the problem of almost periodicity of the integral (primitive) of functions in $AP(R, X)$. The importance of the concept of weak almost periodicity stems from the fact that the primitive is always weakly almost periodic, and only in particular circumstances does it belong to $AP(R, X)$.

An introductory result for the topics discussed in this section is the following.

Proposition 5.10. *If $f(t) \in AP(R, X)$ and*

$$x(t) = \int_0^t f(s)ds, \qquad t \in R, \tag{5.38}$$

then the boundedness of x on R implies $x \in WAP(R, X)$. Furthermore, in order to have $x \in AP(R, X)$, it is necessary and sufficient that $\mathcal{R}_x = \{x(t); t \in R\}$ be relatively compact in X.

Proof. Theorem 5.1 and the definition of $WAP(R, X)$ tell us that the boundedness of $x(t)$ has as a consequence its weakly almost periodicity. We note that for each $\varphi \in X^*$ one has

$$\varphi(x(t)) = \int_0^t \varphi(f(s))ds, \qquad t \in R, \tag{5.39}$$

which justifies the assertion.

For the second statement in Proposition 5.10, the necessity of the condition is obvious. The sufficiency follows directly from Proposition 5.3 above.

Remark 5.6. A pertinent question is whether or not the boundedness of $x(t)$ in X assures its almost periodicity. We shall now provide an example (see Amerio and Prouse [3]) that shows that boundedness only does not suffice for the almost periodicity of the function $x(t) =$ the primitive of $f(t)$. The underlying space would be the Banach space of bounded sequences with complex elements, and the norm is given by $\|x\| = \sup\{|x_n|; \; x_n \in \mathcal{C}, \; n \geq 1\}$, $x = (x_1, x_2, \ldots)$. We shall denote this space by $b(\mathcal{C})$. We consider the following map from R into $\mathcal{C}$:

$$f(t) = \left(\cos t, \frac{1}{2}\cos\frac{t}{2}, \ldots, \frac{1}{n}\cos\frac{t}{n}, \ldots\right). \tag{5.40}$$

One has $f \in AP(R, b(\mathcal{C}))$ because $f(t) = \lim f_n(t)$ as $n \to \infty$ with

$$f_n(t) = \left(\cos t, \frac{1}{2}\cos\frac{t}{2}, \ldots, \frac{1}{n}\cos\frac{t}{n}, 0, 0, \ldots\right).$$

Indeed, $\|f - f_n\| \leq (n+1)^{-1}$, $n > 1$, and each $f_n(t)$ is periodic in t (one period is $2\pi(n!)$). From equation (5.40), we obtain

$$x(t) = \int_0^t f(s)ds = \left(\sin t, \sin \frac{t}{2}, \cdots, \sin \frac{t}{n}, \cdots\right). \tag{5.41}$$

This implies $\|x(t)\| \leq 1$ and hence its boundedness on R.

We shall now prove that $x(t)$ does not belong to $AP(R, b(\mathcal{C}))$. Indeed, assuming $x(t) \in AP(R, b(R))$, we would find that the sequence $\left\{\sin \frac{t}{n}; \ n \geq 1\right\}$ must be equi–almost-periodic. But the sequence of the derivatives $\left\{\frac{1}{n}\cos \frac{t}{n}; \ n \geq 1\right\}$ is uniformly bounded on R, which implies the equicontinuity of $\left\{\sin \frac{t}{n}; \ n \geq 1\right\}$. According to Theorem 3.3 (Section 3.5), the sequence $\left\{\sin \frac{t}{n}; \ n \geq 1\right\}$ should contain a subsequence that converges uniformly on R. It is easy to see that such a subsequence, say $\left\{\sin \frac{t}{n_k}; \ k \geq 1\right\}$, $n_k \to \infty$, cannot exist. The limit should be zero, which is obviously impossible in the case of uniform convergence on R.

Two distinct problems arise in connection with the almost periodicity of the integral of an almost periodic function from $AP(R, X)$.

First is to find conditions on the space X such that the boundedness of the integral suffices for its almost periodicity. We shall see that this problem has a solution, and we will discuss it in some detail.

The second problem is to find conditions on the function in $AP(R, X)$ such that its integral is also almost periodic. We can provide a simple illustration of the fact that such conditions are possible. For instance, if we deal with the space $AP_1(R, X)$, where X stands for an arbitrary Banach space, and assume that

$$f(t) = \sum_{k=1}^{\infty} a_k e^{i\lambda_k t},$$

with $|\lambda_k| \geq \rho > 0$, $k \geq 1$, then the integral $x(t)$ has the Fourier series

$$x(t) = C + \sum_{k=1}^{\infty} a_k \lambda_k^{-1} e^{i\lambda_k t}.$$

In Section 3.1, we presented this situation when $X = \mathcal{C}$. The details are the same.

Concerning the first approach to the almost periodicity of the integral, we shall provide a result due to L. Amerio [2] in which the Banach space X is assumed to be *uniformly convex*.

Definition 5.2. The Banach space X is called *uniformly convex* if for any $\sigma \in (0, 2]$ one can find $\delta(\sigma) > 0$ such that from $x, y \in X$, $\|x\|, \|y\| \leq 1$, and $\|x - y\| \geq \sigma$, there results $\|x + y\| \leq 2[1 - \delta(\sigma)]$.

If $x, y \in X$ are arbitrary, then the condition of uniform convexity is: $\|x - y\| \geq \sigma \max(\|x\|, \|y\|)$ implies $\|x + y\| \geq 2[1 - \delta(\sigma)] \max(\|x\|, \|y\|)$.

A good example of a uniformly convex Banach space is a Hilbert space because of the identity

$$\|x + y\|^2 + \|x - y\|^2 = 2(\|x\|^2 + \|y\|^2). \tag{5.42}$$

Less trivial is the fact that any uniformly convex Banach space is *reflexive* (i.e., $X = X^{**} =$ the Banach space of linear continuous functionals on X^*). For details on these concepts, we send the reader to K. Yosida [103].

The basic result obtained by L. Amerio can be stated as follows.

Theorem 5.4. *Let $f \in \mathrm{AP}(R, X)$ with X a uniformly convex Banach space. Then its primitive (5.38) belongs to $\mathrm{AP}(R, X)$ if and only if it is bounded on R.*

Before we proceed with the proof of Theorem 5.4, we shall establish an auxiliary result.

Proposition 5.11. *Let $x \in \mathrm{WAP}(R, X)$. If there exists a sequence $\{h_n; n \geq 1\} \subset R$ such that $\{x(t + h_n); n \geq 1\}$ is pointwise weakly convergent to $\widetilde{x}(t)$, $t \in R$, then*

$$\sup_{t \in R} \|\widetilde{x}(t)\| = \sup_{t \in R} \|x(t)\|. \tag{5.43}$$

Proof. For $\varphi \in X^*$, $\varphi(x(t)) \in \mathrm{AP}(R, X)$. The sequence $\{\varphi(x(t + h_n)); n \geq 1\} \in \mathrm{AP}(R, X)$ is equicontinuous and equi–almost-periodic. Since it converges pointwise on R, from an argument given in the proof of Theorem 3.3, this sequence converges uniformly on R to $\varphi(\widetilde{x}(t)) \in \mathrm{AP}(R, \mathcal{C})$. But $\varphi \in X^*$ is arbitrary, which implies $\widetilde{x}(t) \in \mathrm{WAP}(R, X)$. Taking into account that $\varphi(x(t + h_n)) \longrightarrow \varphi(\widetilde{x}(t))$ uniformly on R as $n \to \infty$, we obtain $\varphi(\widetilde{x}(t - h_n)) \longrightarrow \varphi(x(t))$ uniformly on R as $n \to \infty$. In other words, due to the arbitrariness of $\varphi \in X^*$, $x(t + h_n) \longrightarrow \widetilde{x}(t)$ and $\widetilde{x}(t - h_n) \longrightarrow x(t)$ as $n \to \infty$, both uniformly on R in the weak topology (i.e., of the space X_w). These facts imply, for sufficiently large n and $\|\varphi\| = 1$,

$$|\varphi(\widetilde{x}(t))| < |\varphi(x(t + h_n))| + \varepsilon \leq \|(t + h_n)\| + \varepsilon \leq \sup_{t \in R} \|x(t)\| + \varepsilon,$$

which amounts to the inequality

$$\|\widetilde{x}(t)\| \leq \sup_{t \in R} \|x(t)\| + \varepsilon,$$

$\varepsilon > 0$ being arbitrary. Therefore

$$\sup_{t \in R} \|\widetilde{x}(t)\| \leq \sup_{t \in R} \|x(t)\|. \tag{5.44}$$

The opposite inequality to (5.44) is obtained by the same argument, which proves the validity of equation (5.43).

Proof of Theorem 5.4. The boundedness is necessary, as seen in Proposition 3.16 of Chapter 3.

We shall now prove the sufficiency by noting first that $x(t) \in WAP(R, X)$ and then proving the relative compactness in X of the range $\mathcal{R}_x$. According to Proposition 5.10 in this section, one has $x(t) \in WAP(R, X)$. There remains to show the relative compactness of the set $\mathcal{R}_x$. We assume the contrary to hold. Then we can find a sequence $\{h_n;\ n \geq 1\} \subset R$ such that

$$\|x(h_j) - x(h_k)\| \geq 2\sigma > 0, \qquad j \neq k. \tag{5.45}$$

This follows from Cauchy's criterion of convergence in complete metric spaces.

Based on Bochner's property and the definition of functions in $AP(R, X)$, we can extract a subsequence of $\{h_n;\ n \geq 1\}$ that we continue to denote in the same way such that

$$\lim_{n \times \infty} f(t + h_n) = \widetilde{f}(t) \tag{5.46}$$

uniformly on R.

Since the sequence $\{x(h_n);\ n \geq 1\}$ is bounded in X and X is reflexive, we can extract a subsequence, still denoted by $\{x(h_n);\ n \geq 1\}$, that is weakly convergent in X:

$$\lim_{n \to \infty} x(h_n) = c \in X \text{ weakly.} \tag{5.47}$$

Now let $t \in R$ be fixed. One has

$$x(t + h_k) = x(h_k) + \int_{h_k}^{t+h_k} f(s)ds = x(h_k) + \int_0^t f(s + h_k)ds.$$

The relationships above and equation (5.46) imply the weak uniform convergence on any compact interval of the sequence $\{x(t + h_n);\ n \geq 1\}$. In other words, we can write

$$\lim_{n \to \infty} x(t + h_n) = c + \int_0^t \widetilde{f}(s)ds = \widetilde{x}(t). \tag{5.48}$$

But $x(t) \in WAP(R, X)$, and Proposition 5.11 above allows us to write

$$\sup_{t \in R} \|\widetilde{x}(t)\| = \sup_{t \in R} \|x(t)\| = M < \infty. \tag{5.49}$$

If $t \in R$ is fixed, then we have

$$\|x(t + h_j) - x(t + h_k)\|$$

$$= \left\| x(h_j) - x(h_k) + \int_0^t [f(s + h_j) - f(s + h_k)]ds \right\|$$

$$\geq \|x(h_j) - x(h_k)\| - \left\| \int_0^t [f(s + h_j) - f(s + h_k)]ds \right\|$$

$$\geq 2\sigma - \left\| \int_0^t [f(s + h_j) - f(s + h_k)]ds \right\|.$$

From equation (5.46), we can write

$$\left\| \int_0^t [f(s + h_j) - f(s + h_k)]ds \right\| < \sigma$$

for $j, k \geq N = N(\sigma, t)$. Combining this with the preceding inequality, we can write

$$\|x(t + h_j) - x(t + h_k)\| \geq 2\sigma - \sigma = \sigma,$$

provided $j, k \geq N$. If we also rely on the definition of uniform convexity and the meaning of M, we obtain from above, for $j, k \geq N$,

$$\|x(t + h_j) - x(t + h_k)\| \geq \sigma M^{-1} \max\{\|x(t + h_j)\|, \|x(t + h_k)\|\}. \tag{5.50}$$

The inequality (5.50) implies, according to the definition of uniform convexity,

$$\|x(t + h_j) + x(t + h_k)\| \leq 2[1 - \delta(\sigma M^{-1})] \max\{\|(t + h_j)\|, \|x(t + h)_k)\|\},$$

from which we get

$$\|x(t + h_j) + x(t + h_k)\| \leq 2M[1 - \delta(\sigma M^{-1})], \qquad j, k \geq N. \tag{5.51}$$

For any $\varphi \in X^*$, $\|\varphi\| = 1$, we have

$$|\varphi(x(t + h_j) + x(t + h_k))| \leq \|x(t + h_j) + x(t + h_k)\| \leq 2M[1 - \delta(\sigma M^{-1})].$$

Letting $j, k \to \infty$, and taking equation (5.48) into account, one gets the inequality

$$\|\widetilde{x}(t)\| \leq M[1 - \delta(\sigma M^{-1})] < M. \tag{5.52}$$

The inequality (5.52) contradicts equation (5.43), and thus Theorem 5.4 is proven.

Remark 5.7. Since the Hilbert spaces are uniformly convex, this result is applicable only in a limited number of situations. We have in mind the function spaces occurring in the theory of partial differential equations, such as certain Sobolev spaces ($W^{k,2}(\Omega)$ for instance).

As we mentioned above in this section, the problem of almost periodicity of the primitive of a function in $\mathrm{AP}(R, X)$, with X a Banach space, has been solved by M. J. Kateds (see, for instance, B. M. Levitan and V. V. Zhikov [63]). The answer to this problem is given in the following theorem.

Theorem 5.5. *The boundedness of the primitive of any function in* $\mathrm{AP}(R, X)$ *is a necessary and sufficient condition for its almost periodicity if and only if* X *does not contain as a subspace the Banach space* $c_0(Z_+, R)$.

The condition is necessary according to Proposition 5.10. The sufficiency is more difficult to prove, requiring further preparation from functional analysis. We will omit the proof of Theorem 5.5 and send the reader to Levitan and Zhikov [63].

The space $c_0(Z_+, R)$ is the space of sequences $x = (x_1, x_2, \ldots, x_n, \ldots)$ with $x_k \in R$, $k \geq 1$, such that $x_n \to 0$ as $n \to \infty$. The norm is $\|x\| = \sup\{|x_n|; n \geq 1\}$. The convergence in c_0 means convergence on each coordinate uniformly with respect to $n \in Z_+$.

There are two basic properties of c_0 that we will need in our construction. First, any linear functional on c_0 has the form $\varphi(x) = \sum_{k=1}^{\infty} a_n x_n$ with $\sum_{n=1}^{\infty} |a_n| < \infty$. Second, a set $S \subset c_0$ is relatively compact if and only if it is uniformly bounded and $x_n \to 0$ as $n \to \infty$ uniformly with respect to $x \in S$. Both properties are easy to prove, and the reader can find the treatment of c_0 in textbooks on functional analysis.

Let $\alpha(t) \in AP(R, R)$ be an arbitrary function whose primitive is bounded on R (and hence also in $AP(R, R)$). We consider the map $f : R \to c_0$ by letting $f_n(t) = n^{-1} \alpha(n^{-1} t)$, $n \geq 1$. Since α is bounded, it is obvious that $f(t) = (f_1(t), f_2(t), \ldots, f_n(t), \ldots) \in c_0$, $t \in R$. Assume now that $A(t)$, $t \in R$, is a primitive of $\alpha(t) : A'(t) = \alpha(t)$, $t \in R$.

First, we note that $f(t)$ is weakly almost periodic. Indeed, according to the general formula of representations of continuous linear functionals on c_0 mentioned above, there results that $(\varphi f)(t)$, $\varphi \in c_0^* = \ell^1$, is almost periodic (as the sum of a series of almost periodic functions that converges uniformly on R). Hence, f is weakly almost periodic.

Second, from $F(t) = \int_0^t f(s) ds = \{A(n^{-1} t) - A(0); n \geq 1\}$, we obtain the weak almost periodicity of $F(t)$ by considering

$$\varphi(F(t)) = \sum_{n=1}^{\infty} a_n [A(n^{-1} t) - A(0)],$$

with $\{a_n; n \geq 1\} \in \ell^1$, and taking into account that $A(t)$ is almost periodic (and hence bounded).

We will now prove that $\mathcal{R}_A \subset c_0$ is not relatively compact, which will show that $F(t)$ cannot belong to $AP(R, c_0)$.

According to the criterion of relative compactness stated above, if $\mathcal{R}_A$ is relatively compact in c_0, then we should have $A(n^{-1} t) - A(0) \longrightarrow 0$ as $n \to \infty$ uniformly with respect to $t \in R$. But this is obviously impossible since $A(t)$ is an almost periodic function in $AP(R, R)$ that does not reduce to a constant, and this fact implies that $A(n^{-1} t)$ takes values as close as we want from both $\sup A(t)$ and $\inf A(t)$ for arbitrarily large values of $t \in R$.

The example examined above, where the Banach space $X = c_0$, clarifies the situation not only for this particular case but also for any Banach space X that contains c_0 as a subspace. Actually, as was shown by M. J. Kadets (Theorem 5.5 above), this feature is characteristic for obtaining a positive answer to the problem of almost periodicity of the primitive if it is bounded.

5.4 Linear Oscillations Described by Ordinary Differential Equations

Having dealt with the simplest equation, $\dot{x}(t) = f(t)$, we shall concentrate now on the more general situation described by linear systems of ordinary differential equations of the form

$$\dot{x}(t) = Ax(t) + f(t), \qquad t \in R, \tag{5.53}$$

where $x : R \longrightarrow C^n$, $A \in \mathcal{L}(C^n, C^n)$, and $f \in E(R, C^n)$, where E stands for one of the spaces of almost periodic functions investigated in the preceding chapters.

We are considering here the case of constant coefficients. The variable coefficient case $A = A(t) \in \mathcal{L}(C^n, C^n)$ for $t \in R$ is more intricate and will be discussed briefly in forthcoming sections.

To the form (5.53) we can reduce higher-order scalar equations of the form

$$y^{(n)} + a_1 y^{(n-1)} + \cdots + a_n y = g(t),$$

where a_k, $1 \leq k \leq n$, are in R or C. Such equations, particularly if $n = 2$, are known to describe linear oscillations in mechanical, electrical, and other types of physical and engineering systems.

It is well known from linear algebra that by means of a linear transformation on x of the form

$$x = Tu, \qquad x, u \in C^n, \tag{5.54}$$

where $T \in \mathcal{L}(C^n, C^n)$ and $\det T \neq 0$, equation (5.53) can be reduced to the special form

$$\dot{u} = Bu + \widetilde{f}(t) \tag{5.55}$$

with an upper-triangular matrix B:

$$B = T^{-1}AT. \tag{5.56}$$

The vector function $\widetilde{f}(t)$ is given by

$$\widetilde{f}(t) = T^{-1}f(t), \qquad t \in R. \tag{5.57}$$

It is also known, but easy to check, that the diagonal elements of B are the characteristic numbers or eigenvalues of the matrix A:

$$B = \begin{bmatrix} \lambda_1 & b_{12} & b_{13} & \cdots & b_{1n} \\ 0 & \lambda_2 & b_{23} & \cdots & b_{2n} \\ \multicolumn{5}{c}{\dotfill} \\ 0 & 0 & 0 & \cdots & \lambda_n \end{bmatrix}. \tag{5.58}$$

The eigenvalues λ_k, $1 \leq k \leq n$, satisfy the algebraic equation

$$\det(\lambda_k I - A) = 0, \qquad k = 1, 2, \ldots, n. \tag{5.59}$$

The identity

$$\det(\lambda I - B) = \det(\lambda I - A)$$

justifies our claim about the nature of the diagonal elements of B.

After this brief recall of some elementary properties of constant matrices, let us move to the investigation of the different meanings of the problem of almost periodicity of the solutions of the equation/system (5.53) under the basic assumption that $f(t)$ on the right-hand side is almost periodic.

We will start with the historically known result, due to Bohr and Neugebauer, related to the case $f \in AP(R, C^n)$.

Proposition 5.12. *If $x(t)$ is a solution of equation (5.53) with $f \in AP(R, C^n)$, then a necessary and sufficient condition for having $x \in AP(R, C^n)$ is $x \in BC(R, C^n)$.*

Proof. The condition is obviously necessary because $AP(R, C^n) \subset BC(R, C^n)$.

To prove the sufficiency of the condition $x \in BC(R, C^n)$, we shall use the "reduced" form of the system (5.53), namely

$$\begin{cases} \dot{u}_1 &= \lambda_1 u_1 + b_{12} u_2 + b_{13} u_3 + \cdots + b_{1n} u_n + \widetilde{f}_1(t), \\ \dot{u}_2 &= \lambda_2 u_2 + b_{23} u_3 + \cdots + b_{2n} u_n + \widetilde{f}_2(t), \\ \cdots\cdots\cdots\cdots\cdots\cdots\cdots\cdots\cdots\cdots\cdots\cdots\cdots\cdots\cdots\cdots \\ \dot{u}_{n-1} &= \lambda_{n-1} u_{n-1} + b_{n-1,n} u_n + \widetilde{f}_{n-1}(t), \\ \dot{u}_n &= \lambda_n u_n + \widetilde{f}_n(t). \end{cases} \tag{5.60}$$

As we can see, the last equation in (5.60) is a scalar equation in u_n with the almost periodic term $\widetilde{f}_n(t)$ on the right-hand side. Therefore, if we prove that $u_n(t) \in AP(R, C)$ and we substitute it in the right-hand side of the equation for $u_{n-1}(t)$, we obtain for $u_{n-1}(t)$ an equation similar to the last equation in equation (5.60),

$$\dot{u} = \lambda u + g(t), \tag{5.61}$$

where $g(t) = b_{n-1,n} u_n(t) + \widetilde{f}_{n-1}(t) \in AP(R, C)$. This process can be continued until we obtain $u_1(t) \in AP(R, C)$, provided any bounded $u(t)$ from equation (5.61) is shown to be in $AP(R, C)$ regardless of the value of $\lambda \in C$ and $g \in AP(R, C)$.

The solution of equation (5.61) is given by the formula

$$u(t) = e^{\lambda t} \left[c + \int_0^t e^{-\lambda s} g(s) ds \right], \qquad t \in R, \tag{5.62}$$

where c stands for an arbitrary complex number.

We need to distinguish three different situations according to the value of $\operatorname{Re} \lambda$: (1) $\operatorname{Re} \lambda < 0$; (2) $\operatorname{Re} \lambda = 0$; (3) $\operatorname{Re} \lambda > 0$.

From our hypothesis on $x(t)$, we know that $u(t)$ must be bounded on R.

Case 1. Since $e^{\lambda t} \longrightarrow \infty$ as $t \longrightarrow -\infty$, the only chance to obtain $u(t)$ from equation (5.62) bounded on R is to have $c + \int_0^{-\infty} e^{-\lambda s} g(s) ds = 0$. The integral always exists due to the boundedness of $g(t)$ on R. This choice for the constant c leads to the following solution for the scalar equation (5.61):

$$u(t) = \int_{-\infty}^t e^{\lambda(t-s)} g(s) ds, \qquad t \in R. \tag{5.63}$$

This expression for $u(t)$ as a convolution in which the first factor is in $L^1(R_-, \mathcal{C})$ defines an almost periodic solution to equation (5.61). Since the homogeneous equation corresponding to equation (5.61), $\dot{u}(t) = \lambda u(t)$, has only the zero solution bounded on R when $\operatorname{Re}\lambda < 0$, equation (5.63) represents the *unique* almost periodic solution of equation (5.61).

It can be easily seen that

$$\sup_{t \in R} |u(t)| \leq \frac{1}{|\operatorname{Re}\lambda|} \sup_{t \in R} |g(t)|. \tag{5.64}$$

Case 2. When $\operatorname{Re}\lambda = 0$, there follows $e^{\lambda t} = e^{i\omega t}$ for some $\omega \neq 0$. But the solution of equation (5.61) is then

$$u(t) = c e^{i\omega t} + e^{i\omega t} \int_0^t e^{-i\omega s} g(s) ds. \tag{5.65}$$

We derive from the above that $u(t)$ is bounded on R if and only if the integral $\int_0^t e^{i\omega s} g(s) ds$ is bounded there. Since the function $e^{-i\omega s} g(s)$ is in $AP(R, \mathcal{C})$, from Theorem 5.1 we find that its integral is also in $AP(R, \mathcal{C})$.

Consequently, all bounded solutions of equation (5.61) are in $AP(R, \mathcal{C})$.

Case 3. This case is completely analogous to Case 1, and for $\operatorname{Re}\lambda > 0$ there exists only one bounded almost periodic solution of equation (5.61). It is given by the formula

$$u(t) = -\int_t^\infty e^{\lambda(t-s)} g(s) ds, \qquad t \in R. \tag{5.66}$$

To summarize, we start with a bounded solution $x = x(t)$ of equation (5.53), which by means of the linear transformation $x = Tu$ generates a bounded solution of equation (5.55). This solution of (5.55) is necessarily in $AP(R, \mathcal{C}^n)$, as seen in the discussion above, when all possible cases for $\operatorname{Re}\lambda$ have been considered and all u_k, $1 \leq k \leq n$, have been found to be almost periodic. Hence, $x(t)$ is also in $AP(R, \mathcal{C}^n)$, which ends the proof of Proposition 5.12.

The case where $\operatorname{Re}\lambda_k \neq 0$, $k = 1, 2, \ldots, n$, is interesting in connection with the uniqueness of the almost periodic solution. The following proposition is basically proven above.

Proposition 5.13. *Consider the system* (5.53) *under the same conditions as in Proposition 5.12 plus* $\operatorname{Re} \lambda_k \neq 0$, $k = 1, 2, \ldots, n$. *Then there exists a unique bounded* (*and hence almost periodic*) *solution* $x(t)$ *for each* $f \in \operatorname{AP}(R, C^n)$. *Moreover, an estimate of the form*

$$|x|_{\text{AP}} \leq K |f|_{\text{AP}} \tag{5.67}$$

is valid with $K > 0$ *independent of* f (*and hence depending on* A *only*).

Proof. As seen in the proof of Proposition 5.12, if $\operatorname{Re} \lambda_k \neq 0$, the corresponding (reduced) equation has only one almost periodic solution in $\operatorname{AP}(R, C^n)$. Since $\operatorname{Re} \lambda_k = 0$ is not a possibility, the uniqueness is proven. It can also be obtained by observing that the homogeneous equation (i.e., when $f(t) = \theta$, the null element in C^n) has no bounded solution on R except $x = 0$ (or $u = 0$).

The estimate (5.67) is a consequence of the estimates on coordinates of the form (5.64).

Remark 5.8. The proposition above tells us that the operator $f \longrightarrow x$ on $\operatorname{AP}(R, C)$ is continuous. By x we mean the unique solution in $\operatorname{AP}(R, C)$ of equation (5.53).

Remark 5.9. The inverse operator of $f \longrightarrow x$ can be expressed as an integral,

$$x(t) = \int_{-\infty}^{t} X_-(t-s)f(s)ds - \int_{t}^{\infty} X_+(t-s)f(s)ds, \tag{5.68}$$

where the matrices X_- and X_+ are such that $X_-(t) + X_+(t) = X(t)$, with $\dot{X}(t) = AX(t)$, $X(0) = I =$ the unit $n \times n$ matrix; moreover, $X_-(t)$ contains only the parts of entries in $X(t)$ with $\operatorname{Re} \lambda_k < 0$, while $X_+(t)$ has a similar meaning related to the terms with $\operatorname{Re} \lambda_k > 0$.

It is a simple exercise to check that $x(t)$ given by equation (5.68) satisfies equation (5.53) and is bounded on R. Hence, it does represent the unique solution of equation (5.53) in $\operatorname{AP}(R, C^n)$.

Remark 5.10. The formula (5.68) can be used to extend the result of Proposition 5.13 to the case where $f \in S(R, C^n)$.

Indeed, we can rewrite equation (5.68) in the convolution form

$$x(t) = \int_{R} K(t-s)f(s)ds, \qquad t \in R, \tag{5.68'}$$

with $K(t) = X_-(t)$ for $t > 0$ and $K(t) = X_+(t)$ for $t \leq 0$.

Since $\|K\| \in L^1(R, R)$, there results $x \in S(R, C^n)$, as seen in Section 3.3, where we defined and investigated the almost periodic function in the sense of Stepanov. In other words, the operator $f \longrightarrow x$, f and x being related by equation (5.53), is defined on $S(R, C^n)$. Of course, it is assumed that $\operatorname{Re} \lambda_k \neq 0$, $k = 1, 2, \ldots, n$.

The result just stated is not the best possible under our assumptions. Actually, the following proposition can be proved.

Proposition 5.14. *Consider the equation* (5.53), *and assume that* $\mathrm{Re}\,\lambda_k \neq 0$, $k = 1, 2, \ldots, n$, *with* λ_k *satisfying* $\det(\lambda_k I - A) = 0$. *Then the operator* $f \longrightarrow x$, *with* f *and* x *connected by equation* (5.53), *is a continuous operator from* $S(R, \mathcal{C}^n)$ *into* $\mathrm{AP}(R, \mathcal{C}^n)$.

Proof. We shall use the convolution (5.68′), representing the connection between f and x. As seen above, the kernel $K(t)$, defined by means of $X_-(t)$ and $X_+(t)$, is in $L^1(R, \mathcal{L}(\mathcal{C}^n, \mathcal{C}^n))$. But taking into account the properties of the matrices $X_-(t)$ and $X_+(t)$, we realize that $K(t)$ is decreasing exponentially at $\pm\infty$. In other words, we can find positive numbers M and α such that

$$\|K(t)\| \leq M e^{-\alpha|t|}, \qquad t \in R. \tag{5.69}$$

Let us now return to the formula (5.68′) expressing the unique solution in $S(R, \mathcal{C}^n)$ of equation (5.53) with $f \in S(R, \mathcal{C}^n)$. Also taking into account the growth of $X_-(t)$ and $X_+(t)$, we can find an estimate of the form

$$|x(t)|_{\mathrm{AP}} \leq N|f(t)|_S, \tag{5.70}$$

with $N > 0$, independent of $f \in S(R, \mathcal{C}^n)$. The inequality (5.70) proves that the solution of equation (5.53) in $S(R, \mathcal{C}^n)$ is actually in $\mathrm{AP}(R, \mathcal{C}^n)$. Moreover, the operator $f \longrightarrow x$, with f and x related by equation (5.53), is continuous from $S(R, \mathcal{C}^n)$ into $\mathrm{AP}(R, \mathcal{C}^n)$.

The details for obtaining the estimate (5.70) from equation (5.68′), when $K(t)$ satisfies formula (5.69), can be summarized as

$$x(t) = \int_{-\infty}^{t} K(t-s)f(s)\,ds + \int_{t}^{\infty} K(t-s)f(s)\,ds = u(t) + v(t),$$

$$|u(t)| \leq M e^{-\alpha t} \int_{-\infty}^{t} e^{\alpha s} |f(s)|\,ds$$

$$\leq e^{-\alpha t}\left(e^{\alpha t}\int_{t-1}^{t} |f(s)|\,ds + e^{\alpha(t-1)}\int_{t-2}^{t-1}|f(s)|\,ds + \cdots \right)$$

$$\leq M\left(1 + e^{-1} + e^{-2} + \cdots\right)|f|_S = \frac{Me}{e-1}\,|f|_S.$$

Similar estimates are valid for $v(t)$, which lead to formula (5.70), and this ends the proof.

Remark 5.11. We can use the convolution formula (5.68′) to obtain a similar result for the space $\mathrm{AP}_1(R, \mathcal{C})$. Indeed, if we take $f \in \mathrm{AP}_1(R, \mathcal{C})$ with

$$f(t) = \sum_{j=1}^{\infty} a_j e^{i\lambda_j t}, \qquad t \in R, \tag{5.71}$$

where $a_j \in \mathcal{C}^n$ and λ_j are real and $\sum_{j=1}^{\infty}|a_j| < \infty$, then one finds with the notation above that

$$x(t) = \int_R K(t-s)f(s)ds = \int_R K(t-s)\left(\sum_{j=1}^{\infty}a_j e^{i\lambda_j s}\right)ds$$

$$= \sum_{j=1}^{\infty}a_j \left(\int_R K(t-s)e^{-i\lambda_j(t-s)}ds\right)e^{i\lambda_j t} = \sum_{j=1}^{\infty}\tilde{a}_j e^{i\lambda_j t},$$

where

$$\tilde{a}_j = a_j \int_R K(u)e^{-i\lambda_j u}du, \qquad j = 1, 2, \ldots.$$

Therefore, $|\tilde{a}_j| \leq |K|_{L^1}|a_j|$, $j = 1, 2, \ldots$, which shows that $\sum_{j=1}^{\infty}|\tilde{a}_j| < \infty$; i.e., $x(t) \in AP_1(R, \mathcal{C}^n)$.

The reader is invited to examine the operator $f \longrightarrow x$ with f and x related by equation (5.53), considering other spaces of almost periodic functions (e.g., Besicovitch spaces $AP_2(R, \mathcal{C}^n)$ and $B(R, \mathcal{C}^n)$). Apparently, these cases have not yet received the attention of researchers.

As seen in Propositions 5.13 and 5.14, the basic hypothesis was $\operatorname{Re}\lambda_k \neq 0$, $k = 1, 2, \ldots, n$, where λ_k are the eigenvalues of the matrix A. It is therefore interesting to examine the problem of almost periodicity of the solutions of equation (5.53) when the matrix A possesses purely imaginary eigenvalues.

For instance, in the case $n = 1$, the equation $\dot{x} = i\alpha x + e^{i\alpha t}$, in which $A = i\alpha = $ the unique eigenvalue ($\alpha = $ real), has the general solution $x(t) = (t + x_0)e^{i\alpha t}$ with arbitrary $x_0 \in \mathcal{C}$, and none of its particular solutions is bounded (and hence cannot be almost periodic).

Of course, this example does not mean that we should not expect almost periodic solutions when A has purely imaginary eigenvalues. Actually, in the introduction, we examined equation (5.4), for which quasi-periodic solutions have been found, under adequate assumptions on the forcing term. But the eigenvalues of the matrix of the system $\dot{x} = \omega y$, $\dot{y} = -\omega x$ are $\pm i\omega$, $\omega \neq 0$.

Instead of dealing with this situation, we shall postpone the investigation until we examine the problem for linear systems of the form

$$\dot{x}(t) = \int_R [dA(s)]x(t-s) + f(t), \tag{5.72}$$

which contain as a very special case the system (5.53).

In concluding this section, we will consider a simple example that occurs in the theory of oscillations described by second-order linear differential equations. Of course, the forcing term will be chosen to be almost periodic, more precisely in $AP(R, \mathcal{C})$.

Consider the second-order linear differential equation

$$\ddot{x} + 2a\dot{x} + bx = f(t), \qquad t \in R, \tag{5.73}$$

with $a, b \in R$ and $f \in AP(R, R)$. Equation (5.73) describes the linear motion of a point attracted or repulsed by the origin that encounters a friction (proportional to its velocity $\dot{x}$) and is subject to an external force represented by $f(t)$.

The linear system equivalent to equation (5.73) is

$$\begin{cases} \dot{x} = y, \\ \dot{y} = -bx - 2ay + f(t), \end{cases} \tag{5.74}$$

and the eigenvalues of the matrix

$$A = \begin{pmatrix} 0 & 1 \\ -b & -2a \end{pmatrix}$$

are $\lambda_1 = -a - \sqrt{a^2 - b}$, $\lambda_2 = -a + \sqrt{a^2 - b}$. If one assumes $ab \neq 0$, then both roots have real parts different from zero when $a^2 - b < 0$ are real and of contrary sign when $a^2 - b > 0$, and both are equal to $-a \neq 0$ when $a^2 - b = 0$.

Therefore, under the main assumption $ab \neq 0$, equation (5.73) has a unique solution in $AP(R, R)$ for any $f \in AP(R, R)$. It is obvious that $\dot{x}$ and $\ddot{x}$ are also in $AP(R, R)$, which means that not only is the space variable $x(t)$ almost periodic but the velocity and the acceleration also.

Based on Proposition 5.14, we can conclude that if $f \in S(R, R)$, the space variable $x(t)$ and the velocity $\dot{x}(t)$ are both in $AP(R, R)$.

We invite the reader to examine the case where either number a, or b, or both, are zero.

5.5 Linear Periodic and Almost Periodic Oscillations in Systems Described by Convolution Equations

In this section, we shall be concerned with periodic or almost periodic solutions of differential equations of convolution type of the form

$$\dot{x}(t) = \int_R [dA(s)]x(t-s) + f(t), \qquad t \in R, \tag{5.75}$$

where $A : R \longrightarrow \mathcal{L}(R^n, R^n)$ is a matrix whose entries are real-valued functions with bounded variation on R always assumed to be left continuous. It is also assumed that $f : R \longrightarrow C^n$, $x : R \longrightarrow C^n$.

It is generally known that the system (5.75) contains as particular cases several types of functional equations/systems that appear in the literature.

For example, the linear equation $\dot{x}(t) = Ax(t) + f(t)$, investigated in Section 5.4, can be obtained from equation (5.75) by letting $A(s) = O =$ the zero matrix of type $n \times n$ for $s \leq 0$,

$$A(s) = (a_{ij}), \qquad i, j = 1, 2, \ldots, n, \text{ for } s > 0.$$

Similarly, if we choose some positive numbers t_k, $k = 1, 2, \ldots, N$, then one obtains from equation (5.75) the system with delays (say $0 < t_1 < t_2 < \cdots < t_N$)

$$\dot{x}(t) = \sum_{k=1}^{N} A_k x(t - t_k) + f(t) \tag{5.76}$$

by choosing $A(s) = 0$ for $s \le t_1$, $A(s) = A_1$ for $t_1 < s \le t_2$, then $A(s) = A_2$ for $t_2 < s \le t_3$, and so on.

We can also obtain more complex systems such as

$$\dot{x}(t) = \sum_{k=1}^{\infty} A_k x(t - t_k) + \int_R B(s) x(t - s) ds \tag{5.77}$$

with $t_k \in R$, $k \ge 1$, and

$$\sum_{k=1}^{\infty} |A_k| < \infty, \qquad \int_R |B(s)| ds < \infty. \tag{5.78}$$

For various properties of systems of the form (5.77) with $t_k \ge 0$ and $B(s) = 0$ for $s > 0$, see the author's book [25].

For the investigation of the system (5.75), we shall need the Fourier–Stieltjes transformation, which is defined by

$$\widetilde{\mathcal{A}}(is) = \int_R [dA(t)] e^{-its}, \qquad s \in R.$$

Under our assumptions on $A(t)$, the Fourier–Stieltjes transform is defined on R. If $A(t)$ is differentiable, which implies $|A'| \in L^1(R, R)$, one obtains the Fourier transform of $A'(s)$ for functions in $L^1(R, R)$. Therefore, we may regard the Fourier transform as a special case of the Fourier–Stieltjes transform. Namely, one integrates in the latter with respect to the measure $dA(t)$ (instead of the Lebesgue measure in the case of the Fourier transform).

We shall now deal with the problem of *periodicity* of solutions of equation (5.75), assuming that the forcing term $f(t)$ is a periodic function of period $\omega > 0$ (i.e., ω is the minimal period for f).

Instead of assuming f to be a continuous periodic function, we shall in fact deal with $f \in L^2([0, \omega], C^n)$ and extend it to the whole real axis R by letting $f(t + \omega) = f(t)$ a.e. In this manner, we allow more generality in the forcing term, including cases where discontinuities are present.

Changing somewhat the notation used in Section 4.6 when discussing the Fourier series corresponding to a periodic function of period $2\pi/\omega$, we shall now write

$$f(t) \sim \sum_{k \in Z} f_k e^{i\omega_k t}, \tag{5.79}$$

where $\omega_k = 2k\pi/\omega$, $k \in Z$.

We note that the Bessel–Parseval formula becomes

$$\sum_{k \in Z} |f_k|^2 = \frac{1}{\omega} \int_0^\omega |f(t)|^2 dt. \tag{5.80}$$

Formula (5.80) is a special case of the Bessel–Parseval formula (see, for instance, the author's book [23] or any book containing the theory of Hilbert spaces).

For equation (5.75), one tries to find periodic solutions of the form

$$x(t) = \sum_{k \in Z} x_k e^{i\omega_k t}, \qquad t \in R. \tag{5.81}$$

We have in mind, of course, solutions in Carathéodory's sense (i.e., satisfying the equation a.e.). Substituting (now formally) $x(t)$ given by equation (5.81) into equation (5.75) and taking into account formula (5.79), we obtain the following infinite system of equations for x_k, $k \in Z$:

$$i\omega_k x_k = \left(\int_R [dA(s)]\, e^{-i\omega_k s} \right) x_k + f_k. \tag{5.82}$$

These equations can also be rewritten in the form

$$[i\omega_k I - \widetilde{A}(i\omega_k)]x_k = f_k, \qquad k \in Z. \tag{5.83}$$

It is obvious from equation (5.83) that the condition

$$\det[isI - \widetilde{A}(is)] \neq 0, \qquad s \in R, \tag{5.84}$$

will suffice for getting a unique x_k, $k \in Z$, which can lead to a unique periodic solution to equation (5.75), provided the series on the right-hand side of equation (5.81) converges in some sense. This would be true for any period $\omega > 0$.

On the other hand, we realize that condition (5.84) is much too strong if we deal with a fixed period ω. As seen from equation (5.83), it will be sufficient to assume that formula (5.84) holds true only for $s = \omega_k$, $k \in Z$.

We shall now prove a proposition that will allow us to formulate an even weaker condition that ensures the unique solvability of equations (5.82) or (5.83).

Proposition 5.15. *Let $s \in R$ be a solution of the equation*

$$\det[isI - \widetilde{A}(is)] = 0. \tag{5.85}$$

Then

$$|s| \leq \int_R |dA(t)| = \gamma < \infty. \tag{5.86}$$

Proof. Let us note that the right-hand side in equation (5.86) represents the total variation of $A(t)$ on R, which we assumed to be finite.

If equation (5.85) holds for some $s \in R$, then the linear system $isv = \widetilde{\mathcal{A}}(is)v$ has nontrivial solution $v \in \mathcal{C}^n$. Hence, $|s||v| \leq |\widetilde{\mathcal{A}}(is)||v|$, where $|\cdot|$ is the Euclidean norm in $\mathcal{C}^n$, while $|\widetilde{\mathcal{A}}(is)|$ is the matrix norm induced by the Euclidean norm for vectors. Since $|v| \neq 0$, one obtains $|s| \leq |\widetilde{\mathcal{A}}(is)|$. But from formula (5.78) one derives

$$|\widetilde{\mathcal{A}}(is)| \leq \int_R |dA(t)|, \tag{5.87}$$

which means that equation (5.86) is true.

Corollary 5.1. *From Proposition 5.15, there results that the only values for ω_k needed in the conditions*

$$\det[i\omega_k I - \widetilde{\mathcal{A}}(i\omega_k)] \neq 0 \tag{5.88}$$

are those for which $\omega_k \leq \gamma < \omega_{k+1}$, as well as their opposite ω_{-k}.

Obviously, there is a maximum $k > 0$ satisfying $\omega_k \leq \gamma < \omega_{k+1}$, which means that we actually involve only a finite number of k's in formulating the solvability condition for the system (5.83).

We now have all the elements necessary to formulate an existence result for periodic solutions to the equation/system (5.75).

Theorem 5.6. *Consider the equation (5.75) under the following assumptions:*

(a) $f \in L^2([0,0], \mathcal{C}^n)$ *and then extended by periodicity to R by $f(t+\omega) = f(t)$ a.e.*

(b) $A(t) \in \mathcal{L}(R^n, R^n)$, $t \in R$, *has bounded variation on R and is left continuous.*

(c) *Condition (5.88) is verified for $|k| \leq p$, with p the greatest integer for which $\omega_p \leq \gamma < \omega_{p+1}$.*

Then there exists a unique solution $x(t)$ of equation (5.75), periodic with period ω, satisfying the equation a.e. such that its Fourier series

$$\sum_{k \in Z} [i\omega_k I - \widetilde{\mathcal{A}}(i\omega_k)]^{-1} f_k e^{i\omega_k t} \tag{5.89}$$

converges uniformly and absolutely.

Proof. The construction of the series (5.83) has been given above, and Corollary 5.1 to Proposition 5.15 allows us to restrict the validity of formula (5.88) to these k such that $|k| \leq p$.

The uniform and absolute convergence of the series (5.89) follows easily from the estimate

$$|[i\omega_k I - \widetilde{\mathcal{A}}(i\omega_k)]^{-1} f_k| \le M(\omega_k^{-2} + |f_k|^2) \qquad (5.90)$$

with $M > 0$ for $|k| \ge p+1$. To obtain formula (5.90), we rely on the fact that $\widetilde{\mathcal{A}}(i\omega_k)/\omega_k \longrightarrow 0$ as $|k| \longrightarrow \infty$. Using equation (5.80), we obtain the uniform and absolute convergence of the series (5.89), which defines a function $x(t)$.

We need to verify the fact that we have obtained indeed a solution of equation (5.75). Using equation (5.80), we have

$$\int_R [dA(s)]x(t-s) = \int_R [dA(s)] \sum_{k \in Z} x_k e^{i\omega_k(t-s)}$$

$$= \sum_{k \in Z} \left(\int_R [dA(s)] e^{-i\omega_k s} \right) x_k e^{i\omega_k t}$$

$$= \sum_{k \in Z} \widetilde{\mathcal{A}}(i\omega_k) x_k e^{i\omega_k t},$$

with x_k, $k \in Z$, given by equation (5.83). Hence,

$$\int_R [dA(s)]x(t-s) + f(t)$$

$$= \sum_{k \in Z} \left\{ \widetilde{\mathcal{A}}(i\omega_k)[i\omega_k I - \widetilde{\mathcal{A}}(i\omega_k)]^{-1} f_k + f_k \right\} e^{i\omega_k t} \qquad (5.91)$$

$$= \sum_{k \in Z} \left\{ i\omega_k [i\omega_k I - \widetilde{\mathcal{A}}(i\omega_k)]^{-1} \right\} f_k e^{i\omega_k t}$$

because

$$\widetilde{\mathcal{A}}(i\omega_k)[i\omega_k I - \widetilde{\mathcal{A}}(i\omega_k)]^{-1} f_k$$

$$= [i\omega_k - i\omega_k + \widetilde{\mathcal{A}}(i\omega_k)][i\omega_k I - \widetilde{\mathcal{A}}(i\omega_k)]^{-1} f_k$$

$$= i\omega_k [i\omega_k I - \widetilde{\mathcal{A}}(i\omega_k)]^{-1} - f_k.$$

The series in the last term of equation (5.91) is the Fourier series of the derivative $\dot{x}(t)$. Hence, we can assert that $x(t)$ given by formula (5.89) satisfies a.e. equation (5.75).

The uniqueness follows from the remark that the difference of two (possible) solutions has a Fourier series that has all coefficients zero.

The proof of Theorem 5.6 is complete.

Remark 5.12. In the case of ordinary differential equations of the form $\dot{x}(t) = Ax(t) + f(t)$, condition (5.88) becomes $\det[i\omega I - A] \ne 0$ for $\omega = \omega_k$. This means that $i\omega_k$ must not coincide with any characteristic root of A. In particular, if $\omega > 0$ is large enough, then formula (5.88) is automatically satisfied. In other words, we can have periodic solutions for the equation $\dot{x}(t) = Ax(t) + f(t)$ when A is such that among its characteristic roots there are some with zero real part, but the period $\omega > 0$ of $f(t)$ does verify the condition $\det[(i\omega_k) I - A] \ne 0$.

Remark 5.13. In case condition (5.88) is violated by some k's, it is obvious from equation (5.83) that we cannot have periodic solutions for equation (5.75), regardless of how we choose $f(t)$. The discussion of such cases, though very important for the theory of oscillations, will be omitted here. See some indications in the author's book [23] and in Schwabik et al. [87].

Using the same approach as in the periodic case, we shall now investigate the problem of almost periodicity for the solutions of equation (5.75). We will maintain the conditions specified for the matrix $A(t)$, but $f(t)$ will be assumed almost periodic in a certain sense. We shall look for solutions to equation (5.75) that belong to the space $AP_1(R, \mathcal{C}^n)$. More general conditions will be examined in Section 5.6.

Let us now assume that $f \in AP(R, \mathcal{C}^n)$, and the Fourier series is

$$f(t) \sim \sum_{k=1}^{\infty} f_k e^{i\lambda_k t}, \tag{5.92}$$

with $f_k \in \mathcal{C}^n$, $k \geq 1$, and $\lambda_k \in R$. It is then natural to seek a solution $x(t)$ to equation (5.75) also in $AP(R, \mathcal{C}^n)$, if any, that will be completely determined by its Fourier series

$$x(t) \sim \sum_{k=1}^{\infty} x_k e^{i\lambda_k t}. \tag{5.93}$$

Substituting formula (5.93) on the right-hand side of formula (5.75), we obtain as above

$$\int_R [dA(s)]x(t-s) + f(t) \sim \sum_{k=1}^{\infty} [\widetilde{\mathcal{A}}(i\lambda_k) + f_k] e^{i\lambda_k t}.$$

It is important to note that the operator

$$x(t) \longrightarrow \int_R [dA(s)]x(t-s)$$

is acting continuously on $AP(R, \mathcal{C}^n)$ due to the conditions imposed on $A(t)$ in this section, namely

$$\left| \int_R dA(s) \left[x(s) - y(s) \right] \right|_{AP} \leq \gamma |x - y|_{AP}.$$

On the other hand, if equation (5.75) has an almost periodic solution, its derivative will also be almost periodic, and therefore

$$\dot{x}(t) \sim \sum_{k=1}^{\infty} i\lambda_k x_k e^{i\lambda_k t}. \tag{5.94}$$

This leads to the system of equations for the coefficients x_k

$$[i\lambda_k I - \widetilde{\mathcal{A}}(i\lambda_k)]x_k = f_k, \qquad k \geq 1, \tag{5.95}$$

similar to the system (5.83). Of course, equation (5.95) must take place if we want $x(t)$ in formula (5.93) to be a solution of equation (5.75).

Based on Proposition 5.15, we are permitted to restrict the discussion of the solvability of equation (5.95) only for those x_k's that verify the condition

$$|\lambda_k| \leq \gamma = \int_R |dA(s)|. \tag{5.96}$$

We shall state and prove a result of the existence of solutions in $AP_1(R, C^n)$ for equation (5.75).

Theorem 5.7. *Consider equation (5.75) under the following assumptions:*

(a) $f \in AP(R, C^n)$ *and formula (5.92) is verified.*
(b) *the same as in Theorem 5.6.*
(c) *For all λ_k's satisfying equation (5.96),*

$$\det[i\lambda_k I - \widetilde{A}(i\lambda_k)] \neq 0. \tag{5.97}$$

(d) *The Fourier exponents of f satisfy*

$$\sum_{k=1}^{\infty} |\lambda_k|^{-2} < \infty. \tag{5.98}$$

Then there exists a unique solution $x(t) \in AP_1(R, C^n)$ to equation (5.75). The Fourier series of this solution is

$$\sum_{k=1}^{\infty} [i\lambda_k I - \widetilde{A}(i\lambda_k)]^{-1} f_k e^{i\lambda_k t}. \tag{5.99}$$

Proof. From equation (5.96) and formula (5.97), the construction of the series (5.99) is assured. We need to prove first its uniform and absolute convergence (on R). The proof is very similar to the proof of Theorem 5.6. We point out that formula (5.98) implies $|\lambda_k| \longrightarrow \infty$ as $k \longrightarrow \infty$. This means that $\widetilde{A}(i\lambda_k)/\lambda_k \longrightarrow 0$ as $k \longrightarrow \infty$, which obviously implies $|iI - \widetilde{A}(i\lambda_k)/\lambda_k| \leq M < \infty$, $k \geq 1$, taking the Euclidean norm for the matrix $iI - \widetilde{A}(i\lambda_k)/\lambda_k$, $k \geq 1$.

We now obtain the inequalities

$$\left| [i\lambda_k I - \widetilde{A}(i\lambda_k)] f_k \right| \leq \frac{M}{2} \left(|\lambda_k|^{-2} + |f_k|^2 \right), \ k \geq 1, \tag{5.100}$$

which imply the validity of the uniform and absolute convergence of the series (5.99) as well as the fact that its sum belongs to $AP_1(R, C^n)$. Of course, we had to rely on the fact that $\sum |f_k|^2 < \infty$, which is a consequence of Bessel's inequality for the Fourier coefficients of functions in $AP(R, C^n)$.

An argument similar to that used in the last part of the proof of Theorem 5.6 will tell us that the sum of the series (5.99) verifies equation (5.75).

The uniqueness can also be proven as in Theorem 5.6, observing that the Fourier coefficients of the difference of two solutions of equation (5.75) must all vanish.

This ends the proof of Theorem 5.7.

Remark 5.14. Based on condition (d) of Theorem 5.7, we can assert that the number of λ_k's for which formula (5.97) must hold is finite. If there is no λ_k satisfying equation (5.96), condition (5.97) is automatically verified, and therefore it can be dropped.

Remark 5.15. Condition (d) in Theorem 5.7 is rather restrictive and diminishes the class of almost periodic functions that are acceptable in the theorem. Fortunately, it is always verified in the case of periodic functions (of any period).

Remark 5.16. An alternate proof of Theorem 5.7, under somewhat more restrictive conditions, is given in the author's paper [26]. It is shown first that if f is a trigonometric polynomial, one also obtains a trigonometric polynomial as the solution, as well as the continuity of the linear map $f \longrightarrow$ unique solution of equation (5.75). Then, the result is extended to the case $f \in AP_1(R, \mathcal{C}^n)$ or $f \in AP(R, \mathcal{C}^n)$.

It is natural to ask whether one can obtain results similar to those in this section when f belongs to spaces of almost periodic functions richer than $AP(R, \mathcal{C}^n)$. We have seen, for instance, that $f \in S(R, \mathcal{C}^n)$ is acceptable for systems/equations of the form (5.53). Let us note that the operator on the right-hand side of equation (5.75) does not leave invariant all spaces of almost periodic functions. Actually, the integral $\int [dA(s)]x(t-s)$ may not make sense in the classical framework. In such cases, one must make recourse to the theory of distributions (or generalized functions), an approach we shall consider in the next section.

We shall now deal with a special case of equation (5.75), namely the one described by equation (5.77). Under adequate conditions, similar to those in Theorem 5.7, we shall be able to prove that choices more general than $f \in AP(R, \mathcal{C}^n)$ are acceptable. For instance, the case $f \in S(R, \mathcal{C}^n)$ can be treated satisfactorily. The condition imposed on the Fourier–Stieltjes transform appearing in equation (5.77) is now formulated in a slightly different way than in the case of equation (5.75), even though it is equivalent to equation (5.84). Let us note that in equation (5.77) the transform has the form

$$\widetilde{\mathcal{A}}(i\omega) = \sum_{k=1}^{\infty} A_k e^{-i\omega t_k} + \int_R B(t) e^{-i\omega t} dt. \tag{5.101}$$

The "characteristic" equation (a special case of formula (5.84)) is

$$\det[i\omega I - \widetilde{\mathcal{A}}(i\omega)] = 0, \qquad \omega \in R,$$

with $\widetilde{\mathcal{A}}(i\omega)$ given by equation (5.101).

The number γ defined by equation (5.86) is now

$$\gamma = \sum_{k=1}^{\infty} |A_k| + \int_R |B(t)| dt. \qquad (5.102)$$

It is always finite if conditions (5.78) are satisfied.

From equation (5.101), we see that $\tilde{A}(i\omega)$ is the sum of an almost periodic function in $AP_1(R, \mathcal{C}^n)$ and the (classical) Fourier transform of the matrix $B(t)$.

It is also worth noting that the operator on the right-hand side of equation (5.77) is defined on the space $S(R, \mathcal{C}^n)$ because both operators

$$x(t) \longrightarrow \sum_{k=1}^{\infty} A_k x(t - t_k)$$

and

$$x(t) \longrightarrow \int_R B(s) x(t - s) ds$$

are defined and continuous on $S(R, \mathcal{C}^n)$. For the first operator, the assertion is obvious, while for the second it has been proven in Section 3.3.

In order to obtain an existence result for equation (5.77), it will be necessary to consider the space $S^2(R, \mathcal{C}^n) \subset S(R, \mathcal{C}^n)$ because for $S(R, \mathcal{C})$ we do not have Bessel's inequality. Since $S^2(R, \mathcal{C}^n) \subset B^2(R, \mathcal{C}^n) = AP_2(R, \mathcal{C}^n)$, we can rely on that inequality for any $f \in S^2(R, \mathcal{C}^n)$.

We shall now state the result for the existence of an almost periodic solution to equation (5.77), which is similar to the result contained in Theorem 5.7.

Theorem 5.8. *Let us consider equation* (5.77) *under the following assumptions:*

(a) *The matrices A_j, $j \geq 1$, and $B(t)$ satisfy the estimates* (5.78).
(b) *$f \in S^2(R, \mathcal{C}^n)$,*

$$f(t) \sim \sum_{k=1}^{\infty} f_k e^{i\lambda_k t},$$

with $\lambda_k \in R$, $k \geq 1$, and such that formula (5.98) *holds.*
(c) *For all λ_k's satisfying equation* (5.96) *with γ given by equation* (5.102), *we have formula* (5.97) *satisfied, where $\tilde{A}(i\lambda_k)$ are obtained from equation* (5.101).

Then, there exists a unique solution $x(t)$ to equation (5.77) *such that $x(t) \in AP_1(R, \mathcal{C}^n)$, while its Fourier series is given by formula* (5.99).

Proof. The proof of Theorem 5.8 goes along the same lines as the proof of Theorem 5.7.

It is not a particular case of the preceding theorem due to the fact that we allow more generality for the free (forcing) term $f(t)$: $f \in S^2(R, C^n)$ instead of $f \in AP(R, C^n)$.

On the other hand, equation (5.77) is a special case of equation (5.75) due to the special choice of $A(t)$.

We have discussed above the specific features occurring when we deal with equation (5.77) instead of equation (5.75), and the key to the proof is the inequality (5.100), which remains valid because Bessel's inequality holds for the elements of $S^2(R, C^n)$.

This ends the proof of Theorem 5.8.

Remark 5.17. The Bessel inequality holds for more general spaces of almost periodic functions (e.g., for $B^2(R, C^n)$). As noted above in this section, the right-hand side of equation (5.77) may not make sense for $x \in B^2(R, C^n)$. We can provide a sense of the convolution product appearing on the right-hand side of equation (5.77) when $x \in B^2(R, C^n)$ if we considerably restrict the condition imposed on $B(t)$.

Namely, it can be proved that $\int_R B(s) x(t - s) ds$ makes sense for each $x \in B^2(R, C^n)$ as soon as $B(t)$ verifies $(1 + t^2)|B(t)| \in L^1(R, R)$. This last condition is much more restrictive than formula (5.78). The reason such a condition works follows from the fact that $B^2(R, C^n)$ is part of a weighted L^1-space, namely for each $x \in B^2$, $(1 + t^2)^{-1}|x(t)|^2 \in L^1(R, R)$.

Some considerations related to this case can be found in our paper [26], where a brief discussion is conducted.

5.6 Further Results on Almost Periodic Oscillations in Linear Time-Invariant Systems

The systems dealt with in the preceding sections belong to the class of time-invariant systems, and all of them can be described by means of convolutions between certain functions (or matrix functions).

Actually, there is a result due to L. Schwartz (see K. Yosida [103]) in the theory of distributions that, roughly speaking, states that any time-invariant operator (to be defined below) is a convolution operator.

We will assume the reader has some acquaintance with the theory of distributions (see L. Schwartz [88] and K. Yosida [103]) in order to better describe the concepts that will appear in this section, especially those related to the *almost periodic distribution* and the Fourier transform of tempered distributions.

The equation to be dealt with is formally the same equation (5.75), which we will rewrite in the (equivalent under some conditions) form

$$x'(t) + (A * x)(t) = f(t), \qquad t \in R, \tag{5.103}$$

where A is a matrix-valued function, $A : R \longrightarrow \mathcal{L}(\mathcal{C}^n, \mathcal{C}^n)$, $x, f : R \longrightarrow \mathcal{C}^n$, and the $*$ sign stands for the convolution of functions.

We will regard equation (5.103) as an equation in distributions, even though A will stand for a finite measure, which is a rather special kind of distribution, and f will usually be assumed to be in $L^\infty(R, \mathcal{C}^n)$ or a subspace of it.

In equation (5.103), $x'(t)$ means the derivative of $x(t)$ in the distributional sense, which implies its existence for any locally integrable $x(t)$.

Since we shall use the Fourier transformation of distributions, it is automatically implied that we mean tempered distributions. This is the case of both *finite measures* and L^∞-functions, a feature that will allow us to take the Fourier transformations of both sides in equation (5.103).

Let us recall that the space of distributions $\mathcal{D}'(R, \mathcal{C})$, as defined by L. Schwartz [88], consists of the linear continuous functionals on the test-function space $C_0^\infty(R, \mathcal{C})$, the latter being the space of infinitely differentiable maps from R into $\mathcal{C}$ with compact support, with the following type of convergence: $x_n \longrightarrow x$ iff there exists a compact $K \subset R$ such that $\operatorname{supp}(x_n) \in K$, $n \geq 1$, while $D^j x_n \longrightarrow D^j x$ uniformly on K for each $j \geq 0$ (here D^j stands for the derivative of order j). The $C_0^\infty(R, \mathcal{C})$ with this sort of convergence/topology is denoted by $\mathcal{D}(R, \mathcal{C})$. The space of distributions $\mathcal{D}'(R, \mathcal{C})$ appears as the dual space of $\mathcal{D}(R, \mathcal{C})$.

According to the definition above, a distribution $T \in \mathcal{D}'(R, \mathcal{C})$ is a linear continuous map: $T : \mathcal{D}(R, \mathcal{C}) \longrightarrow \mathcal{C}$. The continuity means that $x_n \longrightarrow x$ in $\mathcal{D}(R, \mathcal{C})$ implies $T(x_n) \longrightarrow T(x)$ in $\mathcal{C}$. Elementary operations, for distributions such as addition and scalar multiplication, can be routinely defined. The product does not necessarily make sense as well as the convolution product of two distributions.

The derivative of a distribution is defined by

$$T'(x) = -T(x'), \qquad x \in C_0^\infty(R, \mathcal{C}).$$

And since x is indefinitely differentiable, any distribution can be differentiated any number of times, the result of differentiation also being a distribution.

For $f \in L_{\text{loc}}^1(R, \mathcal{C})$, the associated distribution is defined by

$$T_f(x) = \int_R f(t) x(t) dt, \qquad x \in \mathcal{D}(R, \mathcal{C}), \qquad (5.104)$$

which tells us clearly that distributions can be regarded as *generalized functions*.

The Schwartz result we mentioned above states that any linear continuous mapping L from $\mathcal{D}'(R, \mathcal{C})$ into $C^\infty(R, \mathcal{C})$ such that $L(\tau_h x) = \tau_h(Lx)$ for any $h \in R$ and $x \in \mathcal{D}(R, \mathcal{C})$, where $(\tau_h x)(t) = x(t - h)$, can be represented as a convolution, $L(x) = T * x$, where the distribution $T \in \mathcal{D}'(R, \mathcal{C})$ is completely determined by L.

Remark 5.18. The convolution product of a distribution in $\mathcal{D}'(R, \mathcal{C})$ with a function in $\mathcal{D}(R, \mathcal{C})$ is always defined. The convolution product of two distributions is also defined if one of them has compact support (see, for instance, K. Yosida [103]). In the first case, the definition is by the formula

$$(T * x)(t) = T_s(x(t - s)),$$

where T_s means that the distribution T is acting on the test function x, regarded as a function of s.

Remark 5.19. It is known, particularly in the engineering literature, that the so-called time-invariant systems are usually described by means of convolution operators. The result of L. Schwartz above tells us that (in a broad sense) all time-invariant systems can be described by means of operatorsor equations of convolution type in the case of functions as well as distributions (generalized functions).

It is necessary to introduce another class of distributions for which it is possible to define the *Fourier transform*. In the classical literature, the Fourier transformation is usually defined for functions $x \in L^1(R, \mathcal{C})$ as well as for those in $L^2(R, \mathcal{C})$.

For distributions in $\mathcal{D}'(R, \mathcal{C})$, the Fourier transformation is not defined in general. The *tempered* distributions constitute a class for which the Fourier transformation is defined, and it is also a distribution.

We start with the space $S(R, \mathcal{C})$ of test functions consisting of those maps from R into $\mathcal{C}$ differentiable of any order and such that

$$\sup_{t \in R} |t^k D^j x(t)| < \infty, \qquad k, j \geq 0. \tag{5.105}$$

A *tempered* distribution is a linear continuous functional $T : S \longrightarrow \mathcal{C}$.

The seminorms defining the topology on $S(R, \mathcal{C})$ are those appearing in the inequality (5.105) above for the test functions.

The space of tempered distributions is usually denoted by $S'(R, \mathcal{C})$, and it appears as the dual space of $S(R, \mathcal{C})$.

The Fourier transform $\widehat{T}$ of a distribution $T \in S'(R, \mathcal{C})$ is simply defined by the formula

$$\widehat{T}(x) = T(\widehat{x}), \qquad x \in S(R, \mathcal{C}). \tag{5.106}$$

This definition is motivated by the fact that, for each $x \in S(R, \mathcal{C})$, the Fourier transformation

$$\widehat{x}(t) = (2\pi)^{-1/2} \int_R x(s) e^{-its} ds \tag{5.107}$$

is defined and $\widehat{x} \in S(R, \mathcal{C})$. This assertion follows quite easily if we take into account the definition of the space $S(R, \mathcal{C})$ and differentiate (with respect to t) any number of times in (5.107).

Remark 5.20. If $f \in L^1(R, \mathcal{C})$ and T_f is the associated distribution defined by equation (5.104), then $\widehat{T}_f = T_{\widehat{f}}$, where f and $\widehat{f}$ are related by

$$\widehat{f}(t) = (2\pi)^{-1/2} \int_R f(s) e^{-its} ds.$$

This formula shows that the Fourier transform for tempered distributions coincides with the classical transform for functions in $L^1(R, \mathcal{C})$. In the integral $\widehat{T}_f(x) = \int_R f(s)\widehat{x}(s)ds$, one has to change the order of integration.

Remark 5.21. The coefficient $(2\pi)^{-1/2}$ in front of the integral in equation (5.107) is sometimes used because of the symmetry in the inverse formula for equation (5.106):

$$x(t) = (2\pi)^{-1/2} \int_R \widehat{x}(s) e^{its} ds.$$

A similar discussion can be conducted in the case of the classical Fourier transform on the space $L^2(R, \mathcal{C})$, the conclusions being the same: The Fourier transform for tempered distributions coincides with the classical transform of L^2-functions when the distribution is an L^2-function.

The procedure used above to construct the distribution spaces $\mathcal{D}'(R, \mathcal{C})$ and $\mathcal{S}'(R, \mathcal{C})$ can be applied to many other situations. One starts with a test-function space endowed with a topology, and the "distributions" are nothing but the elements of the dual space (i.e., the linear continuous functionals). See the references mentioned above for other examples of distribution spaces.

An important space of particular distributions is the space of *measures*, defined as follows. Let the test-function space consist of all continuous functions in $C(R, \mathcal{C})$ such that each has a compact support; on this space one can define a convergence/topology the same way we proceeded with $C_0^\infty(R, \mathcal{C})$; then, any continuous linear functional will be called a *measure* (on R, with complex values) and similarly for real-valued measures.

If L is such a functional, then we can represent it by the formula

$$L(x) = \int_R x(s) d\mu(s) \tag{5.108}$$

according to a well–known result due to F. Riesz and M. Fréchet. The integral in equation (5.108) is a Stieltjes integral, and μ is a function with bounded variation on any compact interval of R. When μ has bounded variation on R, then we say that μ is a *finite* measure.

Of course, equation (5.108) makes sense when $x \in C_0^\infty(R, \mathcal{C})$, and formula (5.108) can be regarded as describing those distributions that reduce to measures.

It is useful to note that a finite measure is also a tempered distribution, which will allow us to use the Fourier transform for such distributions.

These few facts we have gathered above about distributions will allow us to look at the equation (5.75) or (5.103) as an equation in distributions. We will rewrite these equations in the form

$$x' + \mu * x = f, \tag{5.109}$$

with μ a matrix of type $n \times n$ whose entries are finite measures on R and real valued or complex valued, and $f \in \mathrm{AP}(R, \mathcal{C}^n)$. We shall be interested in solutions to equation (5.109) such that they belong to $\mathrm{AP}_1(R, \mathcal{C}^n)$, $\mathrm{AP}(R, \mathcal{C}^n)$, or another space of almost periodic functions.

Since the Fourier transform method will be used substantially, we need to consider this transform in the distributional sense.

Indeed, the classical Fourier transform cannot be applied to functions in $\mathrm{AP}(R, \mathcal{C}^n)$. For the measure μ, we shall use the Fourier–Stieltjes transform, which coincides with the Fourier transform for tempered distributions.

Let us mention here that a distribution $T \in \mathcal{D}'(R, \mathcal{C})$ is called *almost periodic* iff $T * x \in \mathrm{AP}(R, \mathcal{C})$ for any $x \in \mathcal{D}(R, \mathcal{C})$. This definition recalls the definition of weakly almost periodic functions (see Section 5.2). There is another definition, based on the relative compactness of the set of translates $\{\tau_h T; \, h \in R\}$, where $\tau_h T(x) = T(\tau_{-h} x)$, $h \in R$, with $T \in \mathcal{D}'(R, \mathcal{C})$. Of course, the relative compactness is meant in the sense of convergence in $\mathcal{D}'(R, \mathcal{C})$.

The space of almost periodic distributions will be denoted by $\mathcal{AP}(R, \mathcal{C})$.

We shall also need the concept of a *bounded* distribution, which should be an extension of the usual concept of boundedness for functions.

One says that $T \in \mathcal{D}'(R, \mathcal{C})$ is *bounded* if the set of the translates $\{\tau_h T; \, h \in R\}$ is bounded in $\mathcal{D}'(R, \mathcal{C})$. Let us recall that a set Σ in a topological vector space (and $\mathcal{D}'$ is such a space!) is bounded iff, for each neighborhood $U \ni \theta =$ the zero element, there exists an integer $p > 0$ such that $pU \supset \Sigma$. The space of bounded distributions will be denoted by $\mathcal{B}'(R, \mathcal{C})$.

An equivalent definition is that T is a bounded distribution iff $T * x$ is a bounded function on R for each $x \in \mathcal{D}(R, \mathcal{C})$.

Taking into account the fact that any relatively compact set is also bounded, there results that any almost periodic distribution is bounded.

It can be shown that any *bounded* distribution is also a *tempered* distribution (see L. Schwartz [88]). Therefore, the Fourier transform for distribution can be applied to any bounded distribution and also to any almost periodic distribution.

We can now reformulate the old Bohr–Neugebauer problem, which we have considered in Section 5.4 for ordinary differential equations, for the more general setting where we deal with functional differential equations of the form (5.109). Namely, under what conditions can we assert that a bounded (distributional) solution is almost periodic if f itself is an almost periodic distribution?

This problem will be discussed in the remaining part of this section, following the paper of O. Staffans [93]. We will attempt to be as self-contained as possible, but given the complexity of the topic, it will be necessary to send the reader to other sources (in book form).

We need to say a few words about the *Fourier series* attached to an almost periodic distribution. This concept will allow us to pursue the same procedure as in Section 5.5 to construct almost periodic solutions to equation (5.109); in other words, to obtain from the equation the Fourier series of the solution in terms of the Fourier series of f and the Fourier transform of the measure μ.

Of great help in pursuing the discussion of the problem formulated above (Bohr–Neugebauer, the Fourier series) is a result that will allow us to reduce problems in $\mathcal{B}'(R,\mathcal{C})$ to problems in $\mathrm{UCB}(R,\mathcal{C})$ where $\mathrm{UCB}(R,\mathcal{C})$ is the subspace of $CB(R,\mathcal{C})$ consisting of the uniformly continuous functions on R or reduce problems in $\mathcal{AP}(R,\mathcal{C})$ to problems in $\mathrm{AP}(R,\mathcal{C})$.

Let us consider the auxiliary functions in $L^1(R,\mathcal{C})$

$$e_m(t) = \frac{t^{m-1}}{(m-1)!}\, e^{-t}, \quad t \geq 0 \qquad \text{and } e_m(t) = 0, \quad t < 0. \tag{5.110}$$

The result we have in mind (see L. Schwartz [88]) can be stated as follows.

Proposition 5.16. *For any distribution $x \in \mathcal{B}'(R,\mathcal{C})$, there exists an integer $m \geq 1$ such that $e_m * x \in \mathrm{UCB}(R,\mathcal{C})$. If $x \in \mathcal{B}'(R,\mathcal{C})$, then $x \in \mathcal{AP}(R,\mathcal{C})$ iff $e_m * x \in \mathrm{AP}(R,\mathcal{C})$ for some integer $m \geq 1$.*

Based on Proposition 5.16, we can define the Fourier series for almost periodic distributions. Namely, since for some $m \geq 1$ one has $e_m * x \in \mathrm{AP}(R,\mathcal{C})$ given $x \in \mathcal{AP}(R,\mathcal{C})$, we can write (Section 4.2)

$$e_m * x \sim \sum_{k=1}^{\infty} b_k e^{i\lambda_k t},$$

which leads us to define the Fourier series by

$$x \sim \sum_{k=1}^{\infty} (1 + i\lambda_k)^m b_k e^{i\lambda_k t}. \tag{5.111}$$

The rationale for writing formula (5.111) stems from the fact that $\widehat{e}_m(s) = (1+is)^{-m}$ and $(e_m * x)\widehat{\ } = \widehat{e}_m \widehat{x}$ (the product of a function and a distribution).

If for $x \in \mathcal{AP}(R,\mathcal{C})$ we write

$$x \sim \sum_{k=1}^{\infty} a_k e^{i\lambda_k t}, \tag{5.112}$$

in accordance with formula (5.111), then λ_k, $k \geq 1$, are called the Fourier exponents of x, while a_k, $k \geq 1$, are the Fourier coefficients. The coefficients a_k, $k \geq 1$, do not depend on m, but b_k, $k \geq 1$, do depend on m.

Let us mention that the mean value can also be defined for distributions in $\mathcal{AP}(R,\mathcal{C})$, and the classical approach to defining the Fourier series can be pursued successfully.

It is obvious that what has been said above for the space $\mathcal{AP}(R,\mathcal{C})$ can be extended in the standard manner to the space $\mathcal{AP}(R,\mathcal{C}^n)$, $n \geq 1$, of vector-valued distributions or even to the case of matrix-valued distributions.

We shall now return to equation (5.109) and state the main results of this section. They complete and substantially improve the results contained in Section 5.5. The reason we did not omit the results in Section 5.5 is because they have been obtained by rather elementary procedures, while in this section the presentation requires a lot more concepts and preparation (mainly in connection with distribution theory).

Theorem 5.9. *Consider the functional differential equation* (5.109) *under the following (general) assumptions:*

A_1:*μ is a matrix-valued finite measure on R of type $n \times n$. Equivalently, μ : $R \longrightarrow \mathcal{L}(\mathcal{C}^n, \mathcal{C}^n)$, each entry being a function with bounded variation from R into $\mathcal{C}$.*

A_2:*If $D(s) = isI + \widehat{\mu}(s)$, $s \in R$, where $\widehat{\mu}(s) = \int_R e^{ist}d\mu(t)$ is the Fourier–Stieltjes transform of μ, then the equation*

$$\det D(s) = 0, \qquad s \in R, \tag{5.113}$$

has at most a countable set of solutions.
A_3:*$f \in \mathcal{AP}(R,\mathcal{C}^n)$.*

Then any solution of equation (5.109) *that is a bounded distribution is an almost periodic distribution (i.e., it belongs to $\mathcal{AP}(R,\mathcal{C}^n)$). If $f \in S(R,\mathcal{C}^n)$, then any bounded solution of equation* (5.109)*, in the distributional sense, belongs to* $\mathrm{AP}(R,\mathcal{C})$*. Moreover, if equation* (5.113) *has only a finite number of solutions and $f \in \mathcal{AP}(R,\mathcal{C}^n)$ has the Fourier series*

$$f \sim \sum_{k=1}^{\infty} f_k e^{i\lambda_k t} \tag{5.114}$$

satisfying the condition

$$\sum_k \left| D^{-1}(\lambda_k)f_k \right| < \infty, \qquad \textit{for those } k \textit{ with } \det D(\lambda_k) \neq 0, \tag{5.115}$$

then any bounded distribution solution of equation (5.109) *belongs to* AP_1 $(R,\mathcal{C}^n)$*.*

The proof of Theorem 5.9 is rather lengthy, and we shall present it in several steps.

First, let us dwell on the assumption A_3. Since we are interested in almost periodic solutions to equation (5.109), it is worth noting that the left-hand side of this equation is an almost periodic distribution, any time x is in $\mathcal{AP}(R, \mathcal{C}^n)$. Therefore, if we look for such a solution, one must necessarily accept A_3. Indeed, $x \in \mathcal{AP}(R, \mathcal{C}^n)$ implies $x' \in \mathcal{AP}(R, \mathcal{C}^n)$, and the space $\mathcal{AP}(R, \mathcal{C}^n)$ is invariant (see L. Schwartz [88]) with respect to convolution by a finite matrix-valued measure on R. In other words, if μ satisfies A_1, then $\mu * x \in \mathcal{AP}(R, \mathcal{C}^n)$ for any $x \in \mathcal{AP}(R, \mathcal{C}^n)$. The assertion is well known when $x \in AP(R, \mathcal{C}^n)$. It also holds for $x \in \mathcal{AP}(R, \mathcal{C}^n)$ because $e_m * \mu * x = \mu * e_m * x$. Moreover, if x is as in formula (5.112), then

$$\mu * x = \sum_{k=1}^{\infty} \widehat{\mu}(\lambda_k) a_k e^{i\lambda_k t}. \tag{5.116}$$

Second, in order to carry out the proof of Theorem 5.9, it is convenient to give an equivalent form to equation (5.109). Namely, the following statement holds true.

Proposition 5.17. *Let μ satisfy condition A_1. If $x \in \mathcal{B}'(R, \mathcal{C}^n)$ is a solution of equation* (5.109), *then it verifies the equation*

$$x + \nu * x = e_1 * f \tag{5.117}$$

and vice versa, where

$$\nu = e_1 * \mu - e_1 I. \tag{5.118}$$

Proof. If equation (5.109) is satisfied by $x \in \mathcal{B}'(R, \mathcal{C}^n)$, then necessarily $f \in \mathcal{B}'(R, \mathcal{C}^n)$. Taking the convolution product of both sides of equation (5.109) by e_1, we obtain

$$e_1 * x' + e_1 * \mu * x = e_1 * f,$$

which leads to

$$e_1' * x + e_1 * \mu * x = e_1 * f \tag{5.119}$$

because the derivative of a convolution product is obtained by differentiating either one of the factors. Since $e_1' = \delta - e_1$, with δ the Dirac distributions ($\delta x = x(0)$, $x \in \mathcal{D}$), one obtains equation (5.119) by substituting e_1' in equation (5.118).

Conversely, if $x \in \mathcal{B}'(R, \mathcal{C}^n)$ satisfies equation (5.117), with ν given by equation (5.118), then differentiating both sides of equation (5.117), one obtains

$$x' + \nu' * x = e_1' * f. \tag{5.120}$$

But $e_1' = \delta - e_1$, and $\nu' = e_1' * \mu - e_1' I = \mu - e_1 * \mu - \delta I + e_1 I = \mu - \delta I - \nu$. From equation (5.120), we now get

$$x' + \mu * x - x - \nu * x = f - e_1 * f,$$

and from equation (5.117) one obtains equation (5.109).

Third, let us note that a solution $x \in \mathcal{AP}(R, \mathcal{C}^n)$ of equation (5.109) satisfying formula (5.112) and f satisfying formula (5.114) lead to the identity

$$\sum_{k=1}^{\infty} f_k e^{i\lambda_k t} \equiv \sum_{k=1}^{\infty} D(\lambda_k) a_k e^{i\lambda_k t}. \qquad (5.121)$$

Actually, formula (5.121) should serve for finding the Fourier coefficients of the solution in terms of the Fourier coefficients of f and the Fourier–Stieltjes transform of μ. Obviously, assuming $\det D(\lambda_k) \neq 0$, $k \geq 1$, would suffice for uniquely determining the coefficients a_k, $k \geq 1$. But here we do not attempt to prove the existence of a solution to equation (5.109). All we want now is to show that a solution of equation (5.109), which is a bounded distribution, is in fact in $\mathcal{AP}(R, \mathcal{C}^n)$. From this point of view, another condition proves to be useful. This condition is based on a characterization of functions in $\mathrm{AP}(R, \mathcal{C}^n)$ by means of their Fourier transform (in the distributional sense).

Fourth, we will provide this characterization after introducing the concept of an *asymptotically uniformly continuous function* on R.

We shall say that $x \in L^{\infty}(R, \mathcal{C})$ is asymptotically uniformly continuous if for each $\varepsilon > 0$ there correspond $T(\varepsilon) > 0$ and $\delta(\varepsilon) > 0$ such that $t \geq T(\varepsilon)$ and $|s| < \delta(\varepsilon)$ imply $|x(t + s) - x(t)| < \varepsilon$.

The space of such functions (which is a subspace of L^{∞}) is denoted by $\mathrm{AUCB}(R, \mathcal{C})$. One can easily see that it is invariant with respect to the convolution with a finite measure.

The result we want to state here and prove later follows.

Proposition 5.18. *Let* $x \in \mathrm{AUCB}(R, \mathcal{C})$. *Then* $x \in \mathrm{AP}(R, \mathcal{C})$ *iff the set of points on R where $\hat{x}$ is not almost periodic is countable.*

The proof will be given at the end of this section, the result in Proposition 5.18 being significant in many circumstances.

Fifth, we shall now reduce the case of distribution solutions to the case of functions in $\mathrm{UCB}(R, \mathcal{C}^n)$. In this respect, Proposition 5.16 is the main tool.

Indeed, let us assume with respect to equation (5.109) that $x \in \mathcal{B}'(R, \mathcal{C}^n)$, while $f \in \mathcal{AP}(B, \mathcal{C}^n)$ according to A_3. According to Proposition 5.17, x and f are connected by equation (5.117). If we choose m large enough, then

$$y = e_m * x \in \mathrm{UCB}(R, \mathcal{C}^n) \text{ and } e_{m+1} * f \in \mathrm{AP}(R, \mathcal{C}^n).$$

Therefore, y satisfies the equation

$$y + \nu * y = g \qquad (5.122)$$

with $g = e_{m+1} * f$. As seen in Proposition 5.16, $x \in \mathcal{AP}(R, \mathcal{C}^n)$ if and only if $y \in \mathrm{AP}(R, \mathcal{C}^n)$. Hence we need only prove that for $g \in \mathrm{AP}(R, \mathcal{C})$ every solution $y \in \mathrm{UCB}(R, \mathcal{C}^n)$ is actually in $\mathrm{AP}(R, \mathcal{C}^n)$.

Let us apply the Fourier transform (distributional) to both sides of equation (5.122). One obtains

$$(\delta I + \nu)\hat{\,}\hat{y} = \hat{g} \qquad (5.123)$$

with the matrix-vector product on the left-hand side of equation (5.123). But we can show that

$$(\delta I + \nu)\hat{\,}(s) = (1 + is)^{-1} D(s), \qquad s \in R, \qquad (5.124)$$

by using equation (5.118) and $\hat{e}_1(s) = (1 + is)^{-1}$. Hence, equation (5.123) becomes

$$[(1 + is)^{-1} D(s)]\hat{y}(s) = \hat{g}(s), \qquad s \in R. \qquad (5.125)$$

From equation (5.125), based on the hypothesis of Proposition 5.18 (see formula (5.113) of assumption A_2), we see that $\hat{y}$ may fail, being almost periodic only for a set on R that is at most countable. Proposition 5.18 allows us to conclude that $y(t)$, $t \in R$, is in $AP(R, C^n)$.

This conclusion motivates the first statement in Theorem 5.9.

Sixth, we shall now concentrate on proving the second statement in Theorem 5.9, which claims that $f \in S(R, C^n)$ implies $x \in AP(R, C^n)$.

Considering again equation (5.117), which is equivalent to equation (5.109), we note that for $f \in S(R, C^n)$ one has $e_1 * f \in AP(R, C^n)$. This assertion can be easily proven because

$$|(e_1 * f)(t)| = \left| \int_0^\infty e^{-s} f(t - s) ds \right| \le \sum_{k=0}^\infty e^{-k} \int_k^{k+1} |f(t-s)| ds \le e(e-1)^{-1} |f|_S.$$

In other words, the operator $f \longrightarrow e_1 * f$ is a continuous operator from S into AP (apply the inequality above to $f_\tau - f$).

Therefore, when $f \in S(R, C^n)$, the right-hand side in equation (5.117) belongs to $AP(R, C^n)$.

On the other hand, due to the fact that $e_1 * \mu = \nu \in L^1(R, \mathcal{L}(R^n \times R^n))$, we reduce our problem to proving that any bounded distribution x satisfying the equation

$$x + \nu x = h \qquad (5.126)$$

with ν as above and $h \in AP(R, C^n)$ is actually a function and belongs to the space $AP(R, C^n)$.

Seventh, we shall now prove another auxiliary result, related to equation (5.126), which will allow us to obtain the validity of the second statement in Theorem 5.9. Namely, the following assertion is true with respect to equation (5.126): If x is a bounded distribution in the space $\mathcal{B}'(R, C^n)$ satisfying (5.126), with $f \in UCB(R, C^n)$, then $x \in UCB(R, C^n)$.

Indeed, due to the fact that $\nu \in L^1$, one has $|\hat{\nu}(s)| \longrightarrow 0$ as $|s| \longrightarrow \infty$. This implies that one can find a number $S > 0$ large enough that $I + \hat{\nu}(s)$ is invertible for $|s| \geq S$. By a well-known argument (Wiener; see, for instance, our book [24]), there exists $\mathcal{X}(s) \in L^1$ such that $[I + \widehat{\mathcal{X}}(s)][I + \hat{\nu}(s)] = I$ for

$|s| \geq S$. Now choose $\eta \in L^1$ a scalar function whose Fourier transform has compact support such that $\widehat{\eta}(s) \equiv 1$ for $|s| \leq S + 1$. If we return to equation (5.126), which obviously can be rewritten as $(\delta I + \nu) * x = h$, and take the convolution product of both sides by $\delta - \eta$, we obtain

$$(\delta I + \nu) * (x - \eta * x) = f - \eta * f. \qquad (5.127)$$

The distributional Fourier transform of $x - \eta * x$ vanishes on $(S - 1, S + 1)$, and taking into account the relation $[I + \widehat{\chi}(s)][I + \widehat{\nu}(s)] = I$, $s \in R$, we can write

$$(\delta I + \chi) * (\delta I + \nu) * (x - \eta * x) = x - \eta * x. \qquad (5.128)$$

Taking again the convolution product of equation (5.127) with $\delta I + \chi$, we obtain

$$x - \eta * x = f - \eta * f + \chi * f - \chi * \eta * f. \qquad (5.129)$$

The right-hand side of equation (5.129) belongs to $\mathrm{UCB}(R, C^n)$ because $f \in \mathrm{UCB}(R, C^n)$, which implies $x - \eta * x \in \mathrm{UCB}(R, C^n)$. But the Fourier transform of $\eta * x$ has compact support, which means that $\eta * x \in \mathrm{UCB}(R, C^n)$. Therefore, $x \in \mathrm{UCB}(R, C^n)$, and the assertion is proven.

Eighth, equation (5.126) leads, according to equation (5.125), to the relationship

$$[(1 + is)^{-1} D(s)] \hat{x}(s) = \hat{h}(s), \qquad s \in R, \qquad (5.130)$$

and from the assumption A_2 we obtain that $\hat{x}(s)$ may not be almost periodic only for those values of s that verify equation (5.113); i.e., $D(s) = 0$. Consequently, $\hat{x}(s)$ may not be almost periodic only for a set of at most countable values of s.

Hence, x must be in $\mathrm{AP}(R, C^n)$ according to Proposition 5.18.

Ninth, there remains to show that if equation (5.113) has only a finite number of roots and formula (5.115) holds true, then $x \in \mathrm{AP}_1(R, C^n)$. We already know that any $x \in \mathcal{B}'(R, C^n)$ will be in $\mathrm{AP}(R, C^n)$, as seen above. Since a finite number of terms in the Fourier series of the solution x do not influence the absolute convergence, the condition (5.115) is guaranteeing the validity of the assertion $x \in \mathrm{AP}_1(R, C^n)$.

This ends the proof of Theorem 5.9, but we still have to prove Proposition 5.18. In order to carry out the proof, we need a few more auxiliary concepts and results.

Let $f \in \mathrm{AUCB}(R, C)$ with $\mathrm{AUCB}(R, C)$ as defined just before the statement of Proposition 5.18.

The *limit set* of f, denoted by $\Gamma(f)$, is defined as follows: $g \in \mathrm{UCB}(R, C)$ belongs to $\Gamma(f)$ if there exists a sequence $\{h_m; \ m \geq 1\} \subset R$ with $h_m \longrightarrow \infty$ as $m \longrightarrow \infty$ such that $f(t + h_m) \longrightarrow g(t)$ as $m \longrightarrow \infty$ uniformly on any compact set in R.

It is a simple fact that $f \in \mathrm{AP}(R, C)$ implies $f \in \Gamma(f)$. Indeed, for each $m \in N$, the set of m^{-1}-almost periods of f is relatively dense in R. Therefore, one can find $h_m \geq m$ such that $\sup |f(t + h_m) - f(t)| \leq m^{-1}$, $t \in R$. Hence, $f(t + h_m) \longrightarrow f(t)$ uniformly on R as $m \longrightarrow \infty$, which proves the property $f \in \Gamma(f)$.

This property shows that the set $AP(R, \mathcal{C})$ is closed with respect to a kind of convergence that "a priori" is weaker than that of uniform convergence.

We shall say that the sequence $\{f_n; \ n \geq 1\} \subset AP(R, \mathcal{C})$ *converges asymptotically* to f if for each $\varepsilon > 0$ one can find $N(\varepsilon) > 0$ and $T(\varepsilon) > 0$ with the property

$$|f_n(t) - f(t)| < \varepsilon \qquad \text{for } t \geq T(\varepsilon) \text{ and } n \geq N(\varepsilon). \tag{5.131}$$

The definition of a *Cauchy sequence* with respect to asymptotic convergences is obvious: The sequence $\{f_n(t); \ n \geq 1\} \in AP(R, \mathcal{C})$ is Cauchy if for each $\varepsilon > 0$ one can find $N(\varepsilon) > 0$ and $T(\varepsilon)$ such that

$$|f_n(t) - f_m(t)| < \varepsilon \qquad \text{for } t \geq T(\varepsilon) \text{ and } n, m \geq N(\varepsilon).$$

It is now easy to prove that a Cauchy sequence with respect to asymptotic convergence is uniformly convergent on R. Indeed, since for fixed m, n one has $f_n - f_m \in AP(R, \mathcal{C})$, according to the above-mentioned property ($f \in \Gamma(f)$ for $f \in AP$), we derive $f_n - f_m \in \Gamma(f_n - f_m)$, which implies

$$\sup_{t \in R} |f_n(t) - f_m(t)| = \limsup_{t \longrightarrow \infty} |f_n(t) - f_m(t)|. \tag{5.132}$$

Formula (5.132) shows that a Cauchy sequence with respect to the asymptotic convergence is also a Cauchy sequence with respect to uniform convergence (and vice versa!).

We can therefore conclude that, in the space $AP(R, \mathcal{C})$, the concept of asymptotic convergence (apparently weaker than that of uniform convergence on R) is in fact equivalent to that of uniform convergence.

Let us now deal with a concept that plays an essential role in the proof of Proposition 5.18. We have briefly considered this concept in Section 4.6, but we prefer to repeat its definition here due to the fact that we are now considering the Fourier transform in the distributional sense.

Consider $f \in L^\infty(R, \mathcal{C})$ with $\hat{f}$ its Fourier transform. The point $t_0 \in R$ is called a *point of almost periodicity* for $\hat{f}$ if there exists a neighborhood of t_0 with the property that one can find $g \in AP(R, \mathcal{C})$ such that $\hat{f} = \hat{g}$ in that neighborhood.

Since $\hat{f}$ and $\hat{g}$ are distributions, the definition above requires some explanation. When we say that $\hat{f} = \hat{g}$ in a neighborhood of $t_0 \in R$, we mean that these two distributions take the same value (in $\mathcal{C}$) for every test function with support in that neighborhood.

The following result is an auxiliary one for obtaining Proposition 5.18.

Proposition 5.19. *A function $f \in AUCB(R, \mathcal{C})$ is almost periodic (i.e., in $AP(R, \mathcal{C})$) iff $\hat{f}$ is almost periodic everywhere on R.*

Proof. It is obvious that $f \in AP(R,C)$ implies that $\hat{f}$ is almost periodic everywhere.

Conversely, assuming that $\hat{f}$ is almost periodic everywhere, we need to show that $f \in AP(R,C)$. We distinguish two distinct situations: first, when the spectrum $\sigma(f)$ is compact, and second, when $\sigma(f)$ is not compact. In the first case, we can find a finite number of real intervals I_k, $k = 1, 2, \ldots, N$, and almost periodic functions f_k such that $\bigcup_{k=1}^{N} I_k \supset \sigma(f)$ and $\hat{f} = \hat{f}_k$ in I_k, $k = 1, 2, \ldots, N$.

Let us now choose some test functions $\eta_k \in S(R,C)$ with $\sigma(\eta_k) \subset I_k$, $k = 1, 2, \ldots, N$, and $\Sigma \hat{\eta}_k = 1$ on $\sigma(f)$.

Let us denote $g_k = \eta_k * f$, $k = 1, 2, \ldots, N$. Then $f = (\Sigma \eta_k) * f = \Sigma g_k$. But $\sigma(\eta_k) \subset I_k$, and $\hat{f} = \hat{f}_k$ in I_k, $k = 1, 2, \ldots, N$, which implies $g_k = \eta_k * f = \eta_k * f_k \in AP$, $k = 1, 2, \ldots, N$, from which we obtain $f = \Sigma g_k \in AP$.

Assume now that $\hat{f}$ is everywhere almost periodic but the spectrum $\sigma(f)$ is not compact. Consider as above a sequence $\eta_k \in S$, $k = 1, 2, \ldots$, where η_k form an *approximate identity* with respect to the convolution product (see, for instance, Cora Sadosky [84] or C. Corduneanu [24]) such that $\sigma(\eta_k)$, $k \geq 1$, belong to the same compact set.

If $t_0 \in R$ is fixed and I denotes a neighborhood of t_0, there exists an almost periodic function h such that $\hat{f} = \hat{h}$ in I. Since $\eta_k * h$ is almost periodic, while $(\eta_k * h)\hat{} = \hat{\eta}_k \hat{h} = \hat{\eta}_k \hat{f}$ in I, we obtain $(\eta_k * f)\hat{} \in AP\hat{}$ everywhere. Hence, taking into account that the spectra of $\eta_k * f$ are in a compact set, we obtain the inclusion $\eta_k * f \in AP$, $k \geq 1$. For $k \longrightarrow \infty$, $\eta_k * f \longrightarrow f$ almost everywhere (and asymptotically, too), which implies that $\{\eta_k * f; \ k \geq 1\}$ converges uniformly on R to f (according to the property that in $AP(R,C)$ the uniform and asymptotic convergences (are equivalent) and $f \in AP(R,C)$.

The discussion just carried out proves Proposition 5.19.

Proposition 5.20. *Let $f \in L^{\infty}(R,C)$. Then the set on R where $\hat{f}$ is not almost periodic is perfect.*

Proof. Obviously, the set where $\hat{f}$ is almost periodic is open. Hence, its complement is closed. Assume now that $\hat{f}$ is not almost periodic at some point $t_0 \in R$ but that in some neighborhood of t_0, say I, $\hat{f}$ is almost periodic (i.e., in $I \setminus \{t_0\}$). We choose $\eta \in S$ with $\sigma(\eta)$ compact and $\sigma(\eta) \subset I$, $\hat{\eta} = 1$, in some neighborhood of t_0. Consider the product $g = \eta * f$. Then $\hat{g}$ is almost periodic everywhere, with the possible exception at t_0. Moreover, $\sigma(g)$ is compact. According to a result in Katznelson [54] (Theorem 5.20) applied to the function $e^{it_0 t} g(t)$, $g \in AP(R,C)$. Hence $\hat{f}$ is almost periodic at t_0, which constitutes a contradiction with the fact that the set where $\hat{f}$ is not almost periodic has no isolated points. This ends the proof of Proposition 5.20.

From Propositions 5.19 and 5.20, we derive the validity of Proposition 5.18 due to the fact that any nonempty perfect set is not countable.

Corollary 5.2 (to Proposition 5.18). *If $f \in AUCB(R,C)$ has a countable spectrum, then $f \in AP(R,C)$.*

This is true because in the complement of $\sigma(f)$, $\hat{f}$ is locally zero and hence almost periodic.

We conclude Section 5.6 here. Chapter 5 will conclude with the case of time-varying linear systems.

5.7 Almost Periodic Oscillations in Linear Time-Varying Systems

The simplest type of linear time-varying system is the ordinary differential system

$$\dot{x}(t) = A(t)x(t) + f(t), \qquad t \in R, \tag{5.133}$$

where the given elements are $f \in \mathrm{AP}(R, \mathcal{C}^n)$ and $A \in \mathrm{AP}(R, \mathcal{L}(\mathcal{C}^n, \mathcal{C}^n))$. Of course, we can substitute R^n for $\mathcal{C}^n$ if we want to deal with real-valued functions only.

The special case where A is a constant matrix was discussed in Section 5.4, and the basic result established there was the Bohr–Neugebauer theorem (Proposition 5.12). Regardless of the nature of the characteristic roots of A, any bounded (on R) solution of equation (5.53) is necessarily almost periodic (Bohr). Since Bohr almost periodic functions are bounded, there results that the problem of almost periodicity of solutions is basically the same as the boundedness problem in this case. Of course, the result above does not guarantee the existence of solutions to equation (5.133) bounded on R. Actually, there are examples of almost periodic systems without bounded solutions, a fact that follows from the simple example of the scalar equation $\dot{x}(t) = ix(t) + \exp(it)$, whose general solution is $x(t) = (C + t)\exp(it)$, $C \in \mathcal{C}$ being arbitrary. Since $|x(t)| = |C + t|$, $t \in R$, one sees that this example solves the problem (of nonexistence of bounded solutions) even in the case of constant coefficients.

On the other hand, Proposition 5.13 provides a case where the existence (and also the uniqueness) of a bounded solution is assured (in the absence of characteristic roots with zero real part to the matrix A).

In order to extend such results to the case of systems of the form (5.133), J. Favard [39] has built up a theory of almost periodicity of solutions, which we shall briefly present here.

Before we can enunciate Favard's basic results, we need to define the concept of the *hull* for a function $f \in \mathrm{AP}(R, \mathcal{C})$, which shall be denoted by $H(f)$. By definition, $g \in \mathrm{AP}(R, \mathcal{C})$ is in $H(f)$ if one can find a sequence $\{h_n;\ b \geq 1\} \subset R$ with the property $\lim_{n \to \infty} f(t + h_n) = g(t)$, $t \in R$, uniformly on R.

In the same way, we can define $H(f)$ when $f \in \mathrm{AP}(R, \mathcal{C}^n)$ or, for a matrix-valued $A \in \mathrm{AP}(R, \mathcal{L}(R^n, R^n))$, the hull $H(A)$.

Proposition 5.21. *Consider the system* (5.133), *and assume that each system*

$$\dot{y}(t) = B(t)y(t), \quad B \in H(A), \qquad t \in R, \qquad (5.134)$$

possesses the property that for any bounded nonzero solution of equation (5.134), say $y(t)$, one has

$$\inf_{t \in R} |y(t)| > 0. \qquad (5.135)$$

If the system (5.133) has a bounded solution on R, then there exists an almost periodic solution $x(t)$ to this system, whose module is contained in $\mathrm{mod}(A, f)$: $\mathrm{mod}(x) \subset \mathrm{mod}(A, f)$.

The proof of Proposition 5.21 will follow from that of the similar result in Section 5.2 when we deal with the general (nonlinear) case. The module containment part can be found in the quoted reference by J. Favard or in the book [40] by A. M. Fink.

The definition of the *module*, $\mathrm{mod}(f)$, of a function $f \in AP(R, \mathcal{C})$ is the smallest additive group of reals containing the Fourier exponents of f. For instance, for a periodic function of period $\omega > 0$, the module is the group consisting of all multiples of ω, $\{k\omega; \ k \in Z\}$.

Remark 5.22. In the case of systems with constant coefficients, $A = \mathrm{const.}$, from $\mathrm{Re}\,\lambda_j \neq 0$, $j = 1, 2, \ldots, n$, where λ_j, $1 \leq j \leq n$, are the characteristic values of A, there results the absence of nonzero bounded (on R) solutions to $\dot{x} = Ax$. Hence, condition (5.135) is verified and the existence of a bounded (almost periodic) solution is assured. This fact is known to us from Proposition 5.13.

Remark 5.23. A key condition for the validity of Proposition 5.21 is the separation condition (5.135) for the solutions of the equations in the hull $H(A)$. If one drops this condition, replacing it by its opposite (i.e., there exists at least one $\widetilde{B} \in H(A)$ such that a bounded solution y exists for the equation $\dot{y}(t) = \widetilde{B}(t)y(t)$ with the property $\inf_{t \in R} |y(t)| = 0$), then recent investigations by R. Ortega and M. Tarallo [77] prove that it is necessary to assume further conditions in order to obtain the nonexistence of almost periodic solutions. More precisely, the following statement holds true.

Proposition 5.22. *Consider equation (5.133), and assume that there exists* $\widetilde{B} \in H(A)$ *such that the equation* $\dot{y}(t) = \widetilde{B}(t)y(t)$ *has nonzero bounded solutions all tending to zero as* $|t| \longrightarrow \infty$. *Then, one can find* $f(t) \in AP(R, \mathcal{C}^n)$ *such that* $\mathrm{mod}(f) \subset \mathrm{mod}(A)$ *for which equation (5.133) has bounded solutions but none is almost periodic.*

The proof of Proposition 5.22 is given in the paper by R. Ortega and M. Tarallo quoted above.

We shall now consider the noteworthy case where the matrix A in equation (5.133) is periodic. In other words, we will assume that there exists $T > 0$ such that

$$A(t + T) = A(t), \qquad t \in R. \qquad (5.136)$$

The continuity of $A(t)$ being assumed, there results that $A(t)$ is Bohr almost periodic. Consequently, we are in the framework adopted at the beginning of this section if we preserve the hypothesis $f \in AP(R, \mathcal{C}^n)$.

The periodicity assumption (5.136) will allow us to reduce the problem of almost periodicity of solutions to equation (5.133) to the case of systems with constant coefficients.

Indeed, the so-called Floquet theory of linear systems with periodic coefficients will enable us to find a system equivalent to equation (5.133) but with constant matrix A. This is a sketch of Floquet's reducibility procedure.

Consider the homogeneous system $\dot{x} = A(t)x$ with $A(t)$ satisfying equation (5.136). If $X(t)$ is the fundamental matrix satisfying $X(0) = I$, then $X(t+T)$ is also a fundamental matrix and we can find a nonsingular matrix C of type n by n such that

$$X(t + T) = X(t)C, \qquad t \in R. \tag{5.137}$$

But $\det C \neq 0$ implies the existence of a matrix B also of type n by n such that

$$C = e^{BT}. \tag{5.138}$$

Sometimes the relation (5.138) is written as $B = T^{-1} \log C$, the logarithm of a nonsingular matrix being always defined.

Let us now denote

$$S(t) = X(t)e^{-Bt}, \qquad t \in R, \tag{5.139}$$

and note that

$$S(t + T) = X(t + T)e^{-B(t+T)} = X(t)e^{BT}e^{-Bt}e^{-BT} = X(t)e^{-Bt} = S(t).$$

From equation (5.139), we derive

$$X(t) = S(t)e^{Bt}, \qquad t \in R, \tag{5.140}$$

which constitutes the Floquet representation of the fundamental matrix of the periodic system $\dot{x} = A(t)x$.

If we now operate the change of variable

$$x(t) = S(t)y(t), \qquad t \in R, \tag{5.141}$$

we obtain for y the system

$$\dot{y}(t) = By(t), \qquad t \in R, \tag{5.142}$$

which is a system with constant coefficients. We have more information about the nature of solutions of equation (5.142) than we have initially for the solutions of the time-varying system (5.133). Formula (5.140) is the key to understanding this nature if we keep in mind that $S(t)$ is a periodic (of period T) matrix.

Let us summarize the discussion above in the following.

Proposition 5.23. *Let us consider the system* (5.133) *with* $A(t)$ *continuous and satisfying equation* (5.136). *Then, the fundamental matrix (of solutions)* $X(t)$ *can be represented in the form* (5.140), *where* $S(t)$ *is a periodic nonsingular matrix and* B *is a constant matrix.*

Remark 5.24. The connection between $A(t), S(t)$, and B is given by

$$\dot{S}(t) = A(t)S(t) - S(t)B.$$

We shall now introduce the concepts of the *characteristic multiplier* and *characteristic exponent* of the system (5.133) with $A(t)$ satisfying equation (5.136).

Let us try to find solutions $x(t)$ of $\dot{x}(t) = A(t)x(t)$ such that

$$x(t+T) = \rho x(t), \qquad t \in R, \tag{5.143}$$

for some real or complex ρ. Since any solution of $\dot{x} = A(t)x$ can be represented by $x(t) = X(t)x^0$, where $X(t)$ is the fundamental matrix and $x^0 \in R^n$, there results from equation (5.143) that ρ and x^0 must satisfy the condition

$$[e^{BT} - \rho I]x^0 = \theta, \tag{5.144}$$

which means that ρ must be an eigenvalue of the matrix e^{BT} and x^0 must be an eigenvector corresponding to ρ. Of course, this represents necessary and sufficient conditions for the validity of equation (5.143). Since eigenvalues and eigenvectors do exist, equation (5.143) always has nonzero solutions. More precisely, ρ must be an eigenvalue of the matrix e^{BT}, and these eigenvalues are called *characteristic multipliers* of the system. Each *characteristic exponent* is determined (up to an additive term) from $\rho = e^{\tau T}$, with ρ an eigenvalue of the matrix e^{BT}, and hence τ is an eigenvalue of the matrix B.

We return now to equation (5.133) and assume that $A(t)$ satisfies equation (5.136), while $f(t) \in \mathrm{AP}(R, \mathcal{C}^n)$.

Let $S(t)$ be the periodic matrix described above and operate the change of variable $x(t) = S(t)y(t)$. This will transform the system (5.133) into the system

$$\dot{y}(t) = By(t) + S^{-1}(t)f(t), \qquad t \in R, \tag{5.145}$$

and since $S^{-1}(t)$ is periodic of period T, there results that the product $S^{-1}(t)f(t) \in \mathrm{AP}(R, \mathcal{C}^n)$. In other words, in order to find almost periodic solutions to equation (5.133), one must look for those solutions to the system with constant coefficients

$$\dot{y}(t) = By(t) + g(t), \qquad t \in R, \tag{5.146}$$

where $g(t) = S^{-1}(t)f(t)$. It is obvious that the map $f \longrightarrow S^{-1}(t)f$ of $\mathrm{AP}(R, \mathcal{C}^n)$ into itself is an isomorphism. Therefore, we can consider that $g \in \mathrm{AP}(R, \mathcal{C}^n)$ is arbitrary.

Relying on the discussions carried out in Section 5.4, we can formulate several results in regard to the system (5.133) with periodic $A(t)$. We shall restrict our discussion to two of the possible results to be transposed from the case of constant coefficients to the general almost periodic case (5.133).

Proposition 5.24. *Consider system* (5.133), *and assume that $A(t)$ satisfies equation* (5.136), *while $f \in AP(R, C^n)$. Then any bounded (on R) solution of equation* (5.133) *is in $AP(R, C^n)$. Moreover, if the characteristic exponents of the system $\dot{x} = A(t)x$ (i.e., the eigenvalues of the matrix B) do not lie on the imaginary axis, then the system* (5.133) *has a unique almost periodic solution.*

The problem of almost periodicity of solutions to almost periodic systems of the form (5.133) is far from being clarified. Special types of almost periodic systems of the form (5.133) have been investigated, and criteria have been obtained. We are going to examine two special classes of systems, both time-varying, for which there are some conclusive results in the literature.

The first class is that of systems of the form (5.133) with a triangular matrix. In other words,

$$A(t) = \begin{pmatrix} a_{11}(t) & a_{12}(t) & \cdots & a_{1n}(t) \\ 0 & a_{21}(t) & \cdots & a_{2n}(t) \\ \multicolumn{4}{c}{\dotfill} \\ \multicolumn{4}{c}{\dotfill} \\ 0 & 0 & \cdots & a_{nn}(t) \end{pmatrix}. \tag{5.147}$$

We have chosen the upper-triangular form, but we could as well choose the lower-triangular form. Obviously, the material aspect of the problem would first require us to investigate the scalar equation

$$\dot{y}(t) = a(t)y(t) + b(t), \qquad t \in R, \tag{5.148}$$

with $a, b \in AP(R, C)$.

The following result, due to Cameron and Massera, can be found with a full proof in the book by A. M. Fink [40].

Proposition 5.25. *Consider equation* (5.148), *and assume that $M\{a\} \neq 0$. Then, there is an almost periodic solution $x(t)$ of equation* (5.148) *for any $b \in AP(R, C)$.*

Remark 5.25. In Section 5.4, we have dealt with equation (5.148) for $a(t) = $ const.

Remark 5.26. From Proposition 5.25, we can rather easily get an almost periodicity criterion for the solutions of the triangular system of the form (5.133) with $A(t)$ given by equation (5.147).

Proposition 5.26. *Consider the almost periodic system* (5.133) *with* $A(t)$ *given by equation* (5.147). *If* $M\{a_{ii}(t)\} \neq 0$, $i = 1, 2, \ldots, n$, *then there exists* (*at least*) *one almost periodic solution to equation* (5.133).

The proof is obvious, starting with the last equation of the system (5.133) and applying Proposition 5.25 n times to each preceding equation. First, substitute the almost periodic component $x_n(t)$ into the equation

$$\dot{x}_{n-1}(t) = a_{n-1,n-1}(t)x_{n-1}(t) + a_{n-1,n}^i x_n(t) + b_{n-1}(t),$$

which takes the form (5.148), etc.

The second case we want to present here is related to the concept of exponential dichotomy for the solutions of the homogeneous system

$$\dot{x}(t) = A(t)x(t), \qquad t \in R, \tag{5.149}$$

with $A(t) \in AP(R, \mathcal{L}(\mathcal{C}^n, \mathcal{C}^n))$. The existence of an exponential dichotomy will allow us to the prove the existence of almost periodic solutions for equation (5.133) with $f \in AP(R, \mathcal{C}^n)$.

We shall say that equation (5.149) admits an exponential dichotomy if there exists a projector P of $\mathcal{C}^n$ and some positive constants $K_1, K_2, \alpha_1, \alpha_2$, such that the fundamental matrix $X(t)$ satisfies

$$|X(t)PX^{-1}(s)| \leq K_1 e^{-\alpha_1(t-s)}, \qquad t \geq s,$$

$$|X(t)(I - P)X^{-1}(s)| \leq K_2 e^{-\alpha_2(t-s)}, \qquad t \leq s.$$

It is obvious that $X(t)$ can be any fundamental matrix of equation (5.149).

A good example of an exponential dichotomy is the case of constant coefficients $A(t) = A = $ const., where all the eigenvalues of A have nonzero real parts. We have seen this example in Section 5.4 in Remark 5.9 to Proposition 5.13. An extension of this result to the case of time-varying systems has been given in A. M. Fink's book [40].

The consequence of the existence of an exponential dichotomy for equation (5.149) is the following property.

Proposition 5.27. *Consider the system* (5.133), *and assume that* $A(t)$ *and* $f(t)$ *are almost periodic* (*Bohr*), *while the system* (5.149) *possesses an exponential dichotomy* (*with projector* P). *Then*

$$x(t) = \int_{-\infty}^{t} X(t)PX^{-1}(s)f(s)ds$$

$$- \int_{t}^{\infty} X(t)(I - P)X^{-1}(s)f(s)ds \tag{5.150}$$

represents the unique almost periodic solution of the system (5.133).

Proof. A direct verification of the fact that equation (5.150) is a solution of equation (5.133) can be carried out easily.

Since equation (5.150) can be rewritten as

$$x(t) = \int_R K(t,s)f(s)ds, \qquad t \in R, \qquad (5.151)$$

with $K(t,s)$ satisfying an estimate of the form

$$|K(t,s)| \le K_0 e^{-\alpha|t-s|},$$

one has to apply the result in Section 5.4.

Moreover, one can allow $f \in S(R, C^n)$, still obtaining Bohr almost periodicity for the solution.

6

Almost Periodic Nonlinear Oscillations

6.1 Quasilinear Systems

By quasilinear systems we mean an ordinary differential system of the form

$$\dot{x}(t) = A(t)x(t) + f(t,x), \qquad t \in R, \tag{6.1}$$

in which $f(t,x)$ has some feature of "smallness" so that the linear term in equation (6.1) is the dominant one.

Of course, we want to take advantage of what we already know about linear systems and try to extend this knowledge to the quasilinear case (6.1).

We shall dwell particularly on the case $A(t) = \text{const.}$, where the treatment can be carried out in a satisfactory manner. Proposition 5.13 offers a good example concerned with the linear case that can be used to obtain a similar result for associated quasilinear systems of the form (6.1). More precisely, the following result is obtained.

Proposition 6.1. *Consider the system* (6.1) *under the following assumptions:*

(1) *The matrix* $A(t) = (a_{ij})$, $i, j = 1, 2, \ldots, n$, *and* $a_{ij} \in C$ *has eigenvalues with nonzero real part.*
(2) $f(t,x) = \text{col}(f_1(t, x_1, \ldots, x_n), \ldots, f_n(t, x_1, \ldots, x_n))$ *is a continuous map in the set*

$$t \in R, \ x \in C^n, \qquad |x| \le a, \tag{Δ}$$

and it is almost periodic in t uniformly with respect to x satisfying (Δ).
(3) $f(t,x)$ *is globally Lipschitz in* (Δ); *namely, there exists* $L > 0$ *with the property*

$$|f(t,x) - f(t,y)| \le L|x - y| \tag{6.2}$$

for any $(t, x), (t, y) \in \Delta$.

Then, there exists a unique almost periodic solution $x = x(t)$ *of equation* (6.1) *such that* $|x(t)| \le a$, $t \in R$, *provided* $|f(t, \theta)|$ *and* L *are sufficiently small.*

C. Corduneanu, *Almost Periodic Oscillations and Waves*,
DOI 10.1007/978-0-387-09819-7_6, © Springer Science+Business Media, LLC 2009

Proof. We shall use the Banach contraction method (i.e., successive approximations in a function space framework), taking as the basic (metric) space the ball of radius a centered at the origin in the space $AP(R, C^n)$. The contracting operator, say T, will be defined according to the following scheme: To each $u(t) \in AP(R, C^n)$, $|u| \leq a$, we associate the function $x(t) = (Tu)(t) \in AP(R, C^n)$, defined as the unique solution (according to Proposition 5.13) of the system

$$\dot{x}(t) = Ax(t) + f(t, u(t)), \quad t \in R. \tag{6.3}$$

Equation (6.3), is a linear equation in x with constant matrix. As noted in Remark 5.9 to Proposition 5.13, the unique almost periodic solution of equation (6.3) can be represented by the formula

$$x(t) = \int_{-\infty}^{t} X_-(t-s)f(s, u(s))ds - \int_{t}^{\infty} X_+(t-s)f(s, u(s))ds. \tag{6.4}$$

The significance of $X_-(t)$ and $X_+(t)$ has been explained in Section 5.4. Actually, formula (6.4) provides the representation of the almost periodic solution $x(t) = (Tu)(t)$, and as seen in Section 5.4, one can write for a conveniently chosen $K > 0$, $K = K(A)$,

$$|x|_{AP} \leq K|f|_{AP} \tag{6.5}$$

for any $u \in AP(R, C^n)$ satisfying $|u|_{AP} \leq a$. On the other hand, one has

$$|f|_{AP} \leq L|u|_{AP} + |f(t, \theta)|_{AP}. \tag{6.6}$$

Consequently, formulas (6.5) and (6.6) lead to

$$|x|_{AP} = |Tu|_{AP} \leq KL|u|_{AP} + K|f(t, \theta)|_{AP}, \tag{6.7}$$

which holds true for any $u \in AP(R, C^n)$ with $|u|_{AP} \leq a$.

Now, we shall note that for $u, v \in AP(R, C^n)$, $|u|, |v| \leq a$,

$$|Tu - Tv|_{AP} \leq KL|u - v|_{AP}, \tag{6.8}$$

taking into account assumption (3) of Proposition 6.1. Therefore, the condition

$$KL < 1 \tag{6.9}$$

must be accepted (it shows how small L has to be because $K = K(A)$ is a fixed number). It assures that T will be a contraction.

Returning to inequality (6.7), we note that it implies for each $u \in AP(R, C^n)$ with $|u(t)| \leq a$ on R

$$|Tu|_{AP} \leq KLa + K|f(t, \theta)|_{AP}, \tag{6.10}$$

and the required inclusion by the contraction, $|Tu|_{AP} \leq a$ for any u with $|u|_{AP} \leq a$, takes place as soon as

$$|f(t, \theta)|_{\mathrm{AP}} \leq (1 - KL)K^{-1}a. \tag{6.11}$$

Therefore, we find that $|f(t, \theta)|_{\mathrm{AP}}$ must satisfy the "smallness" condition (6.11) in order to conclude, based on the Banach contraction principle, that the operator T has a unique fixed point in the ball $|x|_{\mathrm{AP}} \leq a$. This fixed point is obviously a solution of the integral equation (derived from equation (6.4))

$$x(t) = \int_{-\infty}^{t} X_-(t - s)f(s, x(s))ds - \int_{t}^{\infty} X_+(t - s)f(s, x(s))ds, \tag{6.12}$$

which leads to equation (6.1) by differentiation. The uniqueness also follows from the Banach principle.

Remark 6.1. The restriction expressed by the inequality (6.11) disappears if we assume $f(t, x)$ to be defined for $t \in R$, $x \in \mathcal{C}^n$, but keeping the Lipschitz condition.

Remark 6.2. The Banach principle implies the uniform convergence on R of the following scheme of successive approximations.

Let $x^0 = \theta$, and define the iterations by means of the equation

$$\dot{x}^{k+1}(t) = Ax^{k+1}(t) + f(t, x^k(t)), \qquad k \geq 0,$$

where $x^{k+1}(t)$ is chosen from above as the unique solution in $\mathrm{AP}(R, \mathcal{C}^n)$. Then

$$x(t) = \lim_{k \to \infty} x^k(t), \qquad t \in R,$$

is the desired solution of equation (6.1).

Remark 6.3. Instead of assuming $A(t) = \mathrm{const.}$ in equation (6.1), we can consider the periodic case; i.e., $A(t + T) = A(t)$, $t \in R$, for some $T > 0$ fixed. The results in Section 5.7 will allow us to reduce the periodic case to the case of systems with constant coefficients. Preserving the notation of that section, we can reduce the system (6.1) by the substitution $x = S(t)y$ to the system

$$\dot{y}(t) = By(t) + S^{-1}(t)f(t, S(t)y(t)). \tag{6.13}$$

Since $S^{-1}(t)$ is periodic of period T the nonlinear term on the right-hand side of equation (6.13) satisfies the assumptions of Proposition 6.1.

Of course, it would be interesting to treat the case where the matrix A in equation (6.1) has at least one eigenvalue with zero real part. Some results are available in the classical treatise [66] of I. G. Malkin.

Remark 6.4. If one starts with a system not necessarily of the form (6.1), then it may be possible to "linearize" it under some circumstances. For instance, when the system has the form

$$\dot{x}(t) = g(t, x(t)), \qquad t \in R, \tag{6.14}$$

with g analytic in x–variables (i.e., developable in power series about the origin $\theta \in R^n$), then equation (6.14) can be rewritten in the form

$$\dot{x}_i(t) = \sum_{j=1}^{n} a_{ij}(t)x_j(t) + [\cdot]_i^{(2)}, \qquad i = 1, 2, \ldots, n, \qquad (6.15)$$

if one also assumes $g(t, \theta) = 0$, $t \in R$. The terms $[\cdot]_i^{(2)}$, $i = 1, 2, \ldots, n$, are of the second order in x-variables, at least. Obviously, the scalar equations (6.15) can be rewritten in concise vector form as

$$\dot{x}(t) = A(t)x(t) + [\cdot]^{(2)}, \qquad (6.16)$$

where the terms in the bracket $[\cdot]^{(2)}$ are at least of the second order. The form (6.16) appears as a variant of equation (6.1).

One question arising from the discussion carried out in regard to the system (6.1) is whether or not a result generalizing that of Proposition 6.1 holds true, provided we can substitute an adequate condition to that concerning the eigenvalues of the matrix A (see assumption (1) in Proposition 6.1). This new assumption on $A(t)$ in equation (6.1) can be formulated as follows.

H. The linear system associated to equation (6.1),

$$\dot{x}(t) = A(t)x(t) + f(t), \qquad t \in R, \qquad (6.17)$$

with both A and f almost periodic, has a unique almost periodic solution for each $f \in AP(R, C^n)$.

It is an elementary fact that condition (1) of Proposition 6.1 implies H, and it is actually equivalent to it.

A result similar to Proposition 6.1, in slightly different formulation, can be stated.

Proposition 6.2. *Consider the system* (6.1) *under hypothesis* H, *and assume* $A(t) \in AP(R, \mathcal{L}(C^n, C^n))$ *with* $f(t, x)$ *such that it is continuous on* $R \times C^n$ *and almost periodic in* t *uniformly with respect to* $x \in M$, *where* M *is any compact subset of* C^n. *Moreover, let* f *be Lipschitz continuous (i.e., satisfying*

$$|f(t, x) - f(t, y)| \leq L|x - y|$$

with $L > 0$ *a constant,* $t \in R$, *and* $x, y \in C^n$ *arbitrary).*

Then equation (6.1) *has a unique almost periodic solution, provided* L *is sufficiently small.*

Proof. We have to prove the continuity (or boundedness) of the map $f \to x_f$, where x_f is the unique solution in $AP(R, C^n)$ of the system (6.17). In other words, this linear map from $AP(R, C^n)$ into itself satisfies an estimate of the form (6.5),

$$|x_f|_{\text{AP}} \le K|f|_{\text{AP}}, \qquad f \in \text{AP}(R, \mathcal{C}^n). \tag{6.18}$$

Such an estimate will enable us to proceed further along exactly the same lines as in the proof of Proposition 6.1, except that we need not impose any restriction on the norm of x ($|x| < \infty$).

The following scheme will lead us to the conclusion that an estimate of the form (6.18) does exist.

First, let us note that the system (6.17) can be rewritten in the form

$$(Tx)(t) = \dot{x}(t) - A(t)x(t) = f(t), \tag{6.19}$$

which tells us that the map $f \longrightarrow x_f$ is actually the inverse map of T, (T^{-1}).

Since any solution in $\text{AP}(R, \mathcal{C}^n)$ of equation (6.1), or $(Tx)(t) = f(t)$, $t \in R$, also has its derivative in $\text{AP}(R, \mathcal{C}^n)$, we shall consider the operator $T :$ $\text{AP}^{(1)}(R, \mathcal{C}^n) \longrightarrow \text{AP}(R, \mathcal{C}^n)$, where $\text{AP}^{(1)}(R, \mathcal{C}^n)$ stands for the Banach space of almost periodic functions with almost periodic derivatives. The norm of this space is

$$|x|_{\text{AP}}^{(1)} = |\dot{x}|_{\text{AP}} + |x|_{\text{AP}}. \tag{6.20}$$

The elementary properties of Bohr almost periodic functions, given in Section 3.2, will lead easily to the conclusion that equation (6.20) defines a norm, and the space is complete when endowed with it.

We can easily prove that the operator T defined by equation (6.19) is a continuous operator from $\text{AP}^{(1)}(R, \mathcal{C}^n)$ into $\text{AP}(R, \mathcal{C}^n)$. Indeed, for $x \in \text{AP}^{(1)}(R, \mathcal{C}^n)$, we have

$$|Tx|_{\text{AP}} = |\dot{x} - A(t)x|_{\text{AP}} \le |\dot{x}|_{\text{AP}} + |A(t)x|_{\text{AP}}.$$

Since $A(t)$ is bounded on R, we get from the equation above the inequality

$$|Tx|_{\text{AP}} \le |\dot{x}|_{\text{AP}} + N|x|_{\text{AP}}, \tag{6.21}$$

where $N = |A(t)|_{\text{AP}}$. Therefore, formula (6.21) leads to

$$|Tx|_{\text{AP}} \le M(|\dot{x}|_{\text{AP}} + |x|_{\text{AP}}) \tag{6.22}$$

with $M = \max(1, N)$. But formula (6.22) means

$$|Tx|_{\text{AP}} \le M|x|_{\text{AP}}^{(1)}, \ x \in \text{AP}^{(1)}, \tag{6.23}$$

which proves the continuity of the operator T.

It remains to note that the hypothesis H assures that the operator T is one-to-one and onto $\text{AP}(R, \mathcal{C}^n)$. According to a well-known result from functional analysis (due to Banach), the inverse operator T^{-1} from $\text{AP}(R, \mathcal{C}^n)$ into $\text{AP}^{(1)}(R, \mathcal{C}^n)$ is also continuous.

Therefore, a formula like (6.18) is valid and from now on the proof follows that of Proposition 6.1. One must take $L < K^{-1}$.

Proposition 6.2 is thereby proven.

Remark 6.5. If we carefully follow the proof of Proposition 6.2, we notice the fact that it is not really material to deal with the system (6.1). We could consider the more general system

$$\dot{x}(t) = (Lx)(t) + f(t; x), \tag{6.24}$$

where $L : AP(R, \mathcal{C}^n) \longrightarrow AP(R, \mathcal{C}^n)$ is a continuous linear mapping. Obviously, in equation (6.1) we have $(Lx)(t) = A(t)x(t)$, and the linearity and continuity are assured.

Instead of hypothesis H as formulated above, we should use a similar one, namely

H'. For each $f \in AP(R, \mathcal{C}^n)$, the linear system

$$\dot{x}(t) = (Lx)(t) + f(t), \qquad t \in R, \tag{6.25}$$

has a unique solution $x = x_f \in AP(R, \mathcal{C}^n)$.

The proof of Proposition 6.2 can be repeated line by line, relying on the hypothesis H' instead of H. We leave the details to the reader.

Finally, let us point out that we could also consider, instead of the nonlinear term $f(t, x(t))$, a more general nonlinearity, say $(fx)(t)$, with $f : AP(R, \mathcal{C}^n) \longrightarrow AP(R, \mathcal{C}^n)$ a nonlinear operator. For instance, one could take

$$(fx)(t) = \int_R k(t - s)g(s, x(s))ds,$$

which corresponds to the Hammerstein type of nonlinearity (or perturbation).

Valuable references for the topic discussed in this section are the books by I. G. Malkin [66] and J. L. Massera and J. J. Schäffer [68]. While the first reference is mostly along the lines of classical analysis, the second takes ample inspiration from the methods of functional analysis, treating the infinite-dimensional case $x \in B = $ a Banach space.

In concluding this section, we note the fact that only the case of a unique almost periodic solution is discussed. This means that the linear homogeneous system $\dot{x}(t) = A(t)x(t)$, or $\dot{x}(t) = (Lx)(t)$, has only the zero solution almost periodic. If the homogeneous system has nonzero almost periodic solutions, then discussion of the problem is considerably more difficult. The references mentioned above contain some elements related to this situation.

6.2 Separated Solutions: Amerio's Theory

We owe to L. Amerio [2] a theory generalizing to the nonlinear case Favard's theory (see Section 5.7) that will allow us to conclude that bounded separated solutions of the nonlinear system

$$\dot{x}(t) = f(t, x(t)), \qquad t \in R, \tag{6.26}$$

are almost periodic (Bohr).

The following conditions will be assumed for the map $f : R \times R^m \to R^m$, $m \geq 1$, throughout this section:

(1) f is continuous on the open set $B \subset R^{m+1}$, where $B = R \times A$, $A \subset R^m$ standing for an open set.

(2) f is almost periodic as a function of t uniformly with respect to $x \in C$, with $C \subset A$ any compact set.

Let $C \subset A$ be a compact set and $\{h_n, ; \; n \geq 1\} \subset R$ a sequence. According to Section 3.6, one can find a subsequence $\{h_{n_k}; \; k \geq 1\} \subset \{h_n; \; n \geq 1\}$ such that $\{f(t + h_{n_k}); \; x\}$ is uniformly convergent on $R \times C$ to a function $g(t, x)$ that is almost periodic in t uniformly with respect to $x \in C$.

We now consider a sequence of compact sets $\{C_n; \; n \geq 1\}$, $C_n \in A$, such that $C_n \subset C_{n+1}$, $\bigcup_{n=1}^{\infty} C_n = A$. If $\{h_n; \; n \geq 1\} \subset R$, then there exists a subsequence $\{h_{1n}; \; n \geq 1\} \subset \{h_n; \; n \geq 1\}$ such that $\{f(t + h_{1n}, x)\}$ is uniformly convergent on $R \times C_1$. From $\{h_{1n}; \; n \geq 1\}$, we extract a subsequence with the property that $\{f(t + h_{2n}, x)\}$ converges uniformly on $R \times C_2$, and so on. From the sequences $\{h_{1n}; \; n \geq 1\}$, $\{h_{2n}; \; n \geq 1\}, \ldots, \{h_{kn}; \; n \geq 1\} \ldots$, we extract the diagonal sequence $\{h_{nn}; \; n \geq 1\}$. This is a subsequence of each $\{h_{kn}, \; n \geq 1\}$, $k \geq 1$, excepting a finite number of terms. Hence, $\{f(t + h_{nn}, x), n \geq 1\}$ is uniformly convergent on each $R \times C_n$, $n \geq 1$, say $f(t + h_{nn}, x) \longrightarrow g(t, x)$ as $n \longrightarrow \infty$, which implies the almost periodicity of the function $g(t, x)$ in t uniformly with respect to $x \in C$, $C \subset A$ being compact. If we choose $C_n = \{(t, x); (t, x) \in R \times A, t^2 + |x|^2 \leq n, \text{dist}[(t, x); \partial(R \times A)] \geq n^{-1}\}$, then $C \subset C_n$ for sufficiently large n.

Let us denote by $H(f)$, as we did in Section 5.7, the set of those $g(t, x)$ that can be obtained as

$$\lim_{n \to \infty} f(t + h_n, x) = g(t, x) \qquad (6.27)$$

uniformly on any set $R \times C$, $C \subset A$ being any compact set, for some $\{h_n; n \geq 1\} \subset R$. In other words, $g \in H(f)$ means that g is in the closure of the set $\{f(t + h, x), h \in R\}$ with respect to the kind of convergence just described. It turns out that each $g \in H(f)$ generates $H(f)$ as described above.

We shall consider in what follows the systems

$$\dot{x}(t) = g(t, x(t)), \; g \in H(f), \qquad (6.28)$$

and formulate conditions for the almost periodicity of solutions for equations (6.26) and (6.28). The relationship existing with this kind of solution is material in investigating the problem of existence of almost periodic solutions.

Let us now define the concept of a *separated solution* of the system (6.26) in a set $D = R \times C$, with $C \subset A$ a compact set. We have in mind only solutions defined on the real line R.

Consider a solution $x(t)$, $t \in R$, of the system (6.26) such that $x(t) \in C$, $t \in R$, which means that the graph of this solution belongs to the set $D \subset R^{m+1}$. We shall say that $x(t)$ is *separated* in D if it is either the only solution of equation (6.26) with its graph in D or if there is another solution $y(t)$, $t \in R$, with the graph in D, for which one can find a number $\delta > 0$ such that $|x(t) - y(t)| \geq \delta$, $t \in R$.

Proposition 6.3. *Consider the system* (6.26) *under assumptions* (1) *and* (2) *formulated above. Then, the following properties hold true for equation* (6.26) *and its associated systems* (6.28):

(a) *The number of separated solutions of the system* (6.26) *with the graph in D is finite.*

(b) *If the system* (6.26) *has a solution $x(t)$ with its graph in D, then each system in $H(f)$ has a solution with its graph in D.*

(c) *If $x(t)$, $t \geq t_0$, is a solution of equation* (6.26) *such that $x(t) \in C$ for $t \geq t_0$, then each system* (6.28) *has a solution defined on R whose graph is in D.*

Proof. (a) The set of all solutions of equation (6.26), with a graph in D is uniformly bounded and equicontinuous. The equicontinuity is a consequence of the fact that the set of their derivatives $\{\dot{x}(t) = f(t, x(t)), t \in R\}$ is also uniformly bounded on D and noting that $|x(t) - x(\tau)| \leq |t - \tau| \sup |f| \leq K|t - \tau|$. Hence, if the set of separated solutions in D were *infinite*, then one could select from this set a sequence that is uniformly convergent on any compact interval of R to a function $\bar{x}(t)$. This solution should belong to D, and it cannot be separated. This contradiction proves the assertion (a).

(b) Let us fix one equation with $g \in H(f)$ and consider a sequence for which equation (6.27) is satisfied. The functions $x_n(t) = x(t + h_n)$, $n \geq 1$, are solutions of $\dot{x}_n(t) = f(t + h_n, x_n(t))$ with their graph in D. As in (a), one can see that the set of functions $\{x_n(t); n \geq 1\}$ is uniformly bounded and equicontinuous on R. Hence, one can determine a subsequence of $\{h_n; n \geq 1\}$, say $\{h_{1n}; n \geq 1\}$, such that the sequence $\{x_{1n}(t); n \geq 1\}$ is uniformly convergent on any compact interval of R. The limit of this sequence is obviously a solution of equation (6.27), whose graph belongs to D.

(c) With the solution $x(t)$, $t \geq t_0$, we can construct a sequence of functions, namely $x_n(t) = x(t + n)$, $n \geq 1$. We note that $x_n(t)$ satisfies $\dot{x}_n(t) = f(t + n, x_n(t))$ on $(t_0 - n, \infty)$. It is obvious that the sequence $\{x_n(t); n \geq 1\}$ is uniformly bounded and equicontinuous on each half-axis (τ, ∞), neglecting (depending on τ) perhaps a finite number of terms not influencing the convergence. Therefore, we can select a subsequence $\{x_{1n}(t); n \geq 1\}$ that converges uniformly on each interval (τ, T), $-\infty < \tau < T < \infty$. Then $\bar{x}(t) = \lim x_{1n}(t)$ as $n \to \infty$ satisfies $\dot{\bar{x}}(t) = \bar{f}(t, \bar{x}(t))$, has its graph in D and $\bar{f} \in H(f)$. Now we have to rely on the proof of (b) above in order to conclude that (c) holds true.

This ends the proof of Proposition 6.3.

We can now state and prove the main result of this section concerning the existence of almost periodic solutions.

Theorem 6.1. *Consider the system* (6.28) *with* $g \in H(f)$ *under conditions* (1) *and* (2), *and assume that each system has its solutions in* D *separated. Then all these solutions are almost periodic.*

Proof. We shall prove first an auxiliary result, namely that there exists a number $\sigma > 0$ independent of $g \in H(f)$ such that for any pair of separated (in D) solutions of (6.28) one has $|x_1(t) - x_2(t)| \geq \sigma$, $t \in R$.

Let us now fix $g \in H(f)$. There exists a largest number $\sigma > 0$ such that two arbitrary solutions of equation (6.26), separated in D, are at a distance $\geq \sigma$ of each other. It is now enough to show that any couple of separated solutions to equation (6.28) are also situated at a distance $\geq \sigma$ from each other.

We have seen in Proposition 6.3, part (a), that when $y(t)$ is a solution of equation (6.26) with its graph in D, then we can select a subsequence of $\{h_n; \ n \geq 1\}$, say $\{h_{1n}; \ n \geq 1\}$, such that $y(t + h_{1n}) \longrightarrow z(t)$, with $z(t)$ being a solution of equation (6.28), also with its graph in D. If $z_1(t)$ and $z_2(t)$ are two distinct solutions of equation (6.26), because they are separated we must have for $t \in R$

$$\inf |z_1(t + h_n) - z_2(t + h_n)| = \inf |z_1(t) - z_2(t)| \geq \sigma.$$

But we can select a subsequence of $\{h_n; \ n \geq 1\}$, say $\{h_{1n}; \ n \geq 1\}$, with the property $z_1(t + h_{1n}) \longrightarrow y_1(t)$, $z_2(t + h_{1n}) \longrightarrow y_2(t)$, both $y_1(t)$ and $y_2(t)$ being solutions of equation (6.28). This implies

$$\inf |y_1(t) - y_2(t)| \geq \sigma, \qquad t \in R.$$

Hence, $y_1(t)$ and $y_2(t)$ are distinct solutions. It means that to distinct solutions of equation (6.26) there correspond distinct solutions of equation (6.28). Therefore, the number of distinct solutions of equation (6.28) is greater than or equal to the number of solutions of equation (6.26). But each element of $H(f)$ generates the whole class, which allows us to conclude that each equation (6.28) has the same number of solutions with graph in D. One can state that the procedure described above leads to the determination of all the separate (in D) solutions of any equation (6.28), starting from equation (6.26). The last inequality above proves the assertion that a number $\sigma > 0$, with $|y_1(t) - y_2(t)| \geq \sigma$, $t \in R$, for any couple of separated solutions of any equation (6.28), does exist.

We now prove that all separated (in D) solutions of any equation (6.28) are almost periodic (Bohr). Let $x(t)$, $t \in R$, be a solution of equation (6.26) separated in D. It would be sufficient to prove that from any sequence $\{x(t + h_n); \ n \geq 1\}$ one can extract a subsequence that converges uniformly on R. According to the Bochner characterization of Bohr almost periodic functions, this will imply the almost periodicity of the separated solution $x(t)$ of equation (6.26).

According to Proposition 6.3, part (c), one can assume that $x(t + h_n) \longrightarrow y(t)$ as $n \to \infty$ uniformly on any compact interval and also

$$\lim_{n \to \infty} f(t + h_n, x) = h(t, x) \tag{6.29}$$

uniformly on D. Moreover,

$$\dot{y}(t) = h(t, y(t)), \qquad t \in R. \tag{6.30}$$

We shall prove that in fact

$$\lim_{n \to \infty} x(t + h_n) = y(t) \tag{6.31}$$

uniformly on R.

We have seen above that there exists a number $\sigma > 0$ with the property that if $x_1(t)$ and $x_2(t)$ are solutions of a system (6.28) whose graphs are in D, then $|x_1(t) - x_2(t)| \geq 2\sigma$, $t \in R$. If $\{x(t + h_n); \ n \geq 1\}$ does not converge uniformly on R to $y(t)$, we shall denote

$$\varphi_{n,p}(t) = |x(t + h_n) - x(t + h_p)|, \qquad n < p, \tag{6.32}$$

$$I_{n,p} = \{t; \ t \in R, \ \varphi_{n,p}(t) \leq \sigma\}. \tag{6.33}$$

One sees that $\varphi_{n,p}(t)$ is continuous on R, while $I_{n,p}$ is a closed set on R (possibly empty). We have

$$\varphi_{n,p}(0) = |x(h_n) - x(h_p)| \leq \sigma, \qquad N \leq n < p,$$

for some $N = N(\sigma) > 0$, due to the convergence of the sequence $\{x(h_n); n \geq 1\}$. Therefore, $0 \in I_{n,p}$, provided $N \leq n < p$, and neglecting some terms if necessary, we can assume without loss of generality that $I_{n,p} \neq \emptyset$. Now denote

$$\delta_{n,p} = \sup \varphi_{n,p}(t), \qquad t \in I_{n,p}, \tag{6.34}$$

which implies $\delta_{n,p} \leq \sigma$ from equation (6.33).

We shall now prove that

$$\overline{\lim} \, \delta_{n,p} = 0 \qquad \text{as } (n, p) \longrightarrow \infty \tag{6.35}$$

cannot take place.

Indeed, if equation (6.35) were true, then for $\varepsilon > 0$, $\varepsilon < \sigma$, we should find $N_\varepsilon = N(\varepsilon)$ with the property that $N_\varepsilon \leq n < p$ implies $\delta_{n,p} < \varepsilon$. According to equation (6.34), we should have $\varphi_{n,p}(t) < \varepsilon$ for $t \in I_{n,p}$, and in the complementary set of $I_{n,p}$, the inequality $\varphi_{n,p}(t) > \sigma$ should be true. Since $\varphi_{n,p}(t)$ is continuous on R, from the above we should get $I_{n,p} = R$, which means $\{x(t + h_n)\}$ is uniformly convergent on R or this contradicts our hypothesis (nonuniform convergence!). The only alternative is

$$\overline{\lim} \, \delta_{n,p} = 2\alpha > 0, \qquad (n, p) \to \infty. \tag{6.36}$$

From equation (6.36), there follows the existence of two sequences of integers $\{n_r; \ r \geq 1\}$ and $\{p_r; \ r \geq 1\}$ such that $\delta_{n_r, p_r} \geq 3\alpha/2$. From equations (6.32),

(6.33), and (6.36), there results the existence of a sequence $\{t_r;\ r \geq 1\} \subset R$ such that $\varphi_{n_r,p_r}(t_r) \geq \alpha$, which means

$$\alpha \leq |x(t_r + h_{n_r}) - x(t_r + h_{p_r})| \leq \sigma. \tag{6.37}$$

From the sequences $\{h_{n_r};\ r \geq 1\}$ and $\{h_{p_r};\ r \geq 1\}$, we shall extract the subsequences $\{\lambda_s;\ s \geq 1\}$ and $\{\mu_s;\ s \geq 1\}$ such that, by letting $t'_s = t_{r_s}$, one has for $s \to \infty$

$$x(t'_s + \lambda_s) \longrightarrow U, \quad x(t'_s + \mu_s) \longrightarrow V, \qquad \text{where } U, V \in R^n. \tag{6.38}$$

Consider now the sequences $\{x(t+t'_s+\lambda_s);\ s \geq 1\}$ and $\{x(t+t'_s+\mu_s);\ s \geq 1\}$. We can extract from each of these sequences some subsequences, which we will denote by $\{x(t + t'_{1s} + \lambda_{1s});\ s \geq 1\}$ and $\{x(t + t'_{1s} + \mu_{1s});\ s \geq 1\}$, that converge uniformly on each compact interval to two solutions $y_1(t)$ and $y_2(t)$ of the equations (6.28)

$$\dot{y}_1(t) = h_1(t, y_1(t)), \qquad \dot{y}_2(t) = h_2(t, y_2(t)), \tag{6.39}$$

with h_1, h_2 in the hull generated by $f(t, x)$; see formula (6.29) above.

We shall prove that $h_1 = h_2$. Indeed, from

$$\lim_{s \to \infty} f(t + t'_{1s} + \lambda_{1s}, y) = h_1(t, y),$$

$$\lim_{s \to \infty} f(t + t'_{1s} + \mu_{1s}, y) = h_2(t, y),$$

and (6.37) and (6.38), which imply $\alpha \leq |U - V| \leq \sigma$, one gets

$$\alpha \leq |y_1(0) - y_2(0)| \leq \sigma. \tag{6.40}$$

But equation (6.31) allows us to write

$$\lim_{s \to \infty} f(t + \lambda_s, y) = \lim_{s \to \infty} f(t + \mu_s, y) = h(t, y) \tag{6.41}$$

because $\{\lambda_s;\ s \geq 1\}$ and $\{\mu_s;\ s \geq 1\}$ are both subsequences of $\{h_n;\ n \geq 1\}$. The convergence in equation (6.41) is the uniform convergence on D. For $s \geq S_1(\varepsilon)$, we can write

$$|f(t + \lambda_{1s}, y) - f(t + \mu_{1s}, y)| < \varepsilon, \qquad (t, y) \in D. \tag{6.42}$$

Further, for $s \geq S(\varepsilon) \geq S_1(\varepsilon)$, we shall have

$$\begin{aligned} |h_1(t, y) - h_2(t, y)| &\leq |h_1(t, y) - f(t + t'_{1s} + \lambda_{1s}, y)| \\ &\quad + |f(t + t'_{1s} + \lambda_{1s}, y) - f(t + t'_{1s} \\ &\quad + \mu_{1s}, y)| + |f(t + t'_{1s} + \mu_{1s}, y) - h_2(t, y)| \\ &< \varepsilon + \varepsilon + \varepsilon = 3\varepsilon \end{aligned}$$

for any $(t, y) \in D$, taking into account the definition of $h_1(t, y)$ and $h_2(t, y)$ as limits as well as formula (6.42).

Therefore, we have shown that $h_1 = h_2$ in D, which means that $y_1(t)$ and $y_2(t)$ are solutions in D of the same equation from the class (6.28). This allows us to write

$$|y_1(t) - y_2(t)| \geq 2\sigma, \qquad t \in R, \qquad (6.43)$$

from what we have proven above at the beginning of the proof of Theorem 6.1.

Since formulas (6.40) and (6.43) are incompatible, we must conclude that the convergence in equation (6.31) is uniform on R. Hence, $y(t)$ is almost periodic.

This ends the proof of Theorem 6.1.

Corollary 6.1. *The existence part in the case of linear time-varying systems (5.134), Proposition 5.21, follows from Theorem 6.1.*

Indeed, it suffices to note that Favard's condition $\inf |y(t)| > 0$, $t \in R$, is nothing but the separability condition of solutions.

Moreover, the assumption of existence of a bounded solution on R can be weakened to the assumption that such a solution does exist on a half-axis $t \geq t_0$.

There are various applications of Theorem 6.1, and in the next section we shall dwell on the so-called second-order equation of nonlinear oscillations, admitting perturbing terms that are almost periodic.

In concluding this section, we note the fact that another proof of Theorem 6.1 can be found in A. M. Fink's book [40], based on another characterization of almost periodic functions (Bochner; pointwise convergence is used instead of uniform convergence). In the quoted book by A. M. Fink, further developments are indicated.

6.3 The Second-Order Equation of Nonlinear Oscillations (Liénard's Type)

In this section, we shall investigate the almost periodicity of solutions to the second-order nonlinear differential equation

$$\frac{d^2 x}{dt^2} + f(x) \frac{dx}{dt} + g(x) = ke(t), \qquad (6.44)$$

where $e(t)$ is an almost periodic real-valued function such that

$$|e(t)| \leq 1, \qquad t \in R, \qquad (6.45)$$

while $k > 0$ is a real parameter.

Throughout this section, we assume that $f(x)$ and $g(x)$ are continuous, real-valued functions on R.

Denote
$$F(x) = \int_0^x f(v)dv, \qquad x \in R,$$

and note that equation (6.44) leads to the system

$$\frac{dx}{dt} = y - F(x), \qquad \frac{dy}{dt} = -g(x) + kp(t), \qquad (6.46)$$

with $p(t)$ any almost periodic function that belongs to the closure of the set $\{e(t+h);\ h \in R\} \subset AP(R,R)$. According to the terminology introduced in Section 5.7, the closure is the *hull* of $e(t)$ in $AP(R,R)$.

If $p(t) = e(t)$, $t \in R$, the system (6.46) is equivalent to equation (6.44). To verify this assertion, one differentiates the first equation (6.46), and then the second equation is taken into consideration.

Throughout this section, we will assume the following conditions:

1. There exist positive numbers a, b, c, d such that $a > b$, $c < d$, $g(c) = k$, $g(-b) = -k$.
2. $k < \min\{[F(d) - F(c)]f(x) + g(-a), [F(-b) - F(-a)]f(x) - g(d)\}$ when $-a \leq x \leq d$.
3. $g(x)$ is continuously differentiable, $g(0) = 0$, $0 < g'(x) \leq \beta$ for $x \neq 0$.
4. $f(x) \geq \alpha > 0$ for $x \in R$.
5. $\beta < \alpha^2$.

In the plane (x, y), we consider a domain Ω whose boundary consists of the following arches of curves:

$$\Gamma_1 : y = F(x) + \beta_1, \qquad -a \leq x \leq c;$$
$$\Gamma_2 : y = F(d), \qquad c \leq x \leq d;$$
$$\Gamma_3 : x = d, \qquad F(d) - \beta_2 \leq y \leq F(d);$$
$$\Gamma_4 : y = F(x) - \beta_2, \qquad -b \leq x \leq d;$$
$$\Gamma_5 : y = F(-a), \qquad -a \leq x \leq -b;$$
$$\Gamma_6 : y = -a, \qquad F(-a) \leq y \leq F(-a) + \beta_1.$$

In the formulas above, $\beta_1 = F(d) - F(c)$, $\beta_2 = F(-b) - F(-a)$. This allows us to assert that the six arches form a simple Jordan curve that coincides with $\partial\Omega$.

We need a simple result before we can proceed with the proof of existence and uniqueness of an almost periodic solution to equation (6.44) with graph in Ω.

Proposition 6.4. *Let $m(t)$ be a continuous real-valued function on R such that $0 < \alpha \leq m(t) \leq M$. Then the equation*

$$\frac{dz}{dt} = z[m(t) - z], \qquad t \in R, \qquad (6.47)$$

has a solution that is defined on R and $\alpha \leq z(t) \leq M$, $t \in R$.

Proof. Let $\{t_n;\ n \geq 1\} \subset R$ with $t_n \to -\infty$ as $n \to \infty$ and $\{\tau_n;\ n \geq 1\} \subset R$ such that $\alpha \leq \tau_n \leq M$. Equation (6.47) is of Bernoulli type, and

$$z_n(t) = \tau_n \exp\left\{\int_{t_n}^{t} m(u)du\right\}\left\{1 + \tau_n \int_{t_n}^{t} \exp\left(\int_{0}^{u} m(\theta)d\theta\right)du\right\}^{-1} \quad (6.48)$$

represents solutions of equation (6.47) satisfying

$$\alpha \leq z_n(t) \leq M, \qquad t \geq t_n,\ n \geq 1. \quad (6.49)$$

The solutions $z_n(t)$, $n \geq 1$, are uniformly bounded and equicontinuous on any half-line $[t_0, \infty) \subset R$. Hence, one can extract from the sequence $\{z_n(t);\ n \geq 1\}$ a subsequence that converges uniformly on any compact interval of R toward a solution $z(t)$ of equation (6.47) that satisfies the inequality $\alpha \leq z(t) \leq M$.

This ends the proof of Proposition 6.4, and we can now establish the main existence result for equation (6.44).

Theorem 6.2. *Consider equation* (6.44), *and assume conditions* (1)–(5) *stated above. Then there exists a unique almost periodic solution $x(t)$ of equation* (6.44) *such that x, $dx/dt \in \mathrm{AP}(R, R)$, and $(x(t), dx(t)/dt) \in \Omega$ for $t \in R$.*

Proof. As we shall see in the proof, each of the systems (6.46) possesses a unique solution with graph in Ω.

The first assertion we shall prove is the property that any solution of the system (6.46) starting at $t_0 \in R$ at some point of Ω remains in Ω for all $t > t_0$. This amounts to the fact that an arbitrary solution of equation (6.46) that passes through a point of Ω cannot leave Ω (i.e., it cannot meet the boundary $\partial\Omega$ for $t > t_0$). This property will be checked by estimating the slope of the tangent to the trajectory (solution) at boundary points of Ω.

On Γ_1, at an arbitrary point, the trajectory of the system (6.46) has the slope

$$\frac{dy}{dx} = \frac{-g(x) + kp(t)}{\beta_1} \leq \frac{-g(-a) + k}{\beta_1}.$$

On behalf of condition (2), there follows

$$\frac{dy}{dx} < f(x) = F'(x).$$

Hence, the slope of the trajectory is less than the slope of the tangent at Γ_1 at the same point. But $dx/dt = \beta_1 > 0$ on Γ_1, which means that the trajectory enters Ω when t increases.

Moving to Γ_2, where $c < x < d$, we obtain $dx/dt = -g(x) + kp(t) < -g(c) + k = 0$. Therefore, the trajectories of equation (6.46) that cross this segment enter Ω if t is increasing.

A similar argument can be applied in the case of Γ_3, for which $F(d) - \beta_2 \leq y < F(d)$, because $dx/dt = y - F(x) < 0$ on Γ_3 and x must decrease.

At the point $(d, F(d))$ on $\partial\Omega$, we have $dx/dt = 0$ and $dy/dt < 0$. Consequently, the trajectory passing through this point is tangent to the line $x = d$. Since $y < F(x)$ under the curve $y = F(x)$, at points of Ω under $y = F(x)$ we have $dy/dt = y - F(x) < 0$. Hence, the trajectory will enter Ω as t increases.

At the point $(c, F(d))$, we'll have $dy/dt = -g(c) + kp(t) \leq 0$, while $dx/dt > 0$. Consequently, the trajectory passing through this point enters Ω or is tangent to the line $y = F(d)$. We notice that about the point $(c, F(d))$ in Ω, and at the right of $x = c$ (since $dx/dt > 0$ shows that x increases with t), $dy/dt < 0$. Therefore, the trajectory will enter Ω.

The remaining parts of the boundary of Ω can be discussed in the same manner, and we shall omit the details.

At this point in the proof, we can apply Proposition 6.3, more precisely the existence part, concluding that each system (6.46) has a solution defined on R with its graph in Ω.

We shall now prove that each system (6.46) has a *unique solution* whose graph is in Ω.

Assume, on the contrary, that equation (6.46) for a given $p(t)$ in the hull of $e(t)$ has at least two solutions, $(x(t), y(t))$ and $(\bar{x}(t), \bar{y}(t))$, defined for $t \in R$ and with their graphs in Ω. The uniqueness theorem for ordinary differential equations tells us that $(x(t), y(t)) \neq (\bar{x}(t), \bar{y}(t))$ for any $t \in R$.

Let us denote $u(t) = x(t) - \bar{x}(t)$, $v(t) = y(t) - \bar{y}(t)$, and note that they verify, from equation (6.46),

$$\frac{du}{dt} = v - m(\bar{x}, u)u, \quad \frac{dv}{dt} = -h(\bar{x}, u)u, \tag{6.50}$$

where

$$m(\bar{x}, u) = \begin{cases} \dfrac{F(\bar{x} + u) - F(\bar{x})}{u}, & u \neq 0, \\ f(\bar{x}), & u = 0, \end{cases} \tag{6.51}$$

$$h(\bar{x}, u) = \begin{cases} \dfrac{g(\bar{x} + u) - g(\bar{x})}{u}, & u \neq 0, \\ g'(\bar{x}), & u = 0. \end{cases} \tag{6.52}$$

From equation (6.51) and condition (4) in Theorem 6.2, we obtain the inequality $\alpha \leq m(\bar{x}, u) \leq M$, $t \in R$, where $M \geq \sup f(\bar{x}(t))$, $t \in R$.

Now applying Proposition 6.4 with $m(t) = m(\bar{x}, u)$, there results a solution $z(t)$ defined on R such that $\alpha \leq z(t) \leq M$, $t \in R$.

Let us now consider the auxiliary function

$$D(t) = \{v^2 + (v - zu)^2\}^{1/2}, \tag{6.53}$$

where $z(t)$ is the solution mentioned above. Since $(u(t), v(t)) \neq (0, 0)$, $t \in R$, we have $D > 0$ on R. Due to the fact that $z(t)$ satisfies equation (6.47) with

$m(t) = m(\bar{x}, u)$, while u and v are defined above in terms of $(x(t), y(t))$ and $(\bar{x}(t), \bar{y}(t))$, we obtain by differentiation

$$DD' = -pu^2 + 2quv - zv^2, \tag{6.54}$$

where $p = z(z^2 - h)$, $q = z^2 - h$, $q^2 - pz = -h(z^2 - h)$.

Then equation (6.54) can be rewritten as

$$DD' = -z\left[\left(v - \frac{q}{z}u\right)^2 + \left(\frac{pz - q^2}{z^2}\right)u^2\right]. \tag{6.55}$$

From equation (6.55), we see that $DD' < 0$ for $(u, v) \neq (0, 0)$ and can vanish only when $(u, v) = (0, 0)$. It is worth noting that $pz - q^2 = h(z^2 - h) \geq 0$, which helps to prove that the right-hand side in equation (6.55) is vanishing only for $h = 0$ (i.e., $u = v = 0$). Therefore, $DD' < 0$ for $(u, v) \neq (0, 0)$ and will vanish and only if $(u, v) = (0, 0)$.

A first consequence of the considerations above is that $D' < 0$ for $(u, v) \neq 0$. Hence, D^2 is a strictly decreasing function of t, which implies $D^2 \longrightarrow D_0^2$, $0 \leq D_0 < \infty$, as $t \to -\infty$. We note that D^2 is bounded on R since u, v, and x are bounded there.

But $D^2 \longrightarrow D_0^2 \neq 0$ as $t \to -\infty$ implies the existence of a sequence $\{t_n;\ n \geq 1\}$, $t_n \to -\infty$ as $n \to \infty$ such that $DD' \longrightarrow 0$ on this sequence. Otherwise, we would have $DD' \leq -\lambda < 0$ for $t \leq \mu$ with $\mu \in R$. The last inequality for D implies $D^2 \longrightarrow +\infty$ as $t \to -\infty$, which contradicts the boundedness of D. Therefore, on the sequence $\{t_n;\ n \geq 1\}$, we must have $v - (q/z) \longrightarrow 0$ and $pz - q^2 \longrightarrow 0$ as $n \to \infty$ because $z \geq \alpha > 0$. This situation can occur only if $u \longrightarrow 0$ and $v \longrightarrow 0$ as $n \to \infty$.

From $(u, v) \longrightarrow (0, 0)$ on the sequence $\{t_n;\ n \geq 1\}$, one derives $D \longrightarrow 0$ as $n \to \infty$, which is incompatible with $D^2 \longrightarrow D_0^2 \neq 0$.

From $v - (q/z)u \longrightarrow 0$ and $pz - q^2 = h(z^2 - h) \longrightarrow 0$ on the sequence above, there results $h \longrightarrow 0$ on the same sequence because $z^2 - h \geq \alpha^2 - \beta > 0$ (see equation (6.52) and condition (3), which imply $h \leq \beta$). If we now admit that u does not tend to zero on that sequence, then we should find a subsequence on which $u \longrightarrow u_0 \neq 0$. Without loss of generality, we can assume that on this subsequence $\bar{x} \longrightarrow \bar{x}_0$. Because $h(\bar{x}, u)$ is continuous, one has $h \longrightarrow h(\bar{x}_0, u_0)$ on the subsequence. But $h \neq 0$ for $u_0 \neq 0$, which leads to a contradiction with the property above, namely $h \longrightarrow 0$ on the sequence initially considered. This implies that $u \longrightarrow 0$ on the sequence, and since $DD' \longrightarrow 0$ on the same sequence, we must have $v \longrightarrow 0$ according to equation (6.54).

Consequently, each system (6.46) has a unique solution in Ω. In particular, equation (6.44) has a unique almost periodic solution $x(t)$ together with its first derivative such that $(x(t), x'(t)) \in \Omega$ for $t \in R$. The almost periodicity is the consequence of Amerio's result, formulated in Theorem 6.1, concerning separated solutions in almost periodic systems.

This ends the proof of Theorem 6.2.

Remark 6.6. It is obvious from equation (6.44) that the second derivative of the (unique) almost periodic solution is also almost periodic.

Remark 6.7. It is easy to prove, following the pattern above, that the almost periodic solution of equation (6.44) under conditions (1)–(5) is *asymptotically stable*. This means that any other solution of equation (6.44), with its graph in Ω for $t \geq t_0$, $t_0 \in R$, tends as $t \to \infty$ to the unique almost periodic solution.

Indeed, if $x(t)$ is the almost periodic solution and $\bar{x}(t)$ is any solution of equation (6.44) with $(t, \bar{x}(t)) \in \Omega$ for $t \geq t_0$, then the distance function $d(t) = |x(t) - \bar{x}(t)|$, $t \geq t_0$, can be treated the same way we have proceeded with $D(t)$ from equation (6.53). One finds that d^2 must be a decreasing function with t, $t \geq t_0$, because $dd' < 0$. The hypothesis $d^2 \longrightarrow d_0^2 > 0$ is not acceptable.

Remark 6.8. Conditions (1) and (2) of Theorem 6.2 are implied by the condition

$$\liminf_{|x| \to \infty} |g(x)| > k. \tag{6.56}$$

Hence, formula (6.56) and conditions (3)–(5) from the statement of Theorem 6.2 suffice for the existence of a unique almost periodic solution to equation (6.44).

Equation (6.44) constitutes a forced perturbation of a classical nonlinear oscillation equation known as Liénard's equation.

6.4 Equations with Monotone Operators

The main feature of differential equations we shall consider in this section consists in the *monotonicity* of the operators on the right-hand side. More precisely, the equations to be dealt with in this section have one of the following forms:

$$\dot{x}(t) = (fx)(t), \qquad t \in R, \tag{6.57}$$

or

$$\ddot{x}(t) = (fx)(t), \qquad t \in R. \tag{6.58}$$

The reduction of equation (6.58) to the form (6.57) can be easily accomplished, but the monotonicity is generally lost.

The main assumption we shall make on the operator f on the right-hand side of either equation (6.57) or (6.58), besides the monotonicity, is that it acts on the space $AP(R, R^n)$ or $AP(R, C^n)$. Of course, other spaces of almost periodic functions are acceptable, and we shall illustrate the case of Stepanov's almost periodic functions; i.e., the elements of the space $S(R, R^n)$.

This section will be dedicated to the results of Bohr–Neugebauer type, which means that the conditions imposed on systems like (6.57) and (6.58) are of such a nature that the property of almost periodicity of solutions is equivalent to that of boundedness on the real line R. In Section 5.4, we dealt with such a case when the differential system was linear. In the present case, we shall cover nonlinear situations.

In dealing with the system (6.57), the following result is helpful.

Proposition 6.5. *Consider the differential inequality*

$$\dot{u}(t) \leq \omega(u(t)), \qquad t \in R, \tag{6.59}$$

under the following assumptions:

(1) $\omega : R_+ \longrightarrow R$ *is continuous.*
(2) $\omega(u) < 0$ *for* $u > \alpha > 0$.
(3) $u : R \longrightarrow R_+$ *is differentiable, satisfies formula* (6.59), *and is bounded on* R.

Then

$$u(t) \leq \alpha, \qquad t \in R. \tag{6.60}$$

Proof. Since $u(t)$ is bounded on R, we can distinguish only two exclusive cases to be examined: first, when $u(t)$ attains its maximum value at some point $t_0 \in R$, and then $u'(t_0) = 0$ and $\omega(u(t_0)) \geq 0$, which means that $u(t) \leq u(t_0) = \sup\{u(t); \ t \in R\}$, and because of condition (2) above, $u(t) \leq u(t_0) \leq \alpha, t \in R$. If $u(t)$ does not attain its maximum value at any $t_0 \in R$, then there exist sequences $\{t_n; \ n \geq 1\} \subset R$, $t_n \longrightarrow \infty$ as $n \to \infty$, with $u(t_n) \longrightarrow \sup\{u(t); \ t \in R\}$ as $n \to \infty$. Or a similar situation occurs with $t_n \longrightarrow -\infty$ as $n \to \infty$.

Let us concentrate on the first case, where $t_n \longrightarrow \infty$ as $n \to \infty$. If $u(t) \longrightarrow \sup\{u(t); \ t \in R\}$ as $t \to \infty$, then one can assume, without loss of generality, that $u'(t_n) \geq 0$ for sufficiently large n. Indeed, in the contrary case it would mean $u'(t) < 0$ for $t \geq T$, which contradicts the fact that $u(t) \uparrow \sup\{u(t); \ t \in R\}$ at least on some intervals (a, b) with sufficiently large a. Hence, on such a sequence $\{t_n; \ n \geq 1\}$ we will have $\omega(u(t_n)) \geq 0$, and these inequalities imply (at the limit) $\omega(\sup\{u(t); \ t \in R\}) \geq 0$. Again, we see that formula (6.60) must be true. If $u(t)$ has no unique limit as $t \to \infty$, we can construct two sequences, $\{t_n; \ n \geq 1\}$ and $\{\bar{t}_n; \ n \geq 1\}$, such that $u(t_n) \longrightarrow \sup\{u(t); \ t \in R\}$, while $u(\bar{t}_n) \longrightarrow U$, $U < \sup\{u(t); \ t \in R\}$. It turns out that in this case one can choose a sequence $\{t_n; \ n \geq 1\}$, $t_n \to \infty$ as $n \to \infty$, such that $\dot{u}(t_n) = 0, n \geq 1$. In other words, $\{t_n; \ n \geq 1\}$ is a sequence of local maxima. For sufficiently large n, the interval $(\bar{t}_n, \bar{t}_{n+1}) \subset R$ must contain a local maximum point, which can be taken as t_n. Again, we obtain $0 \leq \omega(u(t_n))$ for large n, which implies $\sup\{u(t); \ t \in R\} \leq \alpha$.

Remark 6.9. There is a kind of dual result to Proposition 6.5; namely, by the change of variables $t = -\tau$, $u(-\tau) = v(\tau)$, one obtains from formula (6.59) the inequality

$$\frac{dv}{d\tau} \geq -\omega(v(\tau)), \qquad \tau \in R. \tag{6.61}$$

The inequality (6.61) can be discussed in the same way as formula (6.59), the conclusion regarding v being the same: $v(\tau) \leq \alpha$, $\tau \in R$. Only condition (2) has to be changed into $\omega(u) > 0$ for $u > \alpha > 0$.

In order to state and prove the result on almost periodicity of bounded solutions to equation (6.59), we will note that equation (6.59) can be rewritten in the form

$$\dot{x}(t) = (gx)(t) + h(t), \qquad t \in R, \tag{6.62}$$

with

$$(gx)(t) = (fx)(t) - (f\theta)(t), \ h(t) = (f\theta)(t), \tag{6.63}$$

where θ is the zero element in $AP(R, R^n)$. Of course, one can choose, instead of $\theta = \theta(t)$, any given element of $AP(R, R^n)$.

Similarly, equation (6.58) can be rewritten in the form

$$\ddot{x}(t) = (gx)(t) + h(t), \qquad t \in R, \tag{6.64}$$

with the same meaning for g and h.

Let us note that the monotonicity condition on f and g, to be used in what follows, is simultaneously true for the operators f and g. It can be written in the form

$$\langle (gx)(t) - (gy)(t), x(t) - y(t) \rangle \leq -m|x(t) - y(t)|^2, \tag{6.65}$$

with $m > 0$ a constant, and for arbitrary $x, y \in AP(R, R^n)$. On the left-hand side of formula (6.65), we use the scalar product in R^n, denoted by $\langle \cdot, \cdot \rangle$.

We can now formulate the following result related to equation (6.62).

Theorem 6.3. *Consider equation* (6.62) *under the following assumptions:*

(1) *g is a continuous operator on the space $BC(R, R^n)$ satisfying the monotonicity condition* (6.65).
(2) *$h(t) \in AP(R, R^n)$.*

If $x : R \longrightarrow R^n$ is a map verifying equation (6.62) *such that $x \in BC(R, R^n)$, then necessarily $x \in AP(R, R^n)$ together with its derivative.*

Proof. Let $x \in BC(R, R^n)$ be a solution to equation (6.62) and $\tau \in R$ an arbitrary number. Then $x_\tau(t) = x(t + \tau) \in BC(R, R^n)$. From equation (6.62), one easily derives

$$\dot{x}(t + \tau) - \dot{x}(t) = (gx)(t + \tau) - (gx)(t) + h(t + \tau) - h(t), \qquad \tau \in R. \tag{6.66}$$

Multiplying both sides of equation (6.66) scalarly by $x(t+\tau) - x(t)$ and taking formula (6.65) into account, we obtain for $t \in R$

$$\frac{1}{2}\frac{d}{dt}|x(t + \tau) - x(t)|^2$$

$$\leq -m|x(t + \tau) - x(t)|^2 + |h(t + \tau) - h(t)| \, |x(t + \tau) - x(t)|.$$

If we denote $v(t) = |x(t + \tau) - x(t)|^2$, then the inequality above becomes

$$\frac{1}{2}\frac{dv}{dt} \leq -mv + \sqrt{v}\,|h(t + \tau) - h(t)|, \qquad t \in R.$$

Applying Proposition 6.5, and based on the inclusion $h \in \mathrm{AP}(R, R^n)$, we can write for each ε-almost period of h

$$\frac{1}{2}\frac{dv}{dt} \leq -mv + \varepsilon\sqrt{v}, \qquad t \in R. \tag{6.67}$$

Proposition 6.5 allows us to write, for each τ that is an ε-almost period of h, the following inequality:

$$|x(t + \tau) - x(t)|^2 = v \leq \left(\frac{\varepsilon}{m}\right)^2, \qquad t \in R. \tag{6.68}$$

The inequality (6.68) shows that, for each ε-almost period of h, one has

$$|x(t + \tau) - x(t)| \leq \frac{\varepsilon}{m}, \qquad t \in R, \tag{6.69}$$

which means that each τ that is an ε-almost period of h is an (ε/m)-almost period of $x(t)$.

We shall now consider the case of second-order equations of the form (6.58) or equivalently of the form (6.64) when a result similar to Theorem 3.3 can be obtained.

Unlike the case of first-order equations, where the monotonicity type was expressed by condition (6.65), in the second-order case we need the monotonicity property

$$\langle (gx)(t) - (gy)(t), x(t) - y(t)\rangle \geq m|x(t) - y(t)|^2 \tag{6.70}$$

with $m > 0$. Let us note that in the first-order case it is immaterial if we use formula (6.65) or (6.70) due to the fact that by changing t into $-t$ in equation (6.62), g has to be changed in $-g$.

The following result holds true in regard to the equation of comparison (6.71):

$$\ddot{u}(t) \geq \omega(u(t)), \qquad t \in R. \tag{6.71}$$

Proposition 6.6. *Let us consider the differential inequality* (6.71) *under the following assumptions:*

(1) $\omega : R_+ \longrightarrow R$ *is a continuous map.*
(2) $\omega(u) > 0$ *for* $u > \alpha > 0$.
(3) $u : R \longrightarrow R_+$ *is twice continuously differentiable, bounded on* R, *and satisfies formula* (6.71).

Then

$$u(t) \leq \alpha, \qquad t \in R. \tag{6.72}$$

Proof. Again, as in the proof of Proposition 6.5, we distinguish two different situations. First, there is a point $t_0 \in R$ such that $u(t_0) = \sup\{u(t); \, t \in R\}$. In such a case, we obviously have $\ddot{u}(t_0) \leq 0$. Otherwise, t_0 could not be a point of maximum. Hence, $\omega(u(t_0)) \leq 0$, which implies $u(t_0) \leq \alpha$. Second, there is no point $t_0 \in R$ such that the maximum of $u(t)$ is attained at t_0. In this second situation, there must be a sequence $\{t_n; \, n \geq 1\} \subset R$ with $t_n \longrightarrow \infty$ as $n \to \infty$ such that $u(t_n) \longrightarrow \sup\{u(t); \, t \in R\}$ as $n \to \infty$ or a sequence $\{\bar{t}_n; \, n \geq 1\} \subset R$ with $u(\bar{t}_n) \longrightarrow \sup\{u(t); \, t \in R\}$ as $n \to \infty$, while $\bar{t}_n \longrightarrow -\infty$. The second case can be treated in the same way as the first, and we shall deal only with the first situation.

If $\lim u(t) = \sup\{u(t); \, t \in R\}$ as $t \to \infty$, then one can find a sequence $\{t_n; \, n \geq 1\}$, $t_n \to \infty$ as $n \to \infty$, such that $\ddot{u}(t_n) \leq 0$, $n \geq 1$. In the contrary case, we would have $\ddot{u}(t) > 0$ for $t \geq T$, which means that $\dot{u}(t)$ is an increasing map on $[T, \infty)$. We can write the inequality $\dot{u}(t) \geq \dot{u}(T)$ for $t \geq T$, and since $\dot{u}(t)$ is increasing, there must be some $T_1 \geq T$ such that $\dot{u}(T_1) > 0$. Without loss of generality we can consider the case $T_1 = T$. Hence, $\dot{u}(T) > 0$ and $u(t) \geq \dot{u}(T)(t - T) + \text{const.}$ for $t \geq T$. This is impossible because $u(t)$ is bounded on $[T, \infty)$. Therefore, on the sequence $\{t_n; \, n \geq 1\}$, we have $0 \geq \ddot{u}(t_n) \geq \omega(u(t_n))$, and as $n \to \infty$ we obtain $0 \geq \omega(\sup\{u(t); \, t \in R\})$. This inequality leads again to $u(t) \leq \alpha$, $t \in R$.

If $u(t)$ has more than one limit at ∞, reasoning similar to that in the last part of the proof of Proposition 6.5 leads to the desired conclusion: $u(t) \leq \alpha$, $t \in R$.

We shall now return to equation (6.68), which is equivalent to equation (6.64), and prove a result similar to that given in Theorem 6.3 but for second-order differential equations.

Theorem 6.4. *Let us consider equation* (6.64) *under the following assumptions:*

(1) *g is a continuous operator on the space* $\mathrm{BC}(R, R^n)$ *satisfying the monotonicity condition* (6.70).

(2) *$h(t) \in \mathrm{AP}(R, R^n)$.*

If $x : R \longrightarrow R^n$ is a solution of equation (6.64) *such that $x \in \mathrm{BC}(R, R^n)$, then necessarily $x \in \mathrm{AP}(R, R^n)$ together with the derivatives $\dot{x}$ and $\ddot{x}$.*

Proof. Let $x \in \mathrm{BC}(R, R^n)$ be a solution of equation (6.64) and $\tau \in R$ a fixed number. It is known that $x_\tau(t) = x(t + \tau)$, $t \in R$, belongs to $\mathrm{BC}(R, R^n)$. We obtain from equation (6.64) with $t \in R$

$$\ddot{x}(t + \tau) - \ddot{x}(t) = (gx)(t + \tau) - (gx)(t) + h(t + \tau) - h(t).$$

If we multiply both sides by $x(t + \tau) - x(t)$ scalarly and take formula (6.70) into account, we obtain

$$\langle \ddot{x}(t+\tau) - \ddot{x}(t), x(t+\tau) - x(t) \rangle$$
$$\geq m|x(t+\tau) - x(t)|^2 - |h(t+\tau) - h(t)|\,|x(t+\tau) - x(t)|, \qquad t \in R.$$

But

$$\langle \ddot{u}, u \rangle = \frac{d}{dt}\,\langle \dot{u}, u \rangle - |\dot{u}|^2 = \frac{1}{2}\frac{d^2}{dt^2}|u|^2 - |\dot{u}|^2,$$

which allows us to write the inequality

$$\frac{1}{2}\frac{d^2}{dt^2}|x(t+\tau) - x(t)|^2$$
$$\geq m|x(t+\tau) - x(t)|^2 - |h(t+\tau) - h(t)|\,|x(t+\tau) - x(t)|. \tag{6.73}$$

Denote $v(t) = |x(t+\tau) - x(t)|^2$, $t \in R$, and note that formula (6.73) becomes

$$\frac{1}{2}\ddot{v}(t) \geq mv - \sqrt{v}\,|h(t+\tau) - h(t)|, \qquad t \in R. \tag{6.74}$$

Now choose $\tau \in R$ to be an almost period of h corresponding to $\sqrt{\varepsilon/m}$. Then, applying Proposition 6.6 to the inequality

$$\frac{1}{2}\frac{d^2 v}{dt^2} \geq mv - \sqrt{v}\,\sqrt{\varepsilon/m}, \tag{6.75}$$

one obtains

$$v(t) = |x(t+\tau) - x(t)|^2 \leq \varepsilon, \qquad t \in R, \tag{6.76}$$

which proves that $x \in AP(R, R^n)$. The proof of Theorem 6.4 is complete.

Remark 6.10. Both Theorems 6.3 and 6.4 do not state the existence of solutions to equations (6.57) or (6.58) belonging to the space $BC(R, R^n)$. This existence has to be postulated in order to get that solutions in $BC(R, R^n)$ also belong to $AP(R, R^n)$. Within the framework of monotone systems, as seen above, the problems of boundedness and almost periodicity of solutions are equivalent.

In general, this fact is not true even in the case of ordinary differential systems with an almost periodic right-hand side. Already in this chapter, examples of almost periodic systems with bounded solutions but not almost periodic solutions have been provided.

We will now consider the case of ordinary differential equations of the first order, namely

$$\dot{x}(t) = f(t, x(t)), \, t \in R, \tag{6.77}$$

and briefly indicate an existence result that does not require anything but adequate use of the classical methods of successive approximations.

The following hypotheses will be admitted:

(1) $f : R \times R \longrightarrow R$ is a continuous map.
(2) $f(t, x)$ is Bohr almost periodic in t uniformly with respect to x in any bounded set of R.
(3) The derivative $\partial f /] \partial x$ does exist and satisfies an inequality of the form

$$0 < m \le \frac{\partial f}{\partial x} \le M, \qquad (t, x) \in R \times R, \tag{6.78}$$

for some $m, M \in R$, with $0 < m \le M$.

The following existence and uniqueness result is valid.

Proposition 6.7. *Consider equation (6.77) under the preceding conditions (1), (2), and (3). Moreover, we assume that $f(t, 0)$, $t \in R$, is an almost periodic function (i.e., in $\mathrm{AP}(R, R)$). Then there exists a unique solution (6.77), say $x(t)$, such that $x \in \mathrm{AP}(R, R)$. It can be obtained by the iteration procedure described by*

$$\dot{x}_n(t) - M x_n(t) = f(t, x_{n-1}(t)) - M x_{n-1}(t), \tag{6.79}$$

starting, for instance, with $x_0(t) = \theta \in \mathrm{AP}(R, R)$.

Proof. It is useful to note that the linear equation

$$\dot{y}(t) - M y(t) = f(t), \qquad t \in R, \tag{6.80}$$

has a unique bounded solution on R, for each $f \in \mathrm{BC}(R, R)$ that is given by

$$y(t) = - \int_t^\infty e^{M(t-s)} f(s) ds, \tag{6.81}$$

for which the following estimate holds true:

$$|y|_{\mathrm{BC}} \le M^{-1} |f|_{\mathrm{BC}}. \tag{6.82}$$

If one writes the formula as in equation (6.79), changing n into $n+1$, and then subtracts equations side by side, one obtains

$$\frac{d}{dt} [x_{n+1}(t) - x_n(t)] - M[x_{n+1}(t) - x_n(t)]$$
$$= f(t, x_n(t)) - f(t, x_{n-1}(t)) - M[x_n(t) - x_{n-1}(t)].$$

Now applying the estimate (6.82) after elementary operations, one obtains the recurrent inequality

$$|x_{n+1}(t) - x_n(t)|_{\mathrm{BC}} \le \frac{M - m}{M} |x_n(t) - x_{n-1}(t)|_{\mathrm{BC}}, \qquad n \ge 1. \tag{6.83}$$

From formula (6.83), we easily obtain the convergence in $BC(R, R)$ of the sequence $\{x_n(t);\ n \geq 1\}$, and its limit

$$\lim_{n \to \infty} x_n(t) = x(t)$$

represents the unique solution in $BC(R, R)$ to equation (6.80).

Let us note that in each step of the recurrence one chooses $x_n(t)$ as the unique solution in $BC(R, R)$ of the equation (6.79), $n \geq 1$. $x_0(t)$ can be an arbitrary element of $AP(R, R)$.

Remark 6.11. Proposition 6.7 provides a simple example of the existence of a bounded solution that will be almost periodic in $AP(R, R)$ as soon as $f(t, x)$ is almost periodic in t uniformly with respect to x in compacts of R.

For another type of existence result, that involving a generalized monotonicity condition, see the book of Iu. A. Trubnikov and A. I. Perov [96]. The condition (6.65) is substituted by $\langle g(t, x) - g(t, y), x - y \rangle \geq \gamma |x - y|^\alpha$, $\gamma > 0$, $\alpha \geq 2$.

The problem of finding necessary and sufficient conditions for the existence and uniqueness of almost periodic solutions to equation (6.77) has recently been investigated in depth in the paper by M. Bostan [15].

Remark 6.12. As seen above, there exists some parallelism between first- and second-order equations with operators acting on $AP(R, R)$.

6.5 Gradient Type Systems

This section will be dedicated to the investigation of almost periodicity of the solutions to systems of the form

$$\ddot{x}(t) = \operatorname{grad} F(t, x(t)), \qquad t \in R, \tag{6.84}$$

where the grad operator is acting on the x-variables of the (potential) function $F(t, x)$.

It will be assumed in what follows that $x : R \longrightarrow R^n$ is a continuously differentiable map of the second order, while $F(t, x)$, defined on $R \times R^n$, takes the values in R. So, the system (6.84) consists of n differential equations, each of the second order.

In the case $n = 3$, equation (6.84) represents the equations of the motion of a material point (mass $= 1$) in a field of forces that is a gradient (depending on t). In the case $F(t, x) \equiv F(x)$, we obtain the equations of Newton when the field of forces is conservative.

We will pursue the investigation of equation (6.84) in regard to the existence of almost periodic solutions or motions under the hypothesis of almost periodicity for the potential function with respect to the variable t.

As we did at the end of Section 6.4, we shall establish the existence of solutions to equation (6.84) in the space $BC(R, R^n)$. The passage from $BC(R, R^n)$ to $AP(R, R^n)$ is immediate.

We will now consider a rather special case of equation (6.84) that will help us to construct the process of iteration leading to its solution. Namely, we first deal with the linear equation

$$\ddot{x}(t) - Mx(t) = f(t), \qquad t \in R, \tag{6.85}$$

with $M > 0$ a constant. On the right-hand side, $f \in BC(R, R^n)$. It is easily seen that equation (6.85) has a unique solution in $BC(R, R^n)$ given by the formula (see formula (6.81))

$$x(t) = -\frac{1}{2\sqrt{M}} \left\{ e^{-\sqrt{M}\,t} \int_{-\infty}^{t} e^{\sqrt{M}\,s} f(s)ds + e^{\sqrt{M}\,t} \int_{t}^{\infty} e^{-\sqrt{M}\,s} f(s)ds \right\}. \tag{6.86}$$

There follows from equation (6.86), using elementary estimates, that

$$|x|_{BC} \leq M^{-1}|f|_{BC}. \tag{6.87}$$

The iteration process that will ultimately lead to the existence of a bounded solution to equation (6.84) can be constructed as (see also Section 6.4, in which we dealt with the case $n = 1$)

$$\ddot{x}_k(t) - Mx_k(t) = \operatorname{grad} F(t, x_{k-1}(t)) - Mx_{k-1}(t), \tag{6.88}$$

with $k \geq 1$, $M > 0$ a fixed number, and taking for instance $x_0(t) = \theta \in R^n$. At each step in equation (6.88), $x_k(t)$, $k \geq 1$, is the unique solution in $BC(R, R^n)$ of that equation. Of course, we will need some restriction on $\operatorname{grad} F(t, x)$ in order to assure the existence (and uniqueness) of a bounded solution.

Instead of pursuing the proof as sketched above, we prefer to present a fixed-point principle and apply it to prove the existence of a solution to equation (6.84) under suitable hypotheses. In order to also obtain uniqueness, we shall use the *Banach contraction mapping principle*.

Let (S, d) be a complete metric space and $A : S \to S$ a contraction map such that for some α, $0 \leq \alpha < 1$, $d(Ax, Ay) \leq \alpha d(x, y)$ for any $x, y \in S$. Then there exists a unique fixed point for A, $x^* = Ax^*$.

Proof of the principle can be found in most books on functional analysis or its applications; see, for instance, our book [24].

We choose as the underlying space $BC(R, R^n)$ with the usual supremum norm and define an operator on this space according to the following scheme: For each $u \in BC(R, R^n)$, one defines $v = Au$ by means of the equation

$$\ddot{v}(t) - Mv(t) = \operatorname{grad} F(t, u(t)) - Mu(t), \tag{6.89}$$

with $v \in BC(R, R^n)$, while $M > 0$ is a fixed number.

More precisely, we shall rely on the following hypotheses in order to assure the contraction of the operator A defined above.

H_1. The map $F : R \times R^n \longrightarrow R$ is continuously differentiable of the second order in the second argument.

The hypothesis H_1 allows us to define grad $F(t, x)$ by means of the formula

$$\text{grad}\, F(t, x) = \left(\frac{\partial F}{\partial x_1}, \frac{\partial F}{\partial x_2}, \cdots, \frac{\partial F}{\partial x_n} \right) \tag{6.90}$$

and the Hessian matrix of F by

$$H(t, x) = \left(\frac{\partial^2 F}{\partial x_i \partial x_j} \right)_{n \times n}. \tag{6.91}$$

H_2. Let $B_r = \{u;\ u \in R^n,\ |u| \leq r\}$. Assume there exists $K(r) > 0$ such that

$$|\text{grad}\, F(t, x)| \leq K(r), \qquad (t, x) \in R \times B_r. \tag{6.92}$$

H_3. The Hessian matrix (6.91) satisfies the condition

$$mI \leq H(t, x) \leq M(r)I, \qquad (t, x) \in R \times B_r, \tag{6.93}$$

where $m > 0$ and $M(r) > 0$ for $r > 0$. By I we mean the unit matrix of order n.

Let us remind ourselves that for a symmetric matrix C, the inequality $C \geq 0$ stands for the property $\langle Cx, x \rangle \geq 0$ for $x \in R^n$. Hence, $C \geq B$, with B another symmetric matrix, means $\langle Cx, x \rangle \geq \langle Bx, x \rangle$ for $x \in R^n$.

We can now state the following result concerning the solutions, in $\text{BC}(R, R^n)$ or $\text{AP}(R, R^n)$, of equation (6.84).

Theorem 6.5. *Let us consider equation (6.84) under hypotheses* H_1, H_2, *and* H_3. *Then there exists a unique solution* $x \in \text{BC}(R, R^n)$ *of equation (6.84). If* grad $F(t, x)$ *is almost periodic in* t *uniformly with respect to* x *in any compact set of* R^n, *then the unique solution belongs to* $\text{AP}(R, R^n)$.

Proof. We shall consider the operator A defined by equation (6.89), the number M in that formula being in fact one $M(r)$ for sufficiently large r from H_3. We will prove that A is a contraction on a ball $\Sigma_r \subset \text{BC}(R, R^n)$ if r is taken such that $mr \geq K(0)$, with K as in formula (6.92).

From equation (6.84), we obtain, for a couple of functions $u, \bar{u} \in \text{BC}(R, R^n)$, the following equation for $v = Au$, $\bar{v} = A\bar{u}$:

$$\frac{d^2}{dt^2} [v(t) - \bar{v}(t)] - M[v(t) - \bar{v}(t)]$$
$$= \text{grad}\, F(t, u(t)) - \text{grad}\, F(t, \bar{u}(t)) - M[u(t) - \bar{u}(t)].$$

The equation above has the form (6.85), and $v(t) - \bar{v}(t)$ is the only solution in $\text{BC}(R, R^n)$. If we apply the estimate (6.87), we obtain

$$|v(t) - \bar{v}(t)|_{\mathrm{BC}}$$
$$\leq M^{-1}|\mathrm{grad}\, F(t, u(t)) - \mathrm{grad}\, F(t, \bar{v}(t)) - M[u(t) - \bar{u}(t)]|_{\mathrm{BC}}. \quad (6.94)$$

We shall now estimate the difference $\mathrm{grad}\, F(t, u(t)) - \mathrm{grad}\, F(t, \bar{u}(t))$, making use of an integral formula that involves the Hessian matrix $H(t, x)$ associated with $F(t, x)$:

$$\mathrm{grad}\, F(t, u(t)) - \mathrm{grad}\, F(t, \bar{u}(t))$$
$$= \left\{ \int_0^1 H[t, \bar{u}(t) + \tau(u(t) - \bar{u}(t))]d\tau \right\} [u(t) - \bar{u}(t)]. \quad (6.95)$$

Formula (6.95) is a variant of the classical Lagrange formula

$$F(x) - F(x_0) = \int_0^1 F'(x_0 + \theta(x - x_0))d\theta(x - x_0),$$

which holds true for any differentiable operator F with continuous derivative F' on any Banach space (see, for instance, V. Trénoguine [95]).

Let us denote

$$\widetilde{H}(t, u, \bar{u}) = \int_0^1 H(t, \bar{u}(t) + \tau(u(t) - \bar{u}(t))d\tau,$$

and based on inequalities (6.93), note that $\widetilde{H}(t, u, \bar{u})$ satisfies also the inequality

$$mI \leq \widetilde{H}(t, u, \bar{u}) \leq M(r)I \quad (6.96)$$

as soon as $(t, u, \bar{u}) \in R \times B_r \times B_r$. We can therefore derive from formulas (6.94)–(6.96) the estimate

$$|v(t) - \bar{v}(t)|_{\mathrm{BC}} \leq M^{-1}[MI - \widetilde{H}][u(t) - \bar{u}(t)]_{\mathrm{BC}}.$$

But $MI - \widetilde{H}$ is a symmetric matrix, and

$$|MI - \widetilde{H}| = \sup_{|\xi|=1} \left\langle [MI - \widetilde{H}]\xi, \xi \right\rangle \quad (6.97)$$

with the operator matrix norm on the left-hand side. Formula (6.97), together with formula (6.96), leads to the conclusion that

$$|MI - \widetilde{H}| \leq M - m$$

with M as chosen above. This leads to

$$|v(t) - \bar{v}(t)|_{\mathrm{BC}} \leq \frac{M - m}{M}|u(t) - \bar{u}(t)|_{\mathrm{BC}}, \quad (6.98)$$

which proves that the operator $u \longrightarrow v = Au$ is a contraction.

It only remains to show that, for some $r > 0$, one obtains the inclusion

$$A\Sigma_r \subset \Sigma_r. \tag{6.99}$$

We shall again use the estimate (6.87) for the solution in $BC(R, R^n)$ of equation (6.89), taking into account that $\operatorname{grad} F(t, u(t)) - M u(t)$ can be represented as

$$[MI - \widetilde{H}(t, u(t), \theta)]u(t) + \operatorname{grad} F(t, \theta),$$

which leads to the inequality

$$|v(t)|_{\mathrm{BC}} \leq \frac{M(r) - m}{M(r)} r + \frac{K(0)}{M(r)},$$

from which we derive expression (6.99), provided $K(0) \leq mr$.

This ends the proof of Theorem 6.5 concerning existence in $BC(R, R^n)$, but we need to complete the proof of Theorem 6.5 in regard to the almost periodicity of the solution in $BC(R, R^n)$.

We shall take into account the obvious fact that the sequence $\{x_k(t); k \geq 1\}$, defined by equation (6.88), consists of functions in $AP(R, R^n)$. Since it is convergent in $BC(R, R^n)$, whose type of convergence is the same as in $AP(R, R^n)$ (i.e., uniform on R), there results that $x(t) = \lim x_k(t)$ as $k \longrightarrow \infty$ is in $AP(R, R^n)$.

The proof of Theorem 6.5 is now complete.

Remark 6.13. If $M(r)$ from formula (6.93) is bounded (above), say $M(r) = M < \infty$ for $r \geq 0$, then the operator A is acting as a contraction on the whole space $AP(R, R^n)$. Then condition $mr \geq K(0)$ is certainly verified for sufficiently large r.

Remark 6.14. A result similar to Theorem 6.5 can be obtained for the first-order system $\dot{x}(t) = \operatorname{grad} F(t, x(t))$.

Remark 6.15. As shown in a paper by A. R. Aftabizadeh [1], the space $AP(R, R^n)$ can be replaced by the richer space $S(R, R^n)$ of almost periodic functions in the sense of Stepanov. The solution will be in the Bohr space $AP(R, R^n)$.

We shall consider in the second part of this section an equation similar to equation (6.84). Formally, the equation

$$\ddot{x}(t) - \operatorname{grad} V(x(t)) = h(t), \qquad t \in R, \tag{6.100}$$

is a special case of equation (6.84), but the hypotheses we shall use are somewhat different from those stipulated in equation (6.84), as are the method of proof for existence in $BC(R, R^n)$ or almost periodicity of the solutions. Namely, we shall rely on the calculus of variations technique as opposed to the fixed-point principle in the first part of this section.

In order to enunciate the results and provide the adequate hypotheses, we shall write equation (6.100) in a slightly different form. Namely, we shall consider the equation

$$\ddot{x}(t) - Ax(t) - \operatorname{grad} U(x(t)) = h(t), \tag{6.101}$$

which means that we have chosen

$$V(x) = Ax + \operatorname{grad} U(x), \qquad x \in R^n. \tag{6.102}$$

When $A = A^\top$, $2Ax = \operatorname{grad} \langle Ax, x \rangle$, so equation (6.101) is of the form (6.100).

From the calculus of variations (see, for instance, the book by H. Butazzo, M. Giaquinta, and S. Hildebrand [19]), we know that equation (6.101) is the Euler equation associated with the functional

$$J_T(x) = \int_{|t| \leq T} \left\{ \frac{1}{2} |\dot{x}|^2 + \langle Ax, x \rangle + U(x) + \langle h, x \rangle \right\} dt, \tag{6.103}$$

where $T > 0$ is a fixed number.

The following conditions will be imposed on the functions involved in equations (6.101) and (6.103):

(1) $A \in \mathcal{L}(R^n, R^n)$, $n \geq 1$, is symmetric and positive definite

$$\langle Ax, x \rangle \geq \sigma |x|^2, \qquad \sigma > 0, \ x \in R^n.$$

(2) $U : R^n \longrightarrow R$ is continuously differentiable, bounded below, and for some $K > 0$ satisfies the condition

$$\langle \operatorname{grad} U(x), x \rangle \geq 0 \qquad \text{for } |x| \geq K.$$

(3) $h \in \operatorname{BC}(R, R^n)$.

We note first that condition (2) allows us to assume that $U(x) \geq 0$, $x \in R^n$. Indeed, only $\operatorname{grad} U(x)$ appears in equation (6.101), and this does not change if $U(x)$ is substituted by $U(x) + C$, $C = \text{const.}$

We can now state and prove a basic result regarding the solutions in $\operatorname{BC}(R, R^n)$ of equation (6.101).

Theorem 6.6. *Consider equation* (6.101), *and assume the preceding conditions* (1), (2), *and* (3) *are satisfied. Then, there exists a solution* $x = x(t)$, $t \in R$, *of equation* (6.101) *such that* $x \in \operatorname{BC}(R, R^n)$ *as are* $\dot{x}$ *and* $\ddot{x}$.

Proof. We choose as the underlying space for investigating the variational problem related to equation (6.103) the space $H^1([-T, T]; R^n)$ for some fixed $T > 0$. This space is known as a Sobolev space and consists of all maps from $[-T, T]$ into R^n such that $x(t) = \lim x_{0k}(t)$ as $k \to \infty$ with $x_{0k}(t)$, $k \geq 1$, continuously differentiable on $[-T, T]$ with respect to the norm

$$\|u\|^2 = \int_{-T}^{T} \left[|u(t)|^2 + |\dot{u}(t)|^2\right] dt. \tag{6.104}$$

In other words, $H^1([-T,T]; R^n)$ is the completion of the space $C^{(1)}([-T,T]; R^n)$ with respect to the norm (6.104).

The functional $J_T(x)$ defined by equation (6.103), considered on the space $H^1([-T,T]; R^n)$, has some basic properties required in the calculus of variations for the existence of the extremum (minimum in our case): (a) $J_T(x)$ is differentiable; (b) $J_T(x)$ is bounded below (on the whole space!) and its integrand is convex with respect to $\dot{x}$; (c) $J_T(x)$ has superlinear growth in $\dot{x}$.

Let us note that condition (c) is easily checked if one takes into account the term $(Ax, x) \geq \sigma|x|^2$ and $U(x) \geq 0$. Conditions (a) and (b) are obviously verified by $J_T(x)$.

Therefore (see, for instance, the book by G. Butazzo, M. Giaguinta, and S. Hildebrudt [19] quoted above.), there exists an element $v \in H^1([-T,T]; R^n)$ such that $J_T(v) = \min J_T(x)$, $x \in H^1([-T,T]; R^n)$. It satisfies also some boundary value condition, say $\dot{v}(-T) = \dot{v}(T) = \theta \in R^n$.

Let us denote

$$|v|_T = \sup\{v(t); \; |t| \leq T\}, \quad M = |h|_{BC}, \quad C_1 = \max\{K, M\sigma\}.$$

We shall prove now that

$$|v|_T \leq C_1 \qquad \text{for any } T > 0. \tag{6.105}$$

Let us denote $\varphi(t) = |v(t)|^2$, $t \in [-T,T]$, and note that $\dot{\varphi}(t) = 2\langle v(t), \dot{v}(t)\rangle$, $t \in [-T,T]$. This implies, based on the properties of $v(t)$,

$$\dot{\varphi}(-T) = \dot{\varphi}(T) = 0, \tag{6.106}$$

while

$$\begin{aligned}
\ddot{\varphi}(t) &= 2|\dot{v}(t)|^2 + 2\langle v(t), \ddot{v}(t)\rangle \\
&= 2|\dot{v}(t)|^2 + 2\langle Av(t), v(t)\rangle \\
&\quad + 2\langle \operatorname{grad} U(v(t)), v(t)\rangle + 2\langle h(t), v(t)\rangle.
\end{aligned} \tag{6.107}$$

If one assumes, contrary to formula (6.105), that

$$I = \{t; \; |t| \leq T, \; \varphi(t) > C_1^2\} \neq \emptyset, \tag{6.108}$$

then based on conditions (1) and (2) above, we will have from equation (6.107), for every $t \in I$,

$$\ddot{\varphi}(t) \geq 2\sigma\varphi(t) + 2\langle h(t), v(t)\rangle \geq 2\sigma\varphi(t) - 2M|v(t)|,$$

or

$$\ddot{\varphi}(t) \geq 2\sqrt{\varphi(t)}\left[\sigma\sqrt{\varphi(t)} - M\right]. \tag{6.109}$$

At this point, we can exploit the inequality (6.109) to find the estimate equation (6.105). From our assumption (6.108), we have $\sigma\sqrt{\varphi(t)} - M > 0$ on $|t| \leq T$. This means $\ddot{\varphi}(t) > 0$ for $|t| \leq T$, and this implies that $\dot{\varphi}(t)$ is strictly increasing. If $\dot{\varphi}(t_0) \geq 0$ at some t_0, $|t_0| \leq T$, then $[t_0, T] \subset I$ because $\dot{\varphi}(t)$ increases strictly, which means $\dot{\varphi}(T) > 0$, a contradiction with $\varphi(T) = 0$. In a similar way, the assumption $\dot{\varphi}(t_0) < 0$ leads to a contradiction. Therefore, I is the empty set, and this shows that formula (6.105) must be true.

Since $v(t)$ is a solution of the Euler equation (6.101), there results for $\ddot{v}(t)$ the estimate

$$|\ddot{v}|_T \leq \sup_{|t| \leq T} (|Av + \operatorname{grad} U(v)| + M) = C_2, \qquad (6.110)$$

where C_2 is independent of T (i.e., is an absolute constant).

So far, we have found upper bounds for $|v(t)|, |\ddot{v}(t)|$, $t \in [-T, T]$, that do not depend on $T > 0$.

We shall prove that a similar but somewhat different estimate is valid for $\dot{v}(t)$. Indeed, since we have $v(t), \ddot{v}(t) \in C([-T, T], R^n)$, it means that the equation

$$\ddot{v}(t) - v(t) = f(t), \qquad t \in [-T, T], \qquad (6.111)$$

satisfies $f \in C([-T, T], R^n)$. The boundary values for $v(t)$ are, as seen above, $\dot{v}(-T) = \dot{v}(T) = \theta \in R^n$.

From equation (6.111), one obtains the equality

$$\langle \dot{v}, \ddot{v} \rangle = \langle \dot{v}, v \rangle + \langle \dot{v}, f \rangle, \qquad (6.112)$$

which can be rewritten in the form

$$\frac{d}{dt}|\dot{v}(t)|^2 = \frac{d}{dt}|v(t)|^2 + 2\langle \dot{v}(t), f(t)\rangle. \qquad (6.113)$$

By integrating from $-T$ to T, one gets from equation (6.113)

$$|\dot{v}(t)|^2 = |v(t)|^2 - |v(-T)|^2 + 2\int_{-T}^{t} |\dot{v}(t)|\,|f(t)|dt$$

or

$$|\dot{v}(t)|^2 \leq |v(t)|^2 + 2\sup|\dot{v}(t)| \int_{-T}^{T} |f(t)|dt, \qquad t \in [-T, T].$$

Further, based on formula (6.105) and taking the supremum on the left-hand side, one obtains the second-degree inequality

$$|\dot{v}|_T^2 \leq C_1^2 + 2|\dot{v}|_T \int_{-T}^{T} |f(t)|dt, \qquad (6.114)$$

which implies

$$|\dot{v}|_T^2 \leq \int_{-T}^{T} |f(t)|dt + \left\{ \left(\int_{-T}^{T} |f(t)|dt\right)^2 + C_1^2 \right\}^{1/2}. \qquad (6.115)$$

As seen from formula (6.115), $|\dot{v}|$ remains bounded on each interval $[-T, T]$, $T > 0$. Actually, one can obtain an estimate independent of T. But formula (6.115) will suffice to prove an existence result to equation (6.101) using a diagonal procedure.

Before we proceed to the proof of existence, we note that the variational problem of minimizing the functional (6.103) under restrictions $\dot{v}(-T) = \dot{v}(T) = \theta \in R^n$ has a solution for which $|v(t)|$ and $|\ddot{v}(t)|$ are bounded by constants independent of T, while $|\dot{v}(t)|$ has an upper bound depending on T, on each interval $[-T, T]$.

Let us now extend $v(t)$ to the whole real line by putting

$$v_T(t) = \begin{cases} v(-T) & \text{when } t \leq -T, \\ v(t) & \text{when } t \in [-T, T], \\ v(T) & \text{when } t \geq T. \end{cases}$$

We shall now choose a sequence $\{T_k \ k \geq 1\}$ of positive numbers such that $T_k \to \infty$ as $k \to \infty$. Then we consider the sequence $\{v_{T_k}(t); \ k \geq 1\}$ of functions defined on R, all terms being continuously differentiable of the first order and with second derivatives also continuous, except perhaps $\pm T_k$. Now using the diagonal procedure after applying the Ascoli–Arzelà criterion of compactness on each finite interval $[-T, T]$, on which all terms of the sequence $\langle v_{T_k}(t); \ k \geq 1 \rangle$ are defined starting with a certain rank (depending on T), we obtain a function $u(t)$, defined on R, such that a subsequence of $\{v_{T_k}(t); \ k \geq 1\}$ converges uniformly on $[-T, T]$ to $u(t)$, together with the subsequence of the derivatives of the first order. We have to keep in mind the fact that uniform boundedness of the derivatives implies uniform convergence on any compact subset of R. As is known from the calculus of variations, $u(t)$ satisfies equation (6.101) on R (because $T > 0$ is arbitrarily large). So, it also has derivatives of first and second order, and these are continuous on R.

Since $v(t)$, from which we started to construct the sequence $\{v_{T_k}; \ k \geq 1\}$, is in $BC(R, R^n)$, as well as $\{\ddot{v}_{T_k}(t); \ k \geq 1\}$, one obtains that $u(t)$ and $\ddot{u}(t)$ belong to $BC(R, R^n)$. The same is true for $\dot{u}(t)$, as one can see if we consider an equation similar to equation (6.111) but on R,

$$\ddot{u}(t) - u(t) = f(t), \qquad t \in R, \tag{6.116}$$

with $f \in BC(R, R^n)$.

We have seen that equation (6.85), which is the same as equation (6.116), has only one solution in $BC(R, R^n)$. It is given by formula (6.86) with $M = 1$, namely

$$u(t) = -\frac{1}{2}\left\{e^{-t}\int_{-\infty}^{t} e^s f(s)ds + e^t \int_{t}^{\infty} e^{-s} f(s)ds\right\}.$$

One can easily see that $\dot{u}(t) \in BC(R, R^n)$.

To summarize this lengthy proof, we can state that we constructed a solution $u(t)$ to equation (6.101) by means of a variational procedure, and this solution belongs to $BC(R, R^n)$ together with its first and second derivatives.

Theorem 6.6 is now proven.

In connection with the almost periodicity of solutions to equation (6.101), a result similar to Theorem 6.6 is valid, but some changes are necessary in the hypotheses. More precisely, the following result can be obtained.

Proposition 6.8. *Let us consider equation* (6.101), *preserving condition* (1) *from Theorem 6.6 but modifying somewhat conditions* (2) *and* (3). *Namely, it will be assumed that:*

(2′) $U : R^n \to R$ *is continuously differentiable and convex; i.e.,*

$$\langle \operatorname{grad} U(x) - \operatorname{grad} U(y), x - y \rangle \geq 0, \qquad x, y \in R^n.$$

(3′) $h \in AP(R, R^n)$.

Then, the solution whose existence is guaranteed by Theorem 6.6 is the unique almost periodic solution to equation (6.101).

Proof. Since conditions (2′) and (3′) are stronger than conditions (2) and (3) of Theorem 6.6, there results the existence of solution in $BC(R, R^n)$. We need to prove its almost periodicity and uniqueness.

Assume $x(t)$ and $y(t)$ are two solutions of equation (6.101) in $BC(R, R^n)$ of

$$\ddot{y}(t) - Ay(t) - \operatorname{grad} U(y(t)) = g(t), \qquad t \in R,$$

and consider $w(t) = x(t) - y(t)$, $t \in R$. Denote $\psi(t) = |w(t)|^2$, and observe that

$$\ddot{\psi}(t) = 2|\dot{w}(t)|^2 + 2\langle Aw(t), w(t) \rangle$$
$$+ 2\langle \operatorname{grad} U(x) - \operatorname{grad} U(y), x - y \rangle$$
$$+ \langle w(t), h(t) - g(t) \rangle, \qquad t \in R,$$

which obviously implies

$$\ddot{\psi}(t) \geq 2\sigma\psi(t) - |h(t) - g(t)|\sqrt{\psi(t)}, \qquad t \in R. \tag{6.117}$$

Since $\psi(t)$ is bounded on R, one can apply Proposition 6.6 and derive the estimate

$$|x(t) - y(t)|_{BC} \leq \frac{1}{2\sigma}|h(t) - g(t)|_{BC}. \tag{6.118}$$

From the inequality (6.118), one derives the uniqueness of the solution in $BC(R, R^n)$ as well as the Lipschitz continuity of the (unique) solutions to equation (6.101) with respect to the free forcing term h.

Now we consider together with equation (6.101) the equation whose forcing term is $h_\tau(t) = h(t + \tau)$ with $\tau \in R$ a fixed number. Applying the inequality (6.118) for $g(t) = h_\tau(t)$, we obtain

$$|x(t + \tau) - x(t)|_{BC} \leq \frac{1}{2\sigma}|h(t + \tau) - h(t)|, \qquad t \in R,$$

and this shows the almost periodicity of the solution in $BC(R, R^n)$ of equation (6.101).

Further results related to the use of the variational approach to prove the existence of an almost periodic solution in $AP(R, R^n)$ can be found in the paper [20] by C. Carminati. This paper also contains adequate references to the existing literature on this topic.

6.6 Qualitative Differential Inequalities

This section is entirely dedicated to the discussion of some nonlinear differential inequalities having applications not only in regard to almost periodicity of solutions to differential equations but also other qualitative properties of solutions. See the author's paper [27].

We encountered in Section 6.4 two examples of qualitative inequalities and then applied them to establish the almost periodicity of solutions to differential equations with monotone operators. The kind of almost periodicity we had in mind is Bohr's type; i.e., the space $AP(R, R^n)$.

In this section, we shall investigate qualitative inequalities that will allow us to use the spaces $M(R, R^n)$ and $S(R, R^n)$ instead of $BC(R, R^n)$ and $AP(R, R^n)$.

The inequalities we shall consider have the form

$$\dot{x}(t) \geq \lambda(x(t)) - f(t)\mu(x(t)), \qquad t \in R, \tag{6.119}$$

or

$$\ddot{x}(t) \geq \lambda(x(t)) - f(t)\mu(x(t)), \qquad t \in R, \tag{6.120}$$

where λ and μ are maps whose properties will be specified below. We shall look for estimates on the whole real line but not necessarily for solutions that are in $BC(R, R^n)$. The hypothesis $x \in M(R, R^n)$ will also be dealt with.

Before we can investigate the inequalities (6.119) and (6.120) in full generality, we will consider a special case where $\lambda(r) = \lambda r$, $\lambda > 0$. The following result has an auxiliary role.

Proposition 6.9. *Consider the inequalities*

$$\dot{x}(t) \geq \lambda x(t) - f(t), \qquad t \in R, \tag{6.121}$$

and

$$\ddot{x}(t) \geq \lambda x(t) - f(t), \qquad t \in R, \tag{6.122}$$

with $\lambda > 0$ and $f : R \longrightarrow R_+$ such that

$$|f|_M = \sup_{t \in R} \int_t^{t+1} f(s)ds < \infty. \tag{6.123}$$

Let $x : R \longrightarrow R$ be locally absolutely continuous, satisfying the inequality (6.121) a.e. on R. If $x \in M(R, R)$, then there exists a constant $K > 0$ such that

$$x(t) \leq K|f|_M, \qquad t \in R, \tag{6.124}$$

with $K = K(\lambda)$.

Let $x : R \to R$ be such that $\dot{x}$ is locally absolutely continuous and formula (6.122) is satisfied a.e. on R. If $x \in M(R, R)$, then there exists a positive constant $K_1 = K_1(\lambda) > 0$ such that

$$x(t) \leq K_1 |f|_M, \qquad t \in R. \tag{6.125}$$

Proof. Let us multiply both sides of formula (6.121) by $\exp(-\lambda t)$ and integrate from t to T, $T > t$. There results

$$x(t) \leq x(T) \exp\{\lambda(t - T)\} + \int_t^T f(s) \exp\{\lambda(t - s)\} ds.$$

Since $x \in M$, the integral of $x(t) \exp\{-\lambda t\}$ makes sense on any positive semi-axis. Hence, there exists a sequence of values of T, say $\{T_n; \ n \geq 1\}$, with $T_n \to \infty$ as $n \to \infty$, and letting $T = T_n$ in the inequality above, one obtains for $n \to \infty$

$$x(t) \leq \int_t^\infty f(s) \exp\{\lambda(t - s)\} ds, \qquad \forall t \in R. \tag{6.126}$$

Based on the equivalence of the norms in the space M (see, for instance, the proof of Proposition 5.14), we obtain from formula (6.126) the estimate (6.124).

In regard to the second-order inequality (6.122), we first note that it can be rewritten in the form

$$(\dot{x} + \sqrt{\lambda}\, x)^{\cdot} \geq \sqrt{\lambda}(\dot{x} + \sqrt{\lambda}\, x) - f(t), \tag{6.127}$$

which is an inequality of the form (6.121) for $u = \dot{x} + \sqrt{\lambda}\, x$. Consequently, applying the auxiliary inequality from above, one obtains

$$\dot{x}(t) + \sqrt{\lambda}\, x(t) \leq [\dot{x}(T) + \sqrt{\lambda}\, x(T)] \exp\{\sqrt{\lambda}(t - T)\} + \int_t^T f(s) \exp\{\sqrt{\lambda}(t - s)\} ds$$

or

$$\dot{x}(t) + \sqrt{\lambda}\, x(t) \leq K(\lambda) |f|_M, \qquad t \in R, \tag{6.128}$$

if we can get a sequence $\{T_n; \ n \geq 1\}$, $T_n \to \infty$ as $n \to \infty$, for which $[\dot{x}(T_n) + \sqrt{\lambda}\, x(T_n)] \exp\{-\lambda T_n\} \to 0$ as $n \to \infty$.

If one changes t into $-t$ in formula (6.128), one again obtains an inequality of the form (6.121) for $x(t)$, which can be processed as above.

In order for the proof to be complete, we will show that there is a sequence $\{T_n; \ n \geq 1\}$, $T_n \to \infty$ as $n \to \infty$, with the property

$$[\dot{x}(T_n) + \sqrt{\lambda}\, x(T_n)] \exp\{-\lambda T_n\} \longrightarrow 0 \qquad \text{as } n \to \infty.$$

Indeed, otherwise we should have either

$$\liminf_{t \to \infty} [\dot{x}(t) + \sqrt{\lambda}\, x(t)] \exp\{-\sqrt{\lambda}\, t\} = 2\varepsilon_0 > 0$$

or

$$\limsup_{t \to \infty}[\dot{x}(t) + \sqrt{\lambda}\,x(t)]\exp\{-\sqrt{\lambda}\,t\} = -2\varepsilon_0 < 0.$$

These two distinct possibilities can be justified as follows. The set of limit points of $[\dot{x}(t) + \sqrt{\lambda}\,x(t)]\exp\{-\sqrt{\lambda}\,t\}$, as $t \to \infty$, is a connected set on R (i.e., an interval, a semiaxis, or the whole R). One can easily see that this statement is true by using an argument we have given in the proof of Proposition 6.5.

Let us now focus on the first possibility considered above. There results in this case that

$$\dot{x}(t) + \sqrt{\lambda}\,x(t) \geq \varepsilon_0 \exp\{\lambda t\}, \qquad t \geq T_0 > 0.$$

This implies

$$x(t) \geq \varepsilon_0(2\sqrt{\lambda})^{-1}\exp\{\sqrt{\lambda}\,t\} - \eta, \qquad t \geq T_0,$$

for some $\eta > 0$. The inequality above is obviously in contradiction with the assumption $x \in M(R, R)$.

A similar argument is valid in the second case.

We now move to the general inequalities (6.119) and (6.120), where the constant λ is replaced by a function $\lambda(r)$, $r \geq 0$, and $f(t)$ in formula (6.121) or (6.122) is multiplied by a nonlinear term $\mu(x(t))$. The following result is valid.

Theorem 6.7. *Consider the inequalities* (6.119) *and* (6.120) *under the following assumptions on the functions* $\lambda(r)$, $\mu(r)$, *and* $f(t)$ *occurring in both inequalities:*

(1) *The map* $\lambda : R_+ \longrightarrow R_+$ *is continuous and strictly increasing,* $\lambda(0) = 0$, $\lambda(r) \longrightarrow \infty$ *as* $r \to \infty$, *and*

$$\int_{0+} [\lambda(r)]^{-1}dr = +\infty. \tag{6.129}$$

(2) *The function*

$$\frac{d}{dr}\exp\left\{\int_{r_0}^{r}[\lambda(u)]^{-1}du\right\}, \qquad r_0 > 0, \tag{6.130}$$

is nondecreasing on R_+.

(3) *The map* $\mu : R_+ \longrightarrow R_+$ *is continuous and nondecreasing such that* $\mu(r) > 0$ *for* $r > 0$.

(4) *The map* $\nu : R_+ \longrightarrow R_+$, *defined by*

$$\nu(r) = \lambda(r)/\mu(r), \qquad r > 0, \ \nu(0) = 0, \tag{6.131}$$

is strictly increasing, and $\nu(r) \to \infty$ *as* $r \to \infty$.

(5) $f : R \longrightarrow R_+$ *is such that* $f \in M(R, R)$.

Then, if $x : R \longrightarrow R_+$ *is bounded on* R, *locally absolutely continuous, and satisfies inequality (6.119) a.e. on* R, *one can find a positive number* K *such that*

$$x(t) \leq \nu^{-1}(K|f|_M), \qquad t \in R. \tag{6.132}$$

If $x : R \longrightarrow R_+$ *is a bounded solution of formula (6.120) with* $\dot{x}(t)$ *locally absolutely continuous and the function in formula (6.130) is differentiable on* R_+, *then there exists* $K_1 > 0$ *such that*

$$x(t) \leq \nu^{-1}(K_1|f|_M), \qquad t \in R. \tag{6.133}$$

Proof. The proof will be based on a linearization of the inequalities by means of certain adequate transformations of the data.

Let $x(t)$, $t \in R$, be a solution on R of the inequality (6.119) bounded there. We shall make a change of variable, $y = \rho(x)$, with $\rho : R_+ \longrightarrow R_+$ to be determined as a continuously differentiable and increasing function. We will find a "linearized" inequality for y similar to inequality (6.121). Indeed, multiplying both sides of formula (6.119) by $\rho'(x(t))$, which is nonnegative because $\rho(x)$ is increasing, one obtains the inequality

$$\rho'(x(t))\dot{x}(t) \geq \lambda(x(t))\rho'(x(t)) - f(t)\mu(x(t))\rho'(x(t)). \tag{6.134}$$

We will choose ρ such that the first-order differential equation

$$\lambda(r)\rho'(r) = \rho(r), \qquad r > 0, \tag{6.135}$$

is valid, which actually means

$$\rho(r) = \exp\left\{ \int_{r_0}^{r} [\lambda(u)]^{-1} du \right\}, \qquad r_0 > 0, \tag{6.136}$$

with $r_0 > 0$ arbitrarily chosen. Our assumption (1) allows us to define $\rho(0) = 0$. The derivative $\rho'(r)$ does exist for any $r > 0$, and it is obviously continuous and positive for $r > 0$. From assumption (2), we see that $\rho'(r)$ is a nondecreasing function. We can also show that $\rho'(r)$ is bounded in any right neighborhood of the origin $r = 0$, while the right upper Dini derivative is finite at the origin. Indeed, for a continuously differentiable $\lambda(r)$ with $\lambda'(0) = 0$, one has $0 \leq \lambda'(r) \leq 1$ for $0 \leq r \leq r_0$, $r_0 > 0$. That's how we choose r_0 in formula (6.130). From equation (6.135), we then derive, for $0 < r \leq r_0$,

$$\rho'(r) = [\lambda(r)]^{-1}\rho(r) \leq [\lambda(r)]^{-1} \exp\left\{ \int_{r_0}^{r} \lambda'(u)[\lambda(u)]^{-1} du \right\} = [\lambda(r_0)]^{-1}.$$

The inequality above shows the boundedness of $\rho'(r)$ at the right of the origin. In a similar manner, we obtain $\lambda(r) \leq r$ on $[0, r_0]1$, which leads to

$$\rho(r)/r \leq \rho(r)[\lambda(r)]^{-1} \leq [\lambda(r_0)]^{-1}. \tag{6.137}$$

From formula (6.137), we obtain that Dini's number is finite.

Returning to inequality (6.134), we can rewrite it in the form

$$\dot{y}(t) \geq y(t) - f(t)\mu(x(t))\rho'(x(t)), \qquad t \in R, \tag{6.138}$$

which implies

$$\dot{y}(t) \geq y(t) - f(t)\mu(m)\rho'(m), \qquad t \in R, \tag{6.139}$$

where

$$m = \sup x(t), \qquad t \in R. \tag{6.140}$$

But the inequality (6.139) has been dealt with in Proposition 6.9, which means that

$$\sup y(t) = \rho(m) \leq K\rho'(m)\mu(m)|f|_M, \qquad t \in R. \tag{6.141}$$

From equation (6.135), we derive from equation (6.141) that

$$\lambda(m) \leq K\mu(m)|f|_M, \tag{6.142}$$

and dividing both sides by $\mu(m)$, we obtain the inequality (6.132) after taking the inverse of $\nu(r)$.

Concerning the second-order inequality (6.122), we can proceed exactly as in formula (6.121). After multiplying both sides of formula (6.122) by $\rho'(x(t))$, we obtain by means of simple transformations the inequality

$$\frac{d}{dt}[\rho'(x(t))\dot{x}(t)] - \rho''(x(t))[\dot{x}(t)]^2$$
$$\geq \rho'(x(t))\lambda(x(t)) - f(t)\rho'(x(t))\mu(x(t)), \qquad t \in R. \tag{6.143}$$

Due to condition (2), that $\rho'(r)$ has been assumed to be nondecreasing on R_+, there results that $\rho''(r) \geq 0$. Therefore, the inequality (6.143) implies

$$\frac{d^2}{dt^2}\rho(x(t)) \geq \rho(x(t)) - f(t)\rho'(x(t))\mu(x(t)), \tag{6.144}$$

which has as a consequence the inequality

$$\ddot{y}(t) \geq y(t) - f(t)\rho'(m)\mu(m), \qquad t \in R. \tag{6.145}$$

Again, the inequality (6.145) has been discussed in Proposition 6.9, which means that

$$\sup y(t) = \rho(m) \leq K_1\rho'(m)\mu(m)|f|_M, \qquad t \in R. \tag{6.146}$$

As in the case of equation (6.141), from equation (6.146) one obtains

$$\sup x(t) \leq \nu^{-1}(K_1|f|_M), \qquad t \in R, \tag{6.147}$$

which is exactly the inequality (6.133) in Theorem 6.7.

Remark 6.16. If we have a function $\lambda(r)$ satisfying condition (1) in Theorem 6.7, except equation (6.129), we can use with the same result the function

$$\overline{\lambda}(r) = \begin{cases} \displaystyle\int_0^r \lambda(u)\exp\{-u\}du, & r \in [0,\varepsilon_0], \\ [\overline{\lambda}(\varepsilon_0)/\lambda(\varepsilon_0)]\lambda(r), & r > \varepsilon_0, \end{cases}$$

with $\varepsilon_0 > 0$ chosen arbitrarily. One has $\overline{\lambda}(r) < \lambda(r)$ for $r > 0$, $\overline{\lambda}'(0) = 0$, and this implies equation (6.129).

Remark 6.17. One can deal with inequalities of a more general form than formulas (6.119) or (6.120). For instance, we can consider inequalities of the form

$$\dot{x}(t) \geq \lambda(x(t)) - f(t)(\mu x)(t), \qquad t \in R, \tag{6.148}$$

where $\mu : BC(R,R) \longrightarrow BC(R,R)$ is an operator with the properties $\mu(x(t)) \geq 0$, and

$$\sup(\mu x)(t) \leq \widetilde{\mu}(\sup x(t)), \qquad t \in R,$$

where $\widetilde{\mu}(r)$ is a function with the properties of $\mu(r)$ in Theorem 6.8. Then one gets an estimate of the form (6.146), with $\nu(r) = \lambda(r)/\widetilde{\mu}(r)$, for $r > 0$.

A special case of the inequality (6.148) is the integro-differential inequality

$$\dot{x}(t) \geq \lambda(x(t)) - f(t)\int_{-\infty}^t k(t-s)\mu(x(s))ds, \qquad t \in R,$$

with $\lambda(r)$, $\mu(r)$, and $f(t)$ as above, while $k \in L^1(R,R)$. We leave to the reader the task of finding an estimate for the solution of the form (6.147). The second-order inequality can also be discussed.

In concluding this section, we shall consider again, under some modified conditions, the inequality (6.119), but restricting ourselves to the positive half-axis R_+. If we choose the very special case $\mu(r) \equiv 1$ and change t into $-t$, then formula (6.119) becomes

$$\dot{x}(t) + \lambda(x(t)) \leq f(t), \qquad t \in R_+. \tag{6.149}$$

The inequality (6.149) has been used to find estimates for solutions of various equations, including the differential equation in Banach spaces (see, for instance, A. Haraux [48]).

It is interesting to point out a certain relation with stability theory (of solutions of differential equations).

We shall consider the scalar differential equation attached to formula (6.149), namely

$$\dot{y}(t) + \lambda(y(t)) = f(t), \qquad t \in R_+. \tag{6.150}$$

Using the standard terminology from stability theory (see, for instance, N. N. Krasovskii [56]), we note that the solution $y = 0$ of the equation $\dot{y}(t) + \lambda(y(t)) = 0$ is *uniformly asymptotically stable*. This implies the *stability under constantly acting disturbances* from the space $M(R_+, R)$. In particular, there results that any solution of equation (6.150) with $f \in M(R_+, R)$ is bounded on R_+.

Proceeding along the same lines as in the proof of Theorem 6.7, we can obtain an upper estimate for the solution of formula (6.149) satisfying an initial condition of the form $x(0) = x_0 \in R$,

$$\sup x(t) \leq \lambda^{-1}(\lambda(x_0) + K|f|_M),\tag{6.151}$$

where K is a positive constant.

Indeed, if we multiply both sides in formula (6.149) by $\rho'(x(t))$, with $\rho(x)$ chosen as in equation (6.135), then one obtains the inequality

$$\frac{d}{dt}\rho(x(t)) + \rho(x(t)) \leq f(t)\rho'(x(t)), \qquad t \in R_+,\tag{6.152}$$

which leads by integration to

$$\rho(x(t)) \leq \rho(x(0))e^{-t} + \int_0^t e^{-(t-s)}f(s)\rho'(x(s))ds.$$

But $\rho'(x(s)) \leq \rho'(m)$, $m = \sup x(t)$, $\qquad t \in R_+$.

Hence

$$\rho(m) \leq \rho(x_0) + \rho'(m)K|f|_M.\tag{6.153}$$

The inequality (6.151) follows directly from formula (6.153), keeping in mind that $\rho(m) = \rho'(m)\lambda(m)$ and $\rho'(x_0) \leq \rho'(m)$, and formula (6.153) becomes

$$\lambda(m) \leq \lambda(x_0) + K|f|_M.\tag{6.154}$$

Since λ is invertible on R_+, formula (6.154) is equivalent to formula (6.151).

Remark 6.18. Second-order inequalities of the form

$$\ddot{x}(t) \geq \lambda(x(t)) - f(t), \qquad t \in R_+,$$

can be discussed in the manner presented above, and similar estimates can be found for solutions.

Remark 6.19. Comparing Proposition 6.9 and Theorem 6.7, we see that in the latter we have required the boundedness of the solution (its quality!). This is stronger than the requirement in Proposition 6.9 because $BC(R, R) \subset M(R, R)$.

There is an "a priori" explanation for the difference in the requirement: Since the solutions are continuously differentiable (of the same order as the inequality considered), if they are bounded or at least uniformly continuous on R, then they must be in $AP(R, R)$, not only in $M(R, R)$. This is the consequence of Bochner's result on Bohr's almost periodicity of the Stepanov almost periodic functions.

6.7 Further Results on Nonlinear Almost Periodic Oscillations; Perturbed Systems

The amount of results concerning the almost periodicity of solutions to various classes of functional equations is considerable. We mention here only a few sources in book form that contain this type of result.

To reach beyond the framework of this section, the interested reader should consult such references as J. K. Hale [47], A. Halanay [46], I. G. Malkin [66], A. M. Fink [40], M. A. Krasnoselskii, V. S. Burd, and Yu. S. Kolesov [55], and T. Yoshizawa [102]. These references display a wide array of results concerning nonlinear almost periodic oscillations as well as various methods of investigation (analytic methods, use of Fourier series, fixed-point methods, differential inequalities, Lyapunov functions, and others).

We will consider now an application of the results in Section 6.6 to the case of a perturbed monotone system of the form

$$\dot{x}(t) = f(x(t)) + g(t, x(t)), \qquad t \in R, \tag{6.155}$$

or

$$\ddot{x}(t) = f(x(t)) + g(t, x(t)), \qquad t \in R. \tag{6.156}$$

In these systems, f stands for a monotone function from R^n into itself, while g must be regarded as a perturbing term. It is possible to consider the case where f stands for an operator acting on $BC(R^n, R^n)$ as we did in Section 6.5.

Let us deal first with the system (6.155) under the following assumptions.

(1) The map $f : R^n \longrightarrow R^n$ is continuous and monotone; i.e., for any $x, y \in R^n$,

$$\langle f(x) - f(y), x - y \rangle \geq \lambda(|x - y|^2), \tag{6.157}$$

where $\lambda(r)$, $r \in R_+$, is continuous from R_+ into itself and $\lambda(0) = 0$ and increasing for $r > 0$. Moreover, $\lambda(r)$ must be such that the condition (6.129) is satisfied, while $\lambda(r)/\sqrt{r} = \nu(r)$ also verifies that $\nu(0) = 0$ (by definition) and is increasing for $r > 0$ (and hence invertible).

(2) $g : R \times R^n \to R^n$ is continuous and S–almost periodic in the first argument uniformly with respect to $x \in R^n$. Moreover, g is monotone; i.e., for any $t \in R$, $x, y \in R^n$,

$$\langle g(t, x) - g(t, y), x - y \rangle \geq 0. \tag{6.158}$$

We can now state and prove the following result.

Proposition 6.10. *Consider the system* (6.155) *under assumptions* (1) *and* (2) *above. Then any solution in* $BC(R, R^n)$ *is also in* $AP(R, R^n)$.

Proof. Let $x \in BC(R, R^n)$ be a solution of the system (6.155). Then $x(t+\tau) = x_\tau(t)$, $t \in R$, is a solution of the system

$$\dot{x}(t + \tau) = f(x(t + \tau)) + g(t + \tau, x(t + \tau)), \qquad t \in R. \tag{6.159}$$

Subtracting side by side equation (6.155) from equation (6.159) and multiplying both sides scalarly by $x(t + \tau) - x(t)$, one obtains

$$\frac{1}{2}\frac{d}{dt}|x(t+\tau) - x(t)|^2 = \langle f(x(t+\tau)) - f(x(t)), x(t+\tau) - x(t)\rangle$$
$$+ \langle g(t+\tau, x(t+\tau)) - g(t, x(t+\tau)), x(t+\tau) - x(t)\rangle$$
$$+ \langle g(t, x(t+\tau)) - g(t, x(t)), x(t+\tau) - x(t)\rangle.$$

Taking our assumptions into account, we derive from above the inequality

$$\frac{1}{2}\frac{d}{dt}|x(t+\tau) - x(t)|^2 \geq \lambda(|x(t+\tau) - x(t)|^2)$$
$$-|g(t+\tau, \xi) - g(t, \xi)| \, |x(t+\tau) - x(t)|, \tag{6.160}$$

where $\xi = x(t+\tau) \in R^n$. We note that the third term on the right-hand side of the inequality above can be neglected due to the assumption of monotonicity of $g(t, x)$ in respect to $x \in R^n$, which leads to formula (6.160).

The inequality (6.160) above is of the type (6.119), which we investigated in Section 6.6. If one denotes $|x(t + \tau) - x(t)|^2 = u(t)$, the inequality (6.160) becomes

$$\frac{1}{2}\dot{u}(t) \geq \lambda(u(t)) - |g(t+\tau, \xi) - g(t, \xi)|\sqrt{u(t)}, \tag{6.161}$$

which must be verified only for $\xi = x(t + \tau) \in R^n$, $t \in R$. Applying the estimate found for the inequality (6.119), we obtain

$$\sup_{t\in R} |u(t)| \leq \nu^{-1}\left(K \sup_{\xi\in R^n} |g(t+\tau, \xi) - g(t, \xi)|_s \right), \tag{6.162}$$

keeping in mind that the norm in $S(R, R^n)$ is the same as in $M(R, R^n)$. Of course, K is a fixed positive constant.

The inequality (6.162), taking into account that $u(t) = |x(t+\tau) - x(t)|^2$, $t \in R$, proves that $x \in AP(R, R^n)$.

Similar considerations and applications of the estimate found in the inequality (6.120) of Section 6.6 lead to the following result concerning equation (6.156).

Proposition 6.11. *Consider the system* (6.156) *under assumptions* (1) *and* (2) *of Proposition 6.10. Then any solution of this system belonging to* $BC(R, R^n)$ *is also in* $AP(R, R^n)$.

The proof is omitted, being the same as the proof of Proposition 6.10 and relying on the inequality (6.120) instead of formula (6.119).

Remark 6.20. Under the assumptions of Propositions 6.10 and 6.11, the operator appearing on the right-hand side of equations (6.155) or (6.156) is still a "monotone" operator. The case $\lambda(r) = \mu r^2$ is the classical one, also valid in the case of Hilbert spaces (instead of R^n). The major difference between the results of Section 6.4 and those of this section consists in the fact that we deal in the latter with S-almost-periodic functions. The perturbing term g can involve the so-called impulsive terms due to the fact that functions in S, which is a subspace of M, could be unbounded. Hence, renouncing the continuity of the perturbing term, we can deal with much larger classes of perturbations.

Remark 6.21. The results of Propositions 6.10 and 6.11 are of the Bohr–Neugebauer type. In other words, they assume the existence of solutions in $BC(R, R^n)$ and then prove that they are actually in $AP(R, R^n)$. We note that existence has been proven in the case of gradient type systems in Section 6.5. For more general systems, the problem of existence in nonlinear systems is still open. Results for particular types of systems are available, and the references mentioned at the beginning of this section provide useful information.

We will now consider the problem of almost periodicity for some perturbed system of nonlinear equations similar to the systems (6.155) and (6.156). The hypotheses made on the systems will differ to some extent from those already encountered above.

Without any complications with respect to the case of the space R^n, we shall establish the result in the case of functions taking values in a Hilbert space H. The particular case $H = R^n$ is immediate.

Proposition 6.12. Let H be a real Hilbert space and $A : D(A) \longrightarrow H$, $F : R \times H \longrightarrow H$ two maps verifying the following conditions:

(1) A has a type of monotonicity property expressed by the inequality

$$\langle Ax - Ay, x - y \rangle \geq \phi(|x - y|), \qquad x, y \in D(A), \tag{6.163}$$

where $\phi : R_+ \longrightarrow R_+$ is continuous and such that $\phi(0) = 0$, $\phi(r) > 0$, for $r > 0$.

(2) F is continuous and such that

$$\langle F(t, x) - F(t, y), x - y \rangle \geq -\psi(|x - y|) \tag{6.164}$$

for $t \in R$ and $x, y \in H$. The function $\psi : R_+ \longrightarrow R_+$ is supposed continuous, and $\psi(r) \geq 0$, $r \in R_+$.

(3) The map $t \longrightarrow F(t, x)$ from R into H is almost periodic in the sense of Bohr and Bochner for fixed $x \in H$ uniformly on each bounded set in H.

(4) The functions $\phi(r)$ and $\psi(r)$ are such that $\phi(r) = O(r)$ as $r \longrightarrow 0+$ and

$$\underline{\lim} \frac{\phi(r) - \psi(r)}{r} > 0, \tag{6.165}$$

while the largest root $r(\varepsilon)$ of the equation

$$\phi(r) = \psi(r) + \varepsilon r, \qquad \varepsilon > 0, \tag{6.166}$$

verifies the condition

$$\lim r(\varepsilon) = 0 \qquad as \quad \varepsilon \to 0 +.$$

Then any solution $x(t) \in \mathrm{BC}(R, H)$ of the system

$$\dot{x}(t) = Ax(t) + F(t, x(t)), \qquad t \in R, \tag{6.167}$$

or

$$\ddot{x}(t) = Ax(t) + F(t, x(t)), \qquad t \in R, \tag{6.168}$$

also belongs to $\mathrm{AP}(R, H)$.

Proof. The proof relies on the use of qualitative inequalities involving the solution under discussion and follows the same lines as the proofs of Propositions 6.10 and 6.11.

Consider first the case of equation (6.167), and let $x(t) \in \mathrm{BC}(R, H)$, $t \in R$, be a strong solution (i.e., the limit when calculating the derivative $\dot{x}(t)$ is taken in the norm of H). Denote $u(t) = |x(t+h) - x(t)|$, $t \in R$, where $h \in R$ is a fixed number. From equation (6.167) and the similar equation when t is replaced by $t + h$ one obtains by subtraction an equation from which by multiplying both sides by $u(t)$ (scalarly) one obtains the equation

$$\frac{1}{2}\frac{d}{dt}u^2(t) = \langle Ax(t + h) - Ax(t), x(t + h) - x(t) \rangle$$
$$+ \langle F(t + h, x(t + h)) - F(t, x(t)), x(t + h) - x(t) \rangle.$$

Simple transformations allow us to write

$$\frac{1}{2}\frac{d}{dt}u^2(t) = \langle Ax(t + h) - Ax(t), x(t + h) - x(t) \rangle$$
$$+ \langle F(t + h, x(t + h)) - F(t, x(t)), x(t + h) - x(t) \rangle$$
$$+ \langle F(t, x(t + h)) - F(t, x(t)), x(t + h) - x(t) \rangle,$$

which, from our assumptions, leads to the inequality

$$\frac{1}{2}\frac{d}{dt}u^2(t) \geq \phi(u) - \psi(u) - \varepsilon u, \tag{6.169}$$

where $\varepsilon = \sup\{|F(t + h, x(t + h)) - F(t, x(t + h))| : t \in R\} > 0$. The right-hand side is always finite due to our assumption (3) on the almost periodicity. Moreover, ε can be arbitrarily small, provided we choose h among the almost periods of $F(t, \cdot)$.

The inequality (6.169) can be dealt with based on Proposition 6.5 in Section 6.4. One obtains

$$|u(t)| = |x(t + h) - x(t)| < r^2(\varepsilon), \qquad t \in R, \tag{6.170}$$

which shows the almost periodicity of $x(t)$, provided we can show that equation (6.166) has a largest positive root (satisfying $r(\varepsilon) \to 0$ as $\varepsilon \to 0+$, being postulated).

Indeed, in view of the application of Proposition 6.5, we consider, in accordance with the notation used in that proposition, the function

$$\omega(r) = \phi(\sqrt{r}) - \psi(\sqrt{r}) - \varepsilon\sqrt{r} \qquad (6.171)$$

on the semiaxis $r \geq 0$. Let us denote

$$\liminf \frac{\phi(r) - \psi(r)}{r} = \ell > 0 \qquad (6.172)$$

according to our assumption (4). This means

$$\inf_{r \geq K} \frac{\phi(r) - \psi(r)}{r} > \frac{\ell}{2}, \qquad (6.173)$$

where $K > 0$ is sufficiently large. Hence, for $r > K$ we have $\phi(r) - \psi(r) > (\ell/2)r$ and

$$\phi(r) - \psi(r) - \varepsilon r > \left(\frac{\ell}{2} - \varepsilon\right) r > 0$$

if ε is chosen small enough. This shows that the set of zeros of $\phi(r) - \psi(r) - \varepsilon r$ with ε sufficiently small is bounded above. On the other hand, under our assumptions, the set of zeros is not empty on R_+. This fact follows from the inequalities

$$\phi(0) - \psi(0) - \varepsilon \cdot 0 = -\psi(0) \leq 0$$

and

$$\phi(r) - \psi(r) - \varepsilon r > \left(\frac{\ell}{2} - \varepsilon\right) r \qquad r \geq 0 \qquad \text{for } r > K$$

for sufficiently small ε.

Therefore, $r(\varepsilon) > 0$ does exist, and Proposition 6.5 leads to the inequality (6.170) due to the fact that $\omega(r) > 0$ for $\sqrt{r} > r(\varepsilon)$ or $r > r^2(\varepsilon)$.

A similar result can be proven in the case of the second-order equation (6.168). It states the following.

Proposition 6.13. *Under the same hypotheses as in Proposition 6.12, any solution $x(t) \in BC(R, H)$ of equation (6.168) also belongs to $AP(R, H)$.*

Remark 6.22. In both Propositions 6.12 and 6.13, the uniqueness of the almost periodic solution (if any!) is assured by the hypotheses (1)–(4). Indeed, if $x_1(t)$ and $x_2(t)$ are solutions to equation (6.168), then $u(t) = |x_1(t)| - x_2|(t)|$ verifies the inequality

$$\frac{1}{2} \frac{d^2 u^2}{dt^2} \geq \phi(u) - \psi(u), \qquad t \in R,$$

and, of course, the inequality

$$\frac{1}{2}\frac{d^2u^2}{dt^2} \geq \phi(u) - \psi(u) - \varepsilon u \qquad (6.174)$$

for $\varepsilon > 0$ sufficiently small. This fact implies $u(t) = |x_1(t) - x_2(t)| < r^2(\varepsilon)$, $t \in R$, which means $u(t) \equiv 0$ on R.

Remark 6.23. The problem that can be formulated here is whether or not the assumption on the almost periodicity of F can be weakened; for instance, to assume only the fact that $F(t, \cdot)$, $t \in R$, is Stepanov almost periodic. Such a hypothesis has been accepted in Propositions 6.10 and 6.11 in this section, where the monotonicity of F has also been assumed (in respect to x). But the basis of the proof was Theorem 6.7 of Section 6.6.

Remark 6.24. Another problem arising in the context of this section is whether or not the assumption $x(t) \in BC(R, H)$ can be relaxed; for instance, to assume instead $x(t) \in M(R, H)$. As we saw in Section 6.6 when we investigated the inequalities (6.121) and (6.122), such a hypothesis is acceptable, but the inequalities under consideration were linear. For instance, if we admit the linearity of the maps A and F in Proposition 6.12, we may assume that the solution we want to investigate in regard to almost periodicity belongs to the space $M(R, H)$. The conclusion would be that actually the solution is in $AP(R, H)$.

Remark 6.25. Propositions 6.12 and 6.13 can in fact be applied to infinite-dimensional spaces and instead of ordinary differential equations to deal with partial differential equations. There is no need to assume that the operator A, for instance, is a bounded operator on H. We will meet such applications later.

6.8 Oscillations in Discrete Processes

During the last few decades, particularly since the emergence of computers with high performance, the discrete processes have received a lot of attention from researchers. Basic problems such as *stability* and other *asymptotic behavior*, the presence of *oscillations* in such systems, and many other features related to the solutions of *difference equations* have been investigated successfully by many scientists. The applications of such topics to various fields have been a concern of researchers, and a rich body of literature has appeared relatively recently. See, for instance, the book by G. Ladas and I. Gyori [58] and the references therein.

This section will emphasize just a few facts from the theory of oscillations in discrete processes and does not present a systematic treatment of this topic even in regard to the almost periodic aspect.

The basic spaces we shall deal with in this section are the spaces $AP(Z, R)$, $AP(Z, C)$, or their associates, such as $AP(Z, R^n)$, $AP(Z, C^n)$. A brief presentation of these spaces has been provided at the end of Section 3.1.

It is interesting to point out the fact that a certain parallelism takes place between the theory of almost periodic oscillations in the case of continuous time and that in the case of discrete time (as occurs in discrete models).

We shall first illustrate this parallelism in the case of a Bohr–Neugebauer result for linear systems (see Section 5.4, Proposition 5.12).

Proposition 6.14. *Let us consider the linear discrete system*

$$x_{n+1} = Ax_n + b_n, \qquad n \in Z, \tag{6.175}$$

in which $x_n, b_n \in R^m$, $n \in Z$, *and* $A \in \mathcal{L}(R^m, R^m)$ *with* $m \geq 1$ *a fixed integer.*

Assume $\{b_n; \ n \in Z\}$ *is almost periodic. Then a solution* $\{x_n; \ n \in Z\}$ *of equation* (6.175) *is almost periodic iff it is bounded (i.e., it belongs to the sequence space* $b(Z, R^m)$).

Proof. The proof of Proposition 6.14 is the same as that of Proposition 5.12. First, we operate a linear transform on the space R^m, say $x = Ty$, with $\det T \neq 0$ and such that $T^{-1}AT = B$ has the upper-diagonal form

$$B = \begin{bmatrix} \lambda_1 & b_{12} & b_{13} & \cdots & b_{1m} \\ 0 & \lambda_2 & b_{23} & \cdots & b_{2m} \\ \hdotsfor{5} \\ 0 & 0 & 0 & \cdots & \lambda_m \end{bmatrix}.$$

The diagonal entries $\lambda_1, \lambda_2, \ldots, \lambda_m$ are the eigenvalues of B (and of A!), while $b_{ik} \in R$ (or C).

The equation for $y_n = T^{-1}x_n$, $n \in Z$, is, according to equation (6.175),

$$y_{n+1} = By_n + T^{-1}b_n, \qquad n \in Z. \tag{6.176}$$

The sequence (of m-vectors) $c_n = T^{-1}b_n$, $n \in Z$, is itself in $AP(Z, R^m)$. The last equation in equation (6.176) has the form

$$z_{n+1} = \lambda z_n + c_n, \qquad n \in Z, \tag{6.177}$$

which constitutes a scalar equation. As in Proposition 5.14, it suffices to provide the proof of Proposition 6.14, restricting our consideration to the scalar case (6.177). We distinguish the following three (mutually exclusive) situations:

(1) $|\lambda| < 1$;   (2) $|\lambda| > 1$;   (3) $|\lambda| = 1$.

Case 1. In equation (6.177), we substitute n by $n + p$, $p \in Z$ being a fixed integer, and then subtract equation (6.177) term by term from the equation obtained with $n + p$ instead of n. One obtains for $n \in Z$ and fixed $p \in Z$,

$$z_{n+p+1} - z_{n+1} = \lambda(z_{n+p} - z_n) + (c_{n+p} - c_n), \tag{6.178}$$

from which we derive

$$\sup_{n \in Z} |z_{n+p+1} - z_{n+1}| \leq |\lambda| \sup_{n \in Z} |z_{n+p} - z_n| + \sup_{n \in Z} |c_{n+p} - c_n|.$$

Since

$$\sup_{n \in Z} |z_{n+p+1} - z_{n+1}| = \sup_{n \in Z} |z_{n+p} - z_n|,$$

for any $p \geq 1$, there results

$$\sup_{n \in Z} |z_{n+p} - z_n| \leq (1 - |\lambda|)^{-1} \sup_{n \in Z} |c_{n+p} - c_n|. \tag{6.179}$$

But taking into account the fact that $\{c_n;\ n \in Z\} \in \mathrm{AP}(Z, R)$, one sees from formula (6.179) that the sequence $\{z_n;\ n \in Z\} \in \mathrm{AP}(Z, R)$ or $\mathrm{AP}(Z, \mathcal{C})$. Of course, taking the supremum was possible due to the assumption that $\{z_n;\ n \in Z\} \in b(Z, R)$ or $b(Z, \mathcal{C})$.

Case 2. This case can be treated in the same way as Case 1, and one obtains instead of formula (6.179) the inequality

$$\sup_{n \in Z} |z_{n+p} - z_n| \leq (|\lambda| - 1)^{-1} \sup_{n \in Z} |c_{n+p} - c_n|, \tag{6.180}$$

which leads to the same conclusion as the inequality (6.179).

Case 3. In this case, one has $\lambda = \exp(-i\alpha)$ for some $\alpha \in R$. Multiplying both sides of equation (6.177) by $\exp(i(n+1)\alpha)$, one obtains for the sequence $\zeta_n = \exp(in\alpha)z_n,\ n \in Z$, the equation

$$\zeta_{n+1} - \zeta_n = \tilde{c}_n, \qquad n \in Z, \tag{6.181}$$

where $\tilde{c}_n = \exp\{i\alpha\}c_n \exp\{in\alpha\}$, $n \in Z$. We note that $\{\zeta_n,\ n \in Z\}$ is in $B(Z, \mathcal{C})$ iff $\{z_n;\ n \in Z\}$ is in that space.

The detailed proof of almost periodicity of $\{\zeta_n;\ n \in Z\}$ under the boundedness assumption can be found in our book [23] (Theorem I.30).

Remark 6.26. In Cases 1 and 2 discussed above, one obtains the inequality

$$\sup_{n \in Z} |z_n| \leq M \sup_{n \in Z} |\tilde{c}_n|$$

with $M = (1 - |\lambda|)^{-1}$ when $|\lambda| < 1$ and $M = (|\lambda| - 1)^{-1}$ when $|\lambda| > 1$. This inequality implies

$$\sup_{n \in Z} |x_n| \leq M \sup_{n \in Z} |b_n|, \tag{6.182}$$

with the data of the system (6.175), if the eigenvalues of A with zero real part are missing.

The estimate (6.182) is useful in dealing with quasi-linear systems, namely

$$x_{n+1} = Ax_n + f(n, x_n), \tag{6.183}$$

or more general systems in which $f(n, x_n)$ is substituted, for instance, by $f(n, x_n, x_{n-1})$, or even terms such as $f(n; x)$ with $x = \{x_n; \ n \in Z\}$. Results similar to Proposition 6.1 in Section 6.1 can be obtained for the system (6.183) or its generalizations.

Before moving to other cases of almost periodicity in discrete systems, such as systems with monotonicity, it appears useful to establish some results concerning *discrete inequalities*. More precisely, we have in mind the inequalities

$$\Delta y_n \geq \omega(y_n), \qquad n \in Z, \tag{6.184}$$

and

$$\Delta^2 y_n \geq \omega(y_n), \qquad n \in Z, \tag{6.185}$$

where $\Delta y_n = y_{n+1} - y_n$ and $\Delta^2 y_n = \Delta(\Delta y_n) = y_{n+1} - 2y_n + y_{n-1}$, limiting our consideration to the scalar case $y_n \in R, \ n \in Z$, and

$$\omega : R \to R \text{ is continuous with } \omega(y) > 0 \text{ for } y > M > 0 \tag{6.186}$$

for a fixed M.

The following result is a statement analogous to Propositions 6.5 and 6.6 (in the continuous case) but concerned with the discrete time $(t = n \in Z)$.

Proposition 6.15. *Consider the inequality* (6.184) [(6.185)], *with* $\omega : R \to R$ *satisfying formula* (6.186). *If* $\{y_n; \ n \in Z\}$ *is a bounded sequence on R such that formula* (6.184) [(6.185)] *takes place, then necessarily*

$$y_n \leq M, \qquad n \in Z. \tag{6.187}$$

Proof. The proof must be divided into two parts according to whether we deal with inequality (6.184) or (6.185).

In the case of the inequality (6.184), we distinguish several subcases. First, there exists a number $n_0 \in Z$ such that

$$\sup\{y_n; \ n \in Z\} = y_{n_0}. \tag{6.188}$$

In this situation, the inequality (6.184) yields $y_{n_0+1} - y_{n_0} \geq \omega(y_{n_0})$. If we assume $y_{n_0} > M$, then $\omega(y_{n_0}) > 0$. However, the left-hand side in the inequality above, $y_{n_0+1} - y_{n_0} \geq \omega(y_{n_0})$, is negative or zero, which contradicts $\omega(y_{n_0}) > 0$. Hence, $y_{n_0} \leq M$, which proves formula (6.187). If the situation just discussed does not occur, then it may be possible that $y_n \longrightarrow \sup\{y_n; \ n \in Z\} = y_{\max}$ as $n \to \infty$ and, of course, a symmetric situation when $y_n \longrightarrow y_{\max}$ as $n \to -\infty$. If in formula (6.184) we let $n \to \pm\infty$, then we obtain $0 \geq \omega(y_{\max})$. This fact again implies $y_{\max} \leq M$; i.e., formula (6.187) is verified.

Finally, there remains the case where none of the situations above occurs but we can find a subsequence $\{y_{n_k}; \ k \geq 1\} \subset R$ such that $y_{n_k} \longrightarrow y_{\max}$ as $k \to \infty$. Since $y_{n_k+1} - y_{n_k} \geq \omega(y_{n_k})$, one obtains as $k \to \infty$

$$\limsup_{k \longrightarrow \infty} y_{n_k+1} - y_{\max} \geq \omega(y_{\max}). \tag{6.189}$$

But

$$\limsup_{k \longrightarrow \infty} y_{n_k+1} \leq y_{\max}, \tag{6.190}$$

which means that the left-hand side of formula (6.189) is ≤ 0, a fact implying $\omega(y_{\max}) \leq 0$, and also $y_{\max} \leq M$. Therefore, formula (6.187) is proven in all possible cases and also the part of Proposition 6.15 concerning the inequality (6.184).

The inequality (6.185) can be treated in a similar manner. The guidelines are as in the proof of Proposition 6.6. We leave to the reader the task of carrying out the details.

An immediate application of Proposition 6.15 is concerned with the first-order discrete systems of the form

$$\Delta x_n = f(n, x_n), \qquad n \in Z, \tag{6.191}$$

which is obviously equivalent to the system of the form $x_{n+1} = \widetilde{f}(n, x_n)$, $\widetilde{f} = x_n + f$.

The following result corresponding to Theorem 6.3, which relates to the continuous argument, can be stated.

Proposition 6.16. *Consider equation* (6.191) *under the following assumptions:*

(1) $n \longrightarrow f(n, y)$ *is for any* $y \in R^m$ *a map from* Z *into* R^m *such that* $f(n, y) \in AP(Z, R^m)$, *its almost periodicity being uniform with respect to* y *in any bounded set of* R^m.

(2) *The following monotonicity condition is verified:*

$$\langle f(n, x) - f(n, y), x - y \rangle \geq k|x - y|^2, \qquad k > 0, \tag{6.192}$$

for $n \in Z$ *and* $x, y \in R^m$.

(3) $\{x_n; \ n \in Z\} \in b(Z, R^m)$ *satisfies equation* (6.191).

Then $\{x_n; \ n \in Z\} \in AP(Z, R^m)$.

Proof. We write equation (6.191) for $n + p$ instead of n,

$$\Delta x_{n+p} = f(n + p, x_{n+p}), \tag{6.193}$$

and subtract side by side equation (6.191) from formula (6.192),

$$\Delta(x_{n+p} - x_n) = f(n + p, x_{n+p}) - f(n, x_n),$$

which can also be written as

$$\Delta(x_{n+p} - x_n) = f(n + p, x_{n+p}) - f(n, x_{n+p})$$
$$+ f(n, x_{n+p}) - f(n, x_n).$$

We scalarly multiply both sides by $x_{n+p} - x_n$, which leads to

$$\langle x_{n+p} - x_n, \Delta(x_{n+p} - x_n)\rangle = \langle x_{n+p} - x_n, f(n, x_{n+p}) - f(n, x_n)\rangle$$
$$+ \langle x_{n+p} - x_n, f(n + p, x_{n+p}) - f(n, x_{n+p})\rangle.$$

If one takes into account Cauchy's inequality $2\langle u, v\rangle \le 2|u|\,|v| \le |u|^2 + |v|^2$, and $\Delta(x_{n+p} - x_n) = (x_{n+p+1} - x_{n+1}) - (x_{n+p} - x_n)$, then we can write the inequality

$$|x_{n+p+1} - x_{n+1}|^2 - |x_{n+p} - x_n|^2 \ge 2k|x_{n+p} - x_n|^2$$
$$-2|x_{n+p} - x_n|\sup_{n \in Z}|f(n + p, x_{n+p}) - f(n, x_{n+p})|.$$

The last term of the inequality above contains the sup operation, which in this context refers to both n and $x_{n+p} \in$ the ball with center at the origin $\theta \in R^m$ containing all values of $\{x_n;\ n \in Z\}$. This supremum is finite due to our assumption (1) in the statement of Proposition 6.16.

Let us denote $z_n = |x_{n+p} - x_n|^2$, $n \in Z$, for the fixed $p \in Z$. Then, from the last inequality above, we derive

$$\Delta z_n \ge 2kz_n - 2\sqrt{z_n}\sup_{n \in Z}|f(n + p, x_{n+p}) - f(n, x_{n+p})|,$$

which has the form (6.184). Applying the estimate (6.187), one obtains

$$z_n \le k^{-2}\left(\sup_{n \in Z}|f(n + p, x_{n+p}) - f(n, x_{n+p})|\right)^2$$

or

$$|x_{n+p} - x_n| \le k^{-1}\sup_{n \in Z}|f(n + p, x_{n+p}) - f(n, x_{n+p})|. \tag{6.194}$$

The inequality (6.194) shows that any ε-almost period p of $f(n, y)$ is in fact a $k^{-1}\varepsilon$-almost period for $\{x_n;\ n \in Z\}$.

Remark 6.27. A result analogous to Proposition 6.16 can be obtained for the second-order system

$$\Delta^2 x_n = f(n, x_n), \qquad n \in Z.$$

In the remaining part of this section, we shall be concerned with the non-linear system

$$x_{n+1} = f(n, x_n), \qquad n \in Z, \tag{6.195}$$

which is, as noted above, equivalent to the system (6.191). Our purpose is to present a result about the almost periodicity of separated solutions to equation (6.195). Some facts similar to those in Section 6.2 for the continuous case of the argument will be discussed below in a rather succinct fashion.

A general assumption that will be made on the system (6.195) is

$$f(n, x) \in AP(Z, R^m), \qquad n \in Z, \tag{6.196}$$

for each $x \in R^m$, the almost periodicity of $f(n, x)$ being *uniform* with respect to x belonging on any compact subset of R^m.

The concept of a *hull* attached to the map f, say H_f, is defined as follows: $g \in H_f$ iff there exists $\{h_k \ k \geq 1\} \subset Z$ such that $f(n + h_k, x) \longrightarrow g(n, x)$ in R^m uniformly with respect to $(n, x) \in D = Z \times K$ for any compact subset $K \subset R^m$.

Besides equation (6.195), we will consider the related equations

$$x_{n+1} = g(n, x_n), \qquad n \in Z, \tag{6.197}$$

for each $g \in H_f$.

Of course, equation (6.195) is just one of the equations in the class (6.197). We also need the concept of *separated solutions* for equation (6.195) or (6.197). We shall call two solutions x and y separated in D iff their graphs belong to D and if there exists a number $\rho > 0$ with the property $|x - y| \geq \rho$; i.e., $|x_n - y_n| \geq \rho$, $n \in Z$. The number $\rho > 0$ can depend on the solutions involved. If there is only one solution with its graph in D, then we agree to include it in the category of separated solutions.

Before getting the results we are aiming at, one more notation has to be introduced. Namely, for any $x \in AP(Z, R^m)$, we denote by $f_x \in AP(R, R^m)$ the function defined by $f_x(n + u) = x_n + (x_{n+1} - x_n)u$, $n \in Z$, $u \in (0, 1]$. We have dealt with this construction in Section 3.7.

The following result provides the support necessary to establish that Amerio's theory on separated solutions (Section 6.2) in the continuous argument case holds true in a discrete setting.

Proposition 6.17. *Consider the system* (6.195) *with f satisfying* (6.196). *Then the following statements hold true:*

(1) *If $\mathcal{M}$ is a set of solutions of equation* (6.195) *with their graphs in $D = Z \times K$, $K \subset R^m$ being compact, then the set $\mathcal{M}_{a,b} = \{f_x, \ x \in \mathcal{M}\}_{a,b}$ consisting of the restrictions of the functions in $\mathcal{M}$ to the finite interval $[a, b] \subset R$ is relatively compact in $C([a, b], R^m)$.*

(2) *The number of separated solutions of equation* (6.195) *with their graphs in D is finite.*

(3) *If equation* (6.195) *has a solution $x = \{x_n; \ n \in Z\}$ with its graph in D, then each equation* (6.197) *has a solution y whose graph also belongs to Δ.*

(4) *If there exists a solution x of equation* (6.195) *such that $x_n \in K$ is compact for $n \geq n_0$, then equation* (6.197) *has a solution y such that $y_n \in K$ for $n \in Z$.*

(5) *If any system* (6.197) *has separated solutions, then we can find $\sigma > 0$ such that for each $g \in H_f$ and for each couple of solutions of equation* (6.197), *say y and z, we have $|y_n - z_n| \geq \sigma$, $n \in Z$.*

Proof. (1) Define $M = \sup\{|x|; \ x \in K\}$. Then we have $|x_n| \leq M$ and $|x_{n+1} - x_n| \leq 2M$, $n \in Z$. Also, $|f_x| \leq M$ for $t \in R$, and $|f_x(u) - f_x(v)| \leq 2M|u - v|$. Therefore, for $x \in \mathcal{M}$, the set $\{f_x\} \subset AP(Z, R^m)$ is uniformly bounded and equicontinuous on each $[a, b] \subset R$. We now apply the Ascoli–Arzelà criterion of compactness to end the proof of assertion (1) in Proposition 6.17.

(2) If the family of all separated solutions to equation (6.195) were infinite, then according to (1) we could extract a sequence that converges on any compact interval (uniformly!). The limit will also be a solution of equation (6.195), and it could not be separated. The contradiction proves that statement (2) in Proposition 6.17 is true.

(3) Let $g \in H_f$ be given. There exists a sequence $\{h_k; \ k \geq 1\} \subset Z$ such that $f(n + h_k, y) \longrightarrow g(n, y)$ in the sense made precise in the definition of the hull. We consider now the sequence of functions in $AP(R, R^m)$ generated by the terms of the sequence $\{f(n + h_k, \cdot), \ n \in Z, \ k \geq 1\}$ as described in (1) above. This sequence contains a subsequence that converges uniformly on any compact set of $Z \times K$, and the limit of this subsequence is a solution of equation (6.197).

(4) The proof of this assertion is the same as the proof of assertion (b) in Proposition 6.3, which deals with the continuous argument case ($t \in R$ instead of $n \in Z$). If $\{x_n; \ n \geq n_0\}$ is a solution of equation (6.195), then one considers the solutions $\{x_k^j = x_{k+j+n_0}, k + j \geq -n_0\}$, $j \geq 1$, and applies the compactness property on each positive semiaxis of Z. The limit of a convergent subsequence will be a solution of an equation (6.197) on Z. Then one relies on assertion (3).

(5) Let $\sigma > 0$ be the number corresponding to the separation property in equation (6.195). For two solutions z_1 and z_2 of equation (6.195), one can find a common sequence $\{h_k : k \geq 1\} \subset Z$ such that $z_1(n + h_k) \longrightarrow y_1(n)$ as $k \to \infty$ and similarly $z_2(n + h_k) \longrightarrow y_2(n)$ as $k \to \infty$. One has

$$\sigma \leq \inf_{t \in R} |f_{z_1}(t) - f_{z_2}(t)| \leq |y_1 - y_2|,$$

which implies $y_1 \neq y_2$. This means that, for any $g \in H_f$, equations (6.195) and (6.197) have the same number of solutions, while the number σ is the same for equation (6.195) and any equation in (6.197).

We can now state and prove the almost periodicity result, which is similar to the result in Theorem 6.1.

Theorem 6.8. *If all systems* (6.197) *possess separated solutions in* $D{=}Z{\times}K$ *(i.e., for each* $g \in H_f$, *the system* (6.197) *has separated solutions in* D), *then all these solutions are in* $\mathrm{AP}(Z, R^m)$. *In particular, the separated (in* D) *solutions of* (6.195) *are almost periodic.*

Proof. The proof of Theorem 6.8 goes along the same lines as that of Theorem 6.1, which treats the continuous-time situation. Actually, the proof relies on that of Theorem 6.1. The tool is the association to each solution x of equation (6.195) or (6.196), the continuous almost periodic function f_x (see the construction of f_x in terms of x above).

Let x be a solution of equation (6.195) with its graph in D and $\{h_n; n \geq 1\} \subset R$ an arbitrary sequence. It will suffice to show that, from the sequence $\{f_x(t + h_n); n \geq 1\}$, we can extract a subsequence that is convergent on $\mathrm{BC}(R, R^m)$ (i.e., uniformly on R) with respect to the norm of R^m. From Proposition 6.17, we can assume that $\{f_x(t + h_n); n \geq 1\}$ converges uniformly on each compact interval of R.

If the convergence is not uniform on the whole axis R, which we now take as a working hypothesis, then it can be shown that we reach a contradiction. Indeed, since according to Proposition 6.17 we can assume that $\{f_x(t+h_n); n \geq 1\}$ converges uniformly (as $n \to \infty$) on any compact interval of R, we will have a limit $f_z(t)$ such that for some $g \in H_f$, $z_{n+1} = g(n, z_n)$, $n \in Z$. Of course, $z_n = f_z(n)$, $n \in Z$. Again based on Proposition 6.17, we can find $\rho > 0$ such that for any couple of solutions of (6.195), say z and $\bar{z}$, one has $\inf |z_n - \bar{z}_n| \geq 2\rho$, $n \in Z$. Consider for $n < p$ the functions $\phi_{n,p}(t) = |f_x(t + h_n) - f_x(t + h_p)|$ and $I_{n,p} = \phi_{n,p}^{-1}(B_\rho)$, where B_ρ stands for the closed ball of radius ρ centered at the origin $\theta \in R^m$. It is obvious that $\phi_{n,p}(t)$ is continuous on $t \in R$, while $I_{n,p}$ is a closed (on R) set nonempty for large enough n and p (because $\theta \in I_{n,p}$). If $\delta_{n,p} = \sup\{\phi_{n,p}(t); t \in I_{n,p}\}$, then we have $\delta_{n,p} \leq \rho$, with $\limsup \delta_{n,p} \neq 0$ as $(n, p) \longrightarrow (\infty, \infty)$. Otherwise, we would have uniform convergence on R for the sequence $\{f_x(t + h_n), n \geq 1\}$. Hence, $\limsup \delta_{n,p} = 2\alpha > 0$, and therefore one can find sequences of integers $\{n_k; k \geq 1\}$, $n_k \to \infty$, and $\{p_k; k \geq 1\}$, with $p_k \to \infty$, for which $\phi_{n_k,p_k}(t_k) \geq \alpha$ and consequently $\alpha \leq |f_x(t_k + h_{n_k}) - f_x(t_k + h_{p+k})| \leq \rho$. If necessary, we can go to a subsequence such that, in R^m, $f_x(t_k + h_{n_k}) \longrightarrow U$, $f_x(t_k + h_{p_k}) \longrightarrow V$ (without changing notation), from which we derive for U and V the inequalities $\alpha \leq |U - V| \leq \rho$. Proceeding further as in the proof of Theorem 6.1 and passing to subsequences, we can assume that $f_x(t + t_k + h_{n_k}) \longrightarrow z^1(t)$ and $f_x(t + t_k + h_{p_k}) \longrightarrow z^2(t)$ as $k \to \infty$, where $z_{n+1}^i = g_i(n, z_n^i)$, $i = 1, 2$. Since $\alpha \leq |z^2(0) - z^1(0)| \leq \rho$ from above, we can easily derive $g_1 = g_2$. And taking into account that z^1 and z^2 are distinct, one must have $|z^2(n) - z^1(n)| \geq 2\rho$, $n \in Z$, which constitutes a contradiction with $\alpha \leq |z^2(0) - z^1(0)| \leq \rho$.

This ends the proof of Theorem 6.8.

Corollary 6.2. *The linear system*

$$x_{n+1} = A(n)x_n + b_n, \tag{6.198}$$

with $A(n)$ and b_n, $n \in Z$, almost periodic fits into the frame of Theorem 6.8.

In order to achieve the separation property, it suffices to assume that each system $x_{n+1} = B(n)x_n$, $n \in Z$, with $B(t) \in H_{A(t)}$, has only the zero solution to be bounded on Z. Then equation (6.198) has a unique bounded solution (i.e., separated by definition), which implies its almost periodicity.

Many other features encountered in the continuous case can be adapted to the discrete case, as illustrated in this section. More recent results can be found in D. Pennequin [79].

6.9 Oscillations in Systems Governed by Integral Equations

Most of the results in Chapter 6 have been obtained under the main hypothesis that the system is governed by ordinary differential equations (or functional differential equations involving operators acting on the space $AP(R, R^m)$).

In this last section of Chapter 6, dedicated to nonlinear oscillations, we will focus our attention on systems that are governed by integral equations or related functional equations. Let us point out the fact that the existing mathematical/engineering literature, while containing significant results, is by far not as rich as in the case of systems governed by differential equations.

We shall start with a relatively simple approach aimed at oscillating systems governed by integral equations of the form

$$x(t) = h(t) + \int_R k(t - u)(fx)(u)du \tag{6.199}$$

or

$$x(t) = (hx)(t) + \int_R k(t - u)x(u)du. \tag{6.200}$$

We shall explain below the meaning of different functions or operators involved in equations (6.199) and (6.200). In order to simplify the discussion somewhat, we shall restrict ourselves to the scalar case. In other words, each function will be in a space $E(R, C)$ or $E(R, R)$, the operators f and h appearing in the equations acting between such function spaces.

We shall take as the underlying space for our discussions the space $M(R, C)$, defined in Section 6.2. This space will be the richest considered in this section.

The subspace of $M(R, C)$ consisting of Stepanov's almost periodic functions, denoted by $S(R, C)$, was introduced in Section 3.3.

The space $AP(R, C)$ consisting of Bohr's almost periodic functions is part of $S(R, C)$.

Finally, any space of integrable periodic functions of arbitrary period $\omega > 0$, say $P_\omega(R,\mathcal{C})$, is also part of $S(R,\mathcal{C})$, the induced norm being $|x|_\omega = \int_0^\omega |x(s)|ds$. Of course, it makes sense to deal with continuous periodic functions, the meaning of the notation $P_\omega C(R,\mathcal{C})$ being obvious. The norm induced by $AP(R,\mathcal{C})$ is the supremum norm. We are not going to deal with the continuous case, but we note the fact that the inclusion $P_\omega C(R,\mathcal{C}) \subset P(R,\mathcal{C})$ allows us to obtain results similar to the case $P_\omega(R,\mathcal{C}) \subset S(R,\mathcal{C})$, which will be discussed below.

A basic assumption we shall accept in what follows is

(a) $k \in L^1(R,\mathcal{C})$.

The assumption (a), together with its consequences

$$|k * x|_M \leq |k|_{L^1}|x|_M, \qquad x \in M(R,\mathcal{C}), \tag{6.201}$$

$$|k * x|_\omega \leq |k|_{L^1}|x|_\omega, \qquad x \in P_\omega(R,\mathcal{C}), \tag{6.202}$$

leads to the conclusion that $S(R,\mathcal{C})$ and $P_\omega(R,\mathcal{C})$ are invariant subspaces of the convolution linear operator

$$(Kx)(t) = (k * x)(t) = \int_R k(t-s)x(s)ds, \tag{6.203}$$

which is defined and continuous on M (see formula (6.201)).

Moreover, it is easy to prove that the operator K is compact on $P_\omega(R,\mathcal{C})$. One has to apply the inequality (6.202) to the difference $(k*x)(t) - (k*x)(t')$, $t, t' \in R$, to then use the compactness criteria.

Let us consider briefly the linear equation associated to equation (6.199) or (6.200), namely

$$x(t) = h(t) + \int_R k(t-u)x(u)du, \qquad t \in R, \tag{6.204}$$

under assumption (a) for the kernel k. The Fourier transform of $k(t)$ is

$$\widetilde{k}(i\lambda) = \int_R k(t)\exp(-i\lambda t)dt, \qquad \lambda \in R. \tag{6.205}$$

If condition (b),

$$\text{(b)} \qquad \widetilde{k}(i\lambda) \neq 1, \qquad \lambda \in R, \tag{6.206}$$

is satisfied, then $k(t)$ admits a resolvent kernel $k_1(t)$, which is uniquely determined by each of the equations

$$k_1(t) = k(t) + \int_R k(u)k_1(t-u)du \tag{6.207}$$

or

$$\widetilde{k}_1(i\lambda) = \widetilde{k}(i\lambda)[1 - \widetilde{k}(i\lambda)]^{-1}, \qquad \lambda \in R. \tag{6.208}$$

Under assumptions (a) and (b), equation (6.204) has the unique solution $x \in M$ (or S), for each $h \in M(R,\mathcal{C})$ (or $h \in S(R,\mathcal{C})$), given by

$$x(t) = h(t) + \int_R k_1(t-s)h(s)ds, \qquad t \in R. \tag{6.209}$$

Obviously, if $h \in S(R,\mathcal{C})$, then $x \in S(R,\mathcal{C})$. Similarly, when $h \in P_\omega(R,\mathcal{C})$, one has $x \in P_\omega(R,\mathcal{C})$.

For some details about the considerations above, see, for instance, C. Corduneanu [25].

We shall consider next the nonlinear equation (6.199) under the following assumptions guaranteeing the existence and uniqueness of an almost periodic solution.

Theorem 6.9. *Assume that hypothesis* (a) *holds for equation* (6.199). *Moreover, let* $f : S(R,\mathcal{C}) \longrightarrow S(R,\mathcal{C})$ *be Lipschitz continuous; i.e., there exists* $\gamma > 0$ *such that*

$$|fx - fy|_S \leq \gamma|x - y|_S, \qquad x, y \in S. \tag{6.210}$$

Then, for each $h \in S(R,\mathcal{C})$, *equation* (6.199) *has a unique solution* $x \in S(R,\mathcal{C})$, *provided*

$$\gamma|x|_{L^1} < 1. \tag{6.211}$$

The proof is rather standard and consists in the application of the Banach contraction mapping theorem to the operator

$$(Tx)(t) = h(t) + \int_R k(u)(fx)(t-u)du,$$

which acts on the space $S(R,\mathcal{C})$. We omit the details.

Remark 6.28. If we modify somewhat the hypothesis on the operator f, asking only $f : \Sigma_\rho \longrightarrow \Sigma_\rho$, where $\Sigma_\rho = \{x; |x|_s \leq \rho\} \subset S(R,\mathcal{C})$ and formula (6.210), then the unique solution of equation (6.199) in Σ_ρ is assured by the (extra) condition

$$|h|_S + |k|_{L^1}|f\theta|_S \leq \rho(1 - \gamma|k|_{L^1}).$$

The contraction mapping principle can be applied in Σ_ρ.

Remark 6.29. A result similar to that in Theorem 6.9 can be obtained substituting to $S(R,\mathcal{C})$ the space $P_\omega(R,\mathcal{C})$, and again using the contraction mapping principle.

Since the operator K, defined by equation (6.203), is compact in $P_\omega(R,\mathcal{C})$, the Schauder fixed-point theorem can also be applied. In this case, only existence is obtained. One easily obtains, again using the operator T defined above, the following result in $P_\omega(R,\mathcal{C})$:

Equation (6.199) has at least one solution in $P_\omega(R,\mathcal{C})$ if the operator $x \to fx$ satisfies the growth condition defined as follows: Denoting $\alpha(r) = \sup\{|f|_\omega; \ |x|_\omega \le r\}$ for $r > 0$, then $\limsup_{r\to\infty}(\alpha(r)/r) < (|k|_{L^1})^{-1}$.

The Schauder theorem is applied in $P_\omega(R,\mathcal{C})$, choosing as the convex closed set a ball Σ_ρ with ρ satisfying

$$|h|_\omega + |k|_{L^1}\alpha(\rho) \le \rho.$$

We now consider equation (6.200) in view of obtaining an existence result in $S(R,\mathcal{C})$ or $P_\omega(R,\mathcal{C})$. Assumptions (a) and (b) will be made as formulated above in this section; moreover, let $x \to hx$ be Lipschitz continuous on $S(R,\mathcal{C})$ (or on $P_\omega(R,\mathcal{C})$),

$$|hx - hy|_S \le \gamma|x - y|_S, \qquad x,y \in S(R,\mathcal{C}), \tag{6.212}$$

or

$$|hx - hy|_\omega \le \gamma|x - y|_\omega, \qquad x,y \in P_\omega(R,\mathcal{C}). \tag{6.213}$$

Finally, assume that

$$\gamma < (1 + |k_1|_{L^1})^{-1}, \qquad |h\theta|_S \le [(1 + |k_1|)_{L^1}^{-1} - \gamma]\rho, \tag{6.214}$$

where k_1 is the resolvent kernel associated to k. Of course, the second inequality in formula (6.214) must be valid with $|h\theta|_\omega$ instead of $|h\theta|_S$ if we are looking for periodic solutions.

Theorem 6.10. *Consider equation* (6.200) *under assumptions* (a) *and* (b) *and formulas* (6.212) [(6.213)] *and* (6.214) *(its analogue with P_ω instead of S). Then there exists a unique solution of equation* (6.200) *in $\Sigma_\rho \subset S$ [$\Sigma_\rho \subset P_\omega$].*

Proof. Using the resolvent operator associated to k, denoted by k_1, one finds out easily that equation (6.200) is equivalent to

$$x(t) = (hx)(t) + \int_R k_1(u)(hx)(t - u)du. \tag{6.215}$$

We shall now prove that the operator

$$(Tx)(t) = (hx)(t) + \int_R k_1(u)(hx)(t - u)du, \qquad t \in R,$$

is a contraction mapping on $S(R,\mathcal{C})$. Indeed, we obtain, taking also $Ty, y \in S(R,\mathcal{C})$,

$$Tx - Ty = hx - hy + \int_R k_1(u)[hx - hy](t - u)du,$$

which leads to

$$|Tx - Ty|_S \leq (1 + |k_1|_{L^1})|hx - hy|_S \leq \gamma(1 + |k_1|_{L^1})|x - y|_S,$$

showing from (6.214) that T is a contraction mapping on S. The second condition (6.214) is used only if we want the solutions to be in $\Sigma_\rho \subset S$.

Similar considerations are necessary when the solution to equation (6.200) is sought in $P_\omega(r, \mathcal{C})$.

Remark 6.30. Since T in equation (6.215) is completely continuous on $P_\omega(R, \mathcal{C})$, one can also obtain an existence result using Schauder's theorem.

It is worth making a connection between the proven results above for integral equations and some similar results for functional differential equations of the form

$$\dot{x}(t) = Ax(t) + (fx)(t), \qquad t \in R, \tag{6.216}$$

where $A \in \mathcal{L}(R^m, R^m)$ is a constant matrix and $f : S(R, R^m) \longrightarrow S(R, R^m)$ is an operator satisfying conditions to be made precise below. Systems more general than equation (6.216) could be dealt with, but their involvement requires lengthy preparation. For this reason, we confine our discussion to the systems of the form (6.216).

Let us recall the fact that under the main hypothesis we shall accept, namely

$$\det(\lambda I - A) = 0 \Longrightarrow \operatorname{Re} \lambda \neq 0, \tag{6.217}$$

the fundamental matrix $X(t) = \exp(At)$, $t \in R$, of the system $\dot{x} = Ax$ has a decomposition $X(t) = X_-(t) + X_+(t)$, as described in Remark 5.8 to Proposition 5.13 (see also Proposition 5.27 covering the case of time-varying systems in the case of a dichotomy of solutions).

For the linear system associated with equation (6.216), when $(fx)(t) = f(t) \in S(R, R^m)$, the unique almost periodic solution in $AP(R, R^m)$ is given by

$$x(t) = \int_{-\infty}^{t} X_-(t - s)f(s)ds - \int_{t}^{\infty} X_+(t - s)f(s)ds, \tag{6.218}$$

which can also be written as

$$x(t) = \int_{R} K(t - s)f(s)ds, \qquad t \in R, \tag{6.219}$$

where

$$|K(t)| \leq K_0 \exp\{-\alpha|t|\}, \qquad t \in R, \tag{6.220}$$

for some K_0, $\alpha > 0$.

From the representation of the almost periodic solution above given by equation (6.218) or (6.219), one easily sees that equation (6.216) is equivalent, as far as bounded solutions are concerned, with the nonlinear integral equation

$$x(t) = \int_R K(t-s)(fx)(s)ds, \qquad t \in R. \tag{6.221}$$

This equation obviously has the form (6.199). If we take into account $fx = [fx - f\theta] + f\theta$, $\theta \in R^m$ being the null element, and we substitute fx in equation (6.221), then what we obtain has exactly the form (6.199).

It is very important to observe that the estimate (6.220) implies hypothesis (a) on the kernel; i.e., the kernel belongs to $L^1(R, \mathcal{L}(R^m, R^m))$. Hence, any result concerning the existence of almost periodic solutions to equation (6.199), even if they are in $S(R, R^m)$, according to estimate (6.220) will automatically belong to $AP(R, R^m)$.

We shall consider now the scalar integral equation

$$y(t) + \int_0^t k(t-s)\varphi(y(s))ds = u(t), \qquad t \in R_+, \tag{6.222}$$

as well as its "limit equation"

$$x(t) + \int_{-\infty}^t k(t-s)\varphi(x(s))ds = g(t), \qquad t \in R, \tag{6.223}$$

with the scope of providing conditions assuring the existence of an almost periodic solution (Bohr) to equation (6.223) and an asymptotically almost periodic solution to equation (6.222).

Generally speaking, an equation of the form (6.222) is deprived of almost periodic solutions, even though $u(t) \in AP(R, R)$. Fortunately, this is not the case with the so-called "limit equation" (6.223), which admits solutions in $AP(R, R)$.

The result we shall provide below is due to I. W. Sandberg and Gideon J. J. van Zyl [85], and due to the fact that we did not dwell in detail on frequency-domain methods, the (relatively lengthy) proof will be omitted.

The following assumptions will be made on the data occurring in equations (6.222) and (6.223):

(1) $u(t) = u_1(t) + u_2(t)$, with $u_1(t)$ being the restriction to R_+ of a function in $AP(R, R)$, while

$$u_2(t) \in L^\infty(R_+, R) \qquad \text{with } \lim_{t \to \infty} u_2(t) = 0.$$

(2) $k(t) \in L^1_{loc}(R_+, R)$ and

$$(1 + t^2)k(t) \in L^1(R_+, R) \cap L^2(R_+, R).$$

(3) $\varphi : R \longrightarrow R$ is such that $\varphi(0) = 0$ and

$$\alpha \le \frac{\varphi(b) - \varphi(a)}{b - a} \le \beta, \qquad 0 < \alpha < \beta,$$

for $a, b \in R$, $a \ne b$.

(4) The Fourier transform of the kernel $k(t)$, $\widetilde{k}(i\lambda)$, $\lambda \in R$, given by equation (6.205), satisfies the "circle criterion," which means that its locus $\{\widetilde{k}(i\lambda);\ \lambda \in R\}$ in the complex plane is avoiding the disk centered at the point $\left(-\frac{1}{2}\left(\alpha^{-1}+\beta^{-1}\right),0\right)$ with radius $\frac{1}{2}\left(\alpha^{-1}-\beta^{-1}\right)$.

Remark 6.31. Under the preceding conditions (1), (2), and (3), the integral equation (6.222) has a unique solution on R_+ belonging to $L^\infty_{\text{loc}}(R_+, R)$.

This result is easily obtainable if we notice that condition (3) implies a Lipschitz condition. The successive approximations converge in $L^\infty_{\text{loc}}(R_+, R)$.

The following result relates the solutions of equations (6.222) and (6.223) under the extra assumption $g(t) \in AP(R, R)$.

Theorem 6.11. *Consider equations* (6.223) *under conditions* (1)–(4) *and also*

$$g(t) \in AP(R, R). \tag{6.224}$$

Then there is a unique solution $x(t) \in AP(R, R)$ *of equation* (6.223). *Moreover, with* $u_1(t)$ *being the restriction of* $g(t)$ *to* R_+ *and* $y(t)$ *the solution of equation* (6.222) *on* R_+, *one has*

$$\lim_{t \to \infty} |y(t) - x(t)| = 0. \tag{6.225}$$

Since we are not providing the proof of Theorem 6.11, we limit ourselves to noting that the property (6.225) expresses the fact that the solution $y(t)$ of equation (6.222) is asymptotically almost periodic on R_+.

Our last result in this section is due to G. Lellouche [61] and is concerned again with equation (6.222). A sort of oscillation is emphasized for the solution but not almost periodicity. The almost periodicity is certainly an oscillatory property, but it also has a certain regularity property. The meaning of this statement should be understood as follows: The zeros of the solution are "regularly" distributed on the real axis due to the fact that for each $\varepsilon > 0$ there exist infinitely many ε-almost periods of the solution.

We shall rewrite equation (6.222) in the form

$$\sigma(t) = h(t) + \int_0^t k(t-s)\varphi(\sigma(s))ds, \qquad t \in R_+. \tag{6.226}$$

Theorem 6.12. *The following assumptions are accepted for the data involved in equation* (6.226):

(1) $\dot{h}(t), \ddot{h}(t) \in L^1(R_+, R)$, *while* $\int_0^t h(s)ds \in BC(R_+, R)$
(2) $k(t) = k_0(t) - \rho$, $\rho > 0$, $k_0(t), \dot{k}_0(t) \in L^1(R_+, R)$, *and*

$$k_1(t) = -\int_t^\infty k_0(s)ds \in L^1(R_+, R).$$

(3) $\varphi : R \longrightarrow R$ is continuous, and

$$\sigma\varphi(\sigma) > 0 \qquad for \ \sigma \neq 0.$$

(4) There two numbers $q_1, q_2 \in R_+$ such that

$$\mathrm{Re} \left\{ [1 + isq_1 + (is)^{-1}q_2]\widetilde{k}_0(s) - q_1\rho \right\} \leq 0, \qquad s \in R - \{0\},$$

with q_2 verifying the inequality

$$\rho - q_2\widetilde{k}_0(0) > 0.$$

Then all solutions (if any) of equation (6.226) are defined on R_+ and possess the following property: Either

$$\lim_{t \to \infty} \sigma(t) = 0 \qquad\qquad (6.227)$$

or $\sigma(t) \pm \eta$, with $\eta > 0$ arbitrarily small, has infinitely many zeros on any half-axis $[T, \infty)$.

The proof is completely provided in the quoted paper by Lellouche as well as in the book [24] by C. Corduneanu.

Of course, it is interesting to know when one or another case of the alternative is present. For instance, if one takes $q_2 = 0$ in condition (4) above, then equation (6.227) takes place and we have asymptotic stability. See C. Corduneanu [24] for the proof, which is basically due to V. M. Popov. The second case of the alternative can therefore occur only when assumption (4) is verified for $q_2 \neq 0$.

Apparently, it is still an open problem to produce examples in which the oscillatory property is taking place.

It is also worth mentioning, according to Lellouche, that the boundedness of $\sigma(t)$ does not follow from the assumptions of Theorem 6.12, even in the case of linear φ. Hence, oscillations with arbitrarily large amplitudes may occur.

7

Almost Periodic Waves

7.1 Some General Remarks

In the introduction, we presented some elementary considerations related to the classical wave equation

$$u_{tt} = a^2 u_{xx} + f(t, x), \tag{E}$$

where $u = u(t, x)$ stands for the state of the dynamical system whose motion it governs at the moment t and spatial position (one-dimensional) x. For instance, equation (E) can be viewed as describing the motion of a vibrating string, in which case $u(t, x)$ represents the elongation of the point of abscissa x at time t. By $f(t, x)$ one designates the external forces acting on the string during the motion. This example pertains to the category "waves in solids."

A more general equation than equation (E), with several space variables, is the equation

$$u_{tt} = a^2 \Delta u + f(t, x), \tag{E_1}$$

where $x = \mathrm{col}(x_1, x_2, \ldots, x_m)$, $u = u(t, x) = u(t, x_1, x_2, \ldots, x_m)$, and

$$\Delta u = \mathrm{div}(\mathrm{grad}\, u) = u_{x_1 x_1} + u_{x_2 x_2} + \cdots + u_{x_m x_m}$$

(i.e., the Laplacian of u with respect to the variables $x_1, x_2, \ldots, x_m$). The case $m = 3$ is particularly interesting in the physical sciences. It is well known that equations of the form (E_1) appear in continuum mechanics (fluids, elastic solids), electromagnetism, and quantum mechanics, in which fields the notion of a *wave* has significance.

While the concept of a wave is in most circumstances related to a wave equation like equation (E_1), and rather often to the homogeneous equation corresponding to equation (E_1),

$$u_{tt} = a^2 \Delta u, \tag{E_0}$$

one must notice that other types of partial differential equations admit solutions whose physical interpretation is clearly conducive to the concept of

C. Corduneanu, *Almost Periodic Oscillations and Waves*,
DOI 10.1007/978-0-387-09819-7_7, © Springer Science+Business Media, LLC 2009

wave motion. We can mention in the first place the so-called *diffusion equation* encountered in problems concerning heat propagation or diffusion processes in chemistry. This equation has the form

$$u_t = a^2 \Delta u + f(t, x), \tag{E_2}$$

in which the notations have the same significance as above. The unknown $u = u(t, x)$ may be interpreted, for instance, as the temperature in the domain under investigation at the point $x = \mathrm{col}(x_1, x_2, x_3)$ and the moment t. Various interpretations are possible depending on the physical process described by equation (E_2).

Finally, let us mention the fact that nonlinear partial differential equations have appeared in the investigations of various phenomena. One such equation has the form

$$u_t = k(u) \Delta u + f(t, x, u) \tag{E_3}$$

and is encountered in heat conduction processes under special circumstances.

It is a well-known feature of the theory of partial differential equations that a unique solution is determined by prescribing additional conditions, usually initial conditions or boundary value conditions. Such conditions will appear in the subsequent discussion.

Let us now return to the simplest wave equation of the form (E_0) in one space dimension,

$$u_{tt} = a^2 u_{xx}, \qquad t \ge 0, \ 0 \le x \le \ell, \tag{7.1}$$

with the initial conditions

$$u(0, x) = u_0(x), \quad u_t(0, x) = u_1(x), \tag{7.2}$$

on the interval $0 \le x \le \ell$. These are nothing but equation (1.12) and conditions (1.13) from the introduction of this book. The boundary value conditions to be considered are conditions (1.39) from the introduction, namely

$$u(t, 0) = 0, \quad u_x(t, \ell) + h u(t, \ell) = 0, \tag{7.3}$$

where $h > 0$ and $t > 0$.

As seen in the introduction, the solution of equation (7.1) under conditions (7.2) and (7.3) can be represented as a series of the form

$$u(t, x) \sim \sum_{m=1}^{\infty} (A_m \cos \mu_m a t + B_m \sin \mu_m a t) \sin \mu_m x, \tag{7.4}$$

where A_m, B_m, $m \ge 1$, can be determined in terms of the initial conditions (7.2). The numbers μ_m, $m \ge 1$, are the solutions of the numerical equation $\mu h^{-1} = \tan h(\mu \ell)$. There are infinitely many such μ_m, and their asymptotics are given by

$$\mu_m \simeq \left(m + \frac{1}{2}\right) \pi.\ell, \qquad m \to \infty.$$

The numbers $m \geq 1$ serve to determine the frequencies of harmonic oscillations occurring in formula (7.4), namely $2\pi/a\mu_m$, $m \geq 1$, which allows us to conclude that the series (7.4) representing the solution of our problem (7.1), (7.2), (7.3), is defining an almost periodic function in t with values in a function space on $[0, \ell]$, and this function is describing a wave-like process with infinitely many frequencies involved. Depending on the kind of convergence we demand for the series in (7.4), we can obtain classical solutions (i.e., $u(t, x)$ has the derivatives appearing in equation (7.1), and the series can be differentiated term by term) or generalized solutions (as limits in various senses) of classical solutions

$$u_m(t, x) = (A_m \cos \mu_m at + B_m \sin \mu_m at) \sin \mu_m x, \qquad m \geq 1.$$

In any case, one obtains an almost periodic solution that should be regarded as an application from R into the function space $L^2([0, \ell], R)$ or another function space on $[0, \ell]$.

This simple example shows that the almost periodic waves do occur prevalently when considering phenomena described by the wave equation. The periodic case may also occur, for instance when instead of conditions (7.3) we deal with the fixed ends of the string, $u(0, t) = u(\ell, t) = 0$, $t > 0$. In this case, one obtains, instead of the numbers μ_m as defined above, the numbers $m\pi a/\ell$, $m \geq 1$ (see the introduction).

The corresponding frequencies have rational ratios, which characterizes periodic phenomena.

The considerations above regarding the solutions of the wave equation likely led S. Sobolev to publish a series of papers in the *Comptes Rendus de l'Académie des Sciences de l'URSS* (1945) under the title "Sur la presque périodicité des solutions de l'équation des ondes." These contributions, based on the refined use of methods of functional analysis, inaugurated the research work on this topic using modern tools. Embedding theorems for function spaces emphasized the role of compactness in dealing with almost periodicity of solutions to a wave equation. These developments require "hard core" analysis, and we shall only sketch them in what follows.

One more elementary observation we shall make is concerned with the one-dimensional wave equation (E_0)

$$u_{tt} = a^2 u_{xx} \tag{7.5}$$

under initial conditions

$$u(0, x) = \varphi(x), \; u_t(0, x) = \psi(x), \qquad x \in R. \tag{7.6}$$

In terms of the theory of vibrating strings, conditions (7.6) suggest that we are dealing with an "infinite" string.

By means of the change of variables

$$\xi = x + at, \quad \eta = x - at, \tag{7.7}$$

equation (7.5) becomes $u_{\xi\eta} = 0$, which has the general solution

$$u(\xi, \eta) = u_0(\xi) + u_1(\eta), \tag{7.8}$$

where u_0 and u_1 are arbitrary functions of class $C^{(2)}$ on R.

Imposing on

$$u(t, x) = u_0(x + at) + u_1(x - at) \tag{7.9}$$

the initial conditions (7.6), one obtains the well-known formula

$$u(t, x) = \frac{\varphi(x + at) + \varphi(x - at)}{2} + \frac{1}{2a} \int_{x-at}^{x+at} \psi(\tau)d\tau. \tag{7.10}$$

Formula (7.10) allows us to state an elementary result concerning the almost periodicity of solutions to equation (7.5). Namely, one can infer based on formula (7.10) that if the initial data (7.6) are almost periodic functions (Bohr) with ψ such that its primitive is also almost periodic, then the solution of the problem (7.5), (7.6) is also almost periodic (in t and x separately).

Indeed, based on the elementary properties that multiplication or translation of the argument by a constant preserves almost periodicity, there follows from equation (7.10) that both its terms are almost periodic in t (as well as in x). For the second term it is worth observing that, by assumption, $\Psi(\tau) = \int_0^\tau \psi(s)$ is almost periodic, and it can be represented as $(2a)^{-1}[\Psi(x+at) + \Psi(x-at)]$. One may say that all solutions of equation (7.5) are almost periodic if this property holds for the initial functions (determining the position of the string as well as the velocity of each point of the string at the initial moment). Of course, one must choose the physical interpretation of equation (7.5) as describing the vibrations of an elastic string — of large length! — under the action of initial disturbances only.

Equation (7.5) describes the "plane" waves, which means that the front of the wave is a plane surface. In the case of a two-dimensional wave equation, which means we have $\Delta u = u_{x_1x_1} + u_{x_2x_2}$, the wave front has cylindrical shape. The three-dimensional case leads to the wave fronts of the most general nature (i.e., arbitrary surfaces in space).

We have discussed so far the cases of hyperbolic and parabolic equations for which it is well known that they possess solutions having wave-like behavior. These types of partial differential equations are usually investigated in regard to the existence of solutions possessing physical interpretations supporting the terminology for wave motion.

On the other hand, the presence of almost periodic solutions to a partial differential equation describing the dynamics of a physical process, regardless of its particular type, can be naturally interpreted as representing wave-like motion of the system or process. We shall deal below with cases that pertain to this situation.

The literature on almost periodic functions provides many examples of equations with partial derivatives of elliptic type that possess such solutions. Since these equations do not describe dynamical processes (one may say that they describe stationary phenomena), the property of almost periodicity is encountered with respect to one of the space variables.

We will mention here a classical example concerning harmonic functions in two variables. For the details, we send the reader to a book by C. Corduneanu [23] where a complete solution is presented.

The problem we want to discuss can be formulated as follows. Find a harmonic function $u(x, y)$ defined in the half-plane $x > 0$ that is continuous in $x \geq 0$ such that $u(0, y) = f(y)$, $y \in R$, with f an almost periodic function (Bohr).

It is relatively easy to prove that the (unique) answer is given by

$$u(x, y) = \frac{1}{\pi} \int_R \frac{x\, f(s)ds}{x^2 + (y - s)^2}, \qquad x > 0.$$

The solution $u(x, y)$ is almost periodic in y uniformly with respect to $x > 0$.

More complex results are related to the equation $u_{tt} + \Delta u = 0$ with Δu the Laplacian in $n \geq 1$ variables (see C. Corduneanu [29]). This type of problem will also be discussed in this chapter even though our attention will be focused on dynamical equations (in which t will mean time).

In concluding this section, we want to stress again the fact that any time we can prove the existence of almost periodic solutions to a dynamical e-quation, the almost periodicity being valid for the time variable, the process described by such equations can be naturally related to the concept of wave motion.

7.2 Periodic Solitary Waves in One-Dimensional Hyperbolic and Parabolic Structures

This section is dedicated to the discussion of the existence of periodic waves of a certain type in the case of phenomena whose dynamics are described by parabolic or hyperbolic equations. In this approach, we will have the occasion to apply some classical results from the theory of nonlinear oscillations as well as emphasize the concept of solitary (or traveling) waves. This kind of wave represents an important chapter in the theory of wave propagation and its applications in several fields.

We shall provide first the basic result concerning the ordinary differential equation of second order known as the Liénard equation. It is written as

$$y'' + f(y)y' + g(y) = 0, \tag{7.11}$$

where $y' = dy/dx$ with $f(y)$ and $g(y)$ two given functions that are generally nonlinear. The main problem we shall discuss and apply in this section is the existence of a *periodic* solution of Liénard's equation.

The following existence result is a classical one, and it can be found with full proof in the book by F. Brauer and J. A. Nohel [17] or in the book by Coddington and Levinson [22].

Theorem 7.1. *Assume the following conditions are verified by equation* (7.11):

(1) $f(y)$ *and* $g(y)$ *are continuous and differentiable functions from* R *to* R.
(2) $yg(y) > 0$ *for* $y \neq 0$;
(3) *If we denote* $F(y) = \int_0^y f(u)du$, $y \in R$, *then*

$$\lim |F(y)| = +\infty \quad as \quad |y| \to \infty.$$

(4) *There exist real numbers* α *and* β *such that*

$$F(y) < 0 \text{ for } y < -\alpha \quad or \quad 0 < y < \beta$$

and

$$F(y) > 0 \text{ for } -\alpha < y < 0 \quad or \quad y > \beta.$$

Then, there exists a nontrivial periodic solution to equation (7.11).

Remark 7.1. Conditions (1)–(4) do not guarantee the uniqueness of the periodic solution to equation (7.11). If one admits the extra condition that $f(y)$ is symmetric, while $g(y)$ is antisymmetric, then uniqueness holds.

We shall apply Theorem 7.1 in what follows in order to derive the existence of periodic waves in systems described by the nonlinear hyperbolic equation

$$-u_{tt} + u_{xx} + f_0(u)u_x + g_0(u) = 0 \tag{7.12}$$

or by the similar parabolic equation

$$-u_t + u_{xx} + f_0(u)u_x + g_0(u) = 0 \tag{7.13}$$

under assumptions to be specified below for each type of equation.

Let us now consider equation (7.12), and try to find special solutions of the form

$$u = y(x + ct), \qquad c \in R. \tag{7.14}$$

Equation (7.14) leads to the ordinary differential equation

$$(1 - c^2)y'' + f_0(y)y' + g_0(y) = 0, \tag{7.15}$$

which is obviously of the Liénard type.

We shall consider only the case where $|c| < 1$, so $1 - c^2 > 0$. Conditions of existence of a periodic solution can also be formulated in the case $|c| > 1$. Let us note the fact that $|c|$ represents the speed of the wave defined by equation (7.14).

Equation (7.15) can be rewritten in the form (7.11) with

$$f(y) = (1 - c^2)^{-1} f_0(y), \quad g(y) = (1 - c^2)^{-1} g_0(y). \tag{7.16}$$

In order to obtain the existence of a periodic solution to equation (7.12), one has to ensure this property for equation (7.15). The following conditions are sufficient to secure this property.

Theorem 7.2. *Consider equation (7.12) under the following conditions:*

(1) $f_0(y)$ and $g_0(y)$ are continuously differentiable functions from R into R.
(2) $y g_0(y) \neq 0$ for $y \neq 0$.
(3) If we define

$$F(y) = \int_0^y f_0(s) ds, \quad y \in R,$$

then

$$\lim |F(y)| = +\infty \quad as \ |y| \to \infty.$$

(4) Condition (4) in Theorem 7.1 applies.

Then, there exists a nontrivial periodic solution to equation (7.12) of the form $u(t, x) = y(x + ct)$, $|c| < 1$, with $y(x)$ a periodic solution to the nonlinear Liénard equation (7.15).

Remark 7.2. The case $|c| > 1$ also leads to the conclusion expressed in Theorem 7.2, but the case $|c| = 1$ does not lead to the same conclusion. In this case, only the solution of the wave equation $u_{tt} = u_{xx}$ is represented.

A special case of equation (7.12) is the equation

$$-u_{tt} + u_{xx} + g_0(u) = 0 \tag{7.17}$$

corresponding to the choice $f_0(u) \equiv 0$. The associated equation similar to equation (7.15) has the form

$$(1 - c^2) y'' + g_0(y) = 0, \tag{7.18}$$

which is of the pendulum type.

Using Theorem 7.1 from the book of G. Birkhoff and G. C. Rota [11], one obtains the following existence result for periodic solutions to equation (7.17).

Proposition 7.1. *Consider equation (7.17), and assume $g_0(y)$ is continuously differentiable from R to R, antisymmetric, and such that $y g_0(y) > 0$ for $y \neq 0$. Then, for each $c \in R$, $|c| < 1$, there exists a family of periodic solitary waves for equation (7.17) depending on a real parameter.*

An analogous result can be obtained for $|c| > 1$.

Using various results concerning the existence of periodic solutions to Liénard's equation (7.15), one obtains the existence of solitary waves for equation (7.12). Such results can be found in books by G. Sansone and R. Conti [86], E. A. Coddington and N. Levinson [22], and F. Brauer and J. A. Nohel [17].

We shall now consider equation (7.13), known as a *reaction–diffusion* equation. Looking again for solutions of the form (7.14), we find the following ordinary differential equation:

$$y'' + [f_0(y) - c]y' + g_0(y) = 0. \tag{7.19}$$

Compared with equation (7.11), we have the following relations:

$$f(y) = f_0(y) - c, \ g(y) = g_0(y). \tag{7.20}$$

The application of Theorem 7.1 to equation (7.19) leads to the following result for equation (7.13).

Theorem 7.3. *Consider equation (7.13) under the following assumptions.*

(1) *The functions $f_0(y)$ and $g_0(y)$ are continuously differentiable from R into R.*
(2) *The same assumptions as in Theorem 7.2 apply.*
(3) *$f_0(y)$ satisfies the condition*

$$\lim \left| -cy + \int_0^y f_0(s)ds \right| = +\infty \quad as \ |y| \to \infty.$$

(4) *The same assumptions as in Theorem 7.1 apply, with*

$$F(y) = -cy + \int_0^y f_0(s)ds, \qquad y \in R.$$

Then there exists a periodic solution of equation (7.13) that can be represented as $u(t,x) = y(x + ct)$ for any $c \in R$ such that conditions (3) and (4) are verified.

The proof is already provided due to the fact that conditions of Theorem 7.3 and equation (7.20) imply those of Theorem 7.1.

Remark 7.3. The constant c, which represents the wave velocity, can be chosen arbitrarily. Restrictions on the range on which we can choose c may appear if we require more from the periodic solution (e.g., the uniqueness). Let us mention such a situation. It does require another result, one that is similar to Theorem 7.1.

Proposition 7.2. *Consider Liénard's equation (7.11) under the following assumptions:*

(1) $f : R \longrightarrow R$ is continuous, even, and $f(0) < 0$.
(2) $g : R \longrightarrow R$ is locally Lipschitz, odd, and such that $yg(y) > 0$ for $y \neq 0$.
(3) $F(y) = \int_0^y f(s)ds$ has a unique positive zero $y = b$, and it increases at $+\infty$ for $y > b$.

Then there exists a unique periodic solution of equation (7.11). Moreover, it is the only limit cycle of the system equivalent to equation (7.11), $y' = z - F(y)$, $z' = -g(y)$, and it is orbitally stable.

The proof can be found in F. Brauer and J. A. Nohel [17].

From Proposition 7.2, one easily derives the following existence result for equation (7.13).

Theorem 7.4. *Consider equation (7.13) under the following hypotheses:*

(1) $f_0 : R \longrightarrow R$ is continuous, even, and $c > f_0(0)$.
(2) $g_0 : R \longrightarrow R$ is locally Lipschitz, odd, and $yg_0(y) > 0$ for $y \neq 0$.
(3) $F(y) = -cy + \int_0^y f_0(s)ds$ has a unique positive zero $y = b$, and it is increasing to $+\infty$ for $y > b$.

Then, equation (7.13) has a unique periodic solution $u(t, x) = y(x + ct)$, $x, t \in R$, with y the only periodic solution of (7.11), in which f and g are given by equation (7.20).

The proof follows from the fact that the hypotheses of Theorem 7.4 repeat those in Proposition 7.2, and they imply the existence and uniqueness of the periodic solution.

Remark 7.4. The orbital stability of the corresponding plane system $y' = z - F(y)$, $z' = -g(y)$, equivalent to equation (7.11), allows us to conclude that equation (7.11) has solutions that are asymptotically periodic. Of course, there are infinitely many solutions of this kind that generate similar solutions to equation (7.13).

Remark 7.5. The conditions of Theorem 7.4 imply some kind of interrelation between the data, in particular, some restrictions on c. For instance, taking $f_0(y) = 3y^2 - 1$, condition (1) implies $c > -1$. Since $F(y) = y^3 - (1 + c)y$, one sees that a unique positive zero for $F(y)$ is possible only if $1 + c > 0$ or $c > -1$. In other words, the restriction of c to the half-axis $(-1, +\infty)$ is the consequence of the hypotheses of Theorem 7.4.

The procedure used above to derive results for partial differential equations by means of reducing them to ordinary differential equations is widespread in the literature. Let us note that more general equations than equations (7.12) or (7.13) can be reduced to ordinary differential equations when looking for traveling waves. We will mention here the equations

$$-u_{tt} + u_{xx} + f(u, u_x) = 0 \qquad (7.21)$$

and

$$-u_t + u_{xx} + f(u, u_x) = 0. \tag{7.22}$$

Under rather mild assumptions, the corresponding ordinary differential equations, when we let $u(t, x) = y(x + ct)$, are

$$(1 - c^2)y'' + f(y, y') = 0 \tag{7.23}$$

and

$$y'' - cy' + f(y, y') = 0. \tag{7.24}$$

Various properties of solutions to equation (7.21) or (7.22) can be derived from the investigation of equation (7.23) or (7.24). One advantage in dealing with equation (7.23) or (7.24) resides in the fact that each of these equations can be reduced to a first-order equation. There is a classical procedure taking y as the independent variable and equation $y' = z$ as the dependent one. Then $y'' = z\, dz/dy$, and equation (7.23) or (7.24) becomes of first order in z.

7.3 Almost Periodic Heat Waves

In this section, we shall deal with the nonlinear heat (scalar) equation

$$u_t = \Delta u + f(t, x, u), \qquad (t, x) \in R \times \Omega, \tag{7.25}$$

where $u = u(t, x)$, $x = \mathrm{col}(x_1, x_2, \dots, x_m)$, Δ is the classical Laplace operator, $\Delta u = \mathrm{div}(\mathrm{grad}\, u)$, with the grad acting on the x-variables only. Equation (7.25) is also known as the reaction–diffusion equation and describes, besides the process of heat transfer, other physical phenomena such as the diffusion of a substance in a solution or such processes as the spread of a population in a certain area.

In the heat transfer interpretation, the meaning of the variables involved in equation (7.25) is as follows: $u = u(t, x)$ represents the temperature at the moment t, at the point x of a domain $\Omega \subset R^m$. Of course, in this interpretation, one must assume $m \leq 3$.

The function $f(t, x, u)$ takes real values, is defined in the set $R \times \overline{\Omega} \times R$, and is usually interpreted as the density of heat sources in Ω. It says that the output of these sources depends on the temperature as well as their location and time.

The domain $\Omega \subset R^m$ will be assumed to be *bounded*, and its boundary $\partial\Omega$ should be a smooth enough manifold in R^m such that the Dirichlet problem related to equation (7.25) under the boundary value condition

$$u|_{\partial\Omega} = \varphi, \qquad t \in R, \tag{7.26}$$

where φ is a given function on $\partial\Omega$ (real valued), is solvable. Other boundary value conditions are encountered. Condition (7.26) means that the temperature is known on the boundary $\partial\Omega$ at any moment. If instead of equation

(7.26) one assigns the value of the normal derivative $\partial u/\partial n$ on $\partial\Omega$, this will be understood as being the exchange of heat between Ω and its external region. Besides equation (7.25), other equations will be discussed that are usually more general than equation (7.25).

The results we are going to establish are of the Bohr–Neugebauer type. This means that the conditions imposed on the data must be of such a nature that the existence of a *bounded* solution (assumed) in the sense to be considered in the problem is also *almost periodic*, such as a map from R into $L^2(\Omega) : t \longrightarrow u(t,x)$, $t \in R$, $x \in \Omega$, or into another function space on Ω.

We shall also prove that, under our conditions, the uniqueness of the almost periodic solution is assured.

The following basic assumptions will be accepted in regard to equation (7.25):

(a) The domain $\Omega \subset R^m$ is bounded, and $\partial\Omega$ is such that the Dirichlet problem

$$\Delta u + \lambda u = 0 \ \text{ in } \Omega, \quad u|_{\partial\Omega} = 0, \tag{7.27}$$

has a sequence of eigenvalues $\{\lambda_k; \ k \geq 1\}$, $0 < \lambda_1 < \lambda_2 < \cdots < \lambda_k < \lambda_{k+1} < \cdots$, $\lambda_k \to \infty$ as $k \to \infty$.

(b) The map $f : R \times \overline{\Omega} \times R \to R$ is continuous and $L^2(\Omega)$–almost-periodic in t uniformly with respect to $u \in R$; i.e., to each $\varepsilon > 0$, one can associate a number $\ell = \ell(\varepsilon) > 0$ such that any interval $(\alpha, \alpha + \ell) \subset R$ contains a point τ with the property

$$\int_\Omega |f(t + \tau, x, u) - f(t, x, u)|^2 dx < \varepsilon^2 \tag{7.28}$$

for all $t, u \in R$.

(c) The map f is of class $C^{(1)}$ with respect to u, and $f_u = \partial f/\partial u$ satisfies *one* of the following inequalities:

$$f_u \leq \mu < \lambda_1 \ \text{ in } R \times \overline{\Omega} \times R \tag{7.29}$$

or

$$\lambda_k < p_k \leq f_u \leq q_k < \lambda_{k+1} \ \text{ in } R \times \overline{\Omega} \times R, \tag{7.30}$$

where $k \geq 1$ and μ, p_k, q_k are real numbers as seen in formulas (7.29), and (7.30).

(d) For each k, there exists $m = m(k)$, $p_k < m < q_k$, such that

$$\max\{(m - \lambda_k)^{-1}(m - p_k)^{-1}, (m - \lambda_k)^{-1}(q_k - m)^{-1},$$
$$(\lambda_{k+1} - m)^{-1}(m - p_k)^{-1}, (\lambda_{k+1} - m)^{-1}(q_k - m)^{-1}\} < 1.$$

Remark 7.6. Taking into account condition (a), we see that conditions (7.30) are mutually exclusive for different values of the integer k. Condition (c) postulates the existence of the numbers μ, p_k, q_k, $k \geq 1$, such that formulas (7.29) and (7.30) are verified.

We shall now establish our first result of this section, stating the almost periodicity of solutions of equation (7.25) under the assumption that they are $L^2(\Omega)$-bounded, i.e., a condition of the form

$$\int_\Omega u^2(t,x)dx \le M < \infty, \qquad t \in R, \tag{7.31}$$

takes place for some $M > 0$.

Theorem 7.5. *Consider equation* (7.25) *under assumptions* (a), (b), (c), *and* (d). *If* $u = u(t,x)$ *is a solution of equation* (7.25) *such that* $u|_{\partial\Omega} = 0$ *and formula* (7.31) *is verified, then* $u(t,x)$ *is almost periodic in* t *as a map from* R *into* $L^2(\Omega)$. *When* f *is periodic in* t, *so is* $u(t,x)$.

Proof. We are distinguishing two complementary situations. First is when f_u satisfies condition (7.29) and second when one of conditions (7.30) takes place for a given k. The methods of proof will be different for each case.

Case 1. Let $u(t,x)$ be a solution of equation (7.25) satisfying the condition stipulated in Theorem 7.5. For the function $u(t+\tau,x)$ with fixed $\tau \in R$, we have the equation

$$u_t(t+\tau,x) = \Delta u(t+\tau,x) + f(t+\tau,x,u(t+\tau,x)).$$

Let us denote $v(t,x) = u(t+\tau,x) - u(t,x)$ for fixed $\tau \in R$, and subtracting equation (7.25) from the above, side by side, we obtain for $v(t,x)$ the equation (similar to equation (7.25))

$$v_t = \Delta v + f(t+\tau,x,u(t+\tau,x)) - f(t,x,u(t,x)), \tag{7.32}$$

which together with the boundary condition

$$v|_{\partial\Omega} = 0, \qquad t \in R, \tag{7.33}$$

leads to a qualitative inequality for

$$V(t) = \int_\Omega v^2(t,x)dx, \qquad t \in R, \tag{7.34}$$

namely

$$\frac{1}{2}\frac{dV}{dt} \le -(\lambda_1 - \mu)V + \sqrt{V}\left(\sup\int_\Omega |f(t+\tau,x,u) - f(t,x,u)|^2 dx\right)^{1/2}, \tag{7.35}$$

in which the supremum is taken for $t, u \in R$.

Indeed, multiplying both sides of equation (7.32) by v and integrating both sides on Ω, one obtains

$$\frac{1}{2}\frac{d}{dt}\int v^2(t,x)dx = \int_\Omega v\Delta v\, dx$$
$$+ \int_\Omega v[f(t+\tau,x,u(t+\tau,x)) - f(t,x,u(t,x))]dx.$$

Applying Green's formula to the first integral on the right-hand side, one derives

$$\frac{1}{2}\frac{d}{dt}\int_\Omega v^2(t,x)dx$$
$$= -\int_\Omega |\text{grad}\, v|^2 dx + \int_\Omega v[f(t+\tau,x,u(t+\tau)) - f(t+\tau,x,u(t,x))]dx$$
$$+ \int_\Omega v[f(t+\tau,x,u(t,x)) - f(t,x,u(t,x))]dx.$$

Poincaré's inequality

$$\lambda_1 \int_\Omega v^2(t,x)dx \leq \int_\Omega |\text{grad}\, v(t,x)|^2 dx,$$

with λ_1 the smallest eigenvalue of the Laplace operator (see condition (a) of Theorem 7.5), combined with the preceding equation, allows us to write the inequality

$$\frac{1}{2}\frac{d}{dt}\int_\Omega v^2(t,x)dx \leq -\lambda_1 \int_\Omega v^2(t,x)dx + \int_\Omega \left(\frac{\partial f}{\partial u}\right)^* v^2(t,x)dx$$
$$+ \left(\int_\Omega v^2(t,x)dx\right)^{1/2}\left(\sup_{t,u}\int_\Omega |f(t+\tau,x,u) - f(t,x,u)|^2 dx\right)^{1/2}, \tag{7.36}$$

where $\left(\frac{\partial f}{\partial u}\right)^*$ stands for an intermediate value of $\partial f/\partial u$ between $u(t,x)$ and $u(t+\tau,x)$. Taking into account condition (7.29) and the notation (7.34), we see that inequality (7.36) is exactly inequality (7.35) for $V(t)$.

The qualitative inequality (7.35) can be treated using Proposition 6.5, leading to the estimate

$$V(t) \leq (\lambda_1 - \mu)^{-2}\sup\int_\Omega |f(t+\tau,x,u) - f(t,x,u)|^2 dx,$$

the supremum being taken with respect to $t,u \in R$. In other words, we can write the inequality

$$\sup\int_\Omega |u(t+\tau,x) - u(t,x)|^2 dx$$
$$\leq (\lambda_1 - \mu)^{-2}\sup\int_\Omega |f(t+\tau,x,u) - f(t,x,u)|^2 dx, \tag{7.37}$$

with the supremum on the left-hand side taken for $t \in R$.

The inequality (7.37) shows that the solution $u(t, x)$ of equation (7.25) satisfying $u|_{\partial\Omega} = 0$ and formula (7.31) is in $AP(R, L^2(\Omega))$. From formula (7.37), we also derive that if $f(t, x, u)$ is periodic in t, then $u(t, x)$ is also periodic in t with the same period.

This ends the proof of Theorem 7.5 under assumption (7.29) for f_u.

Case 2. In this case, f_u will satisfy an inequality of the form indicated in, formula (7.30). In what follows, the integer k will be fixed. An auxiliary result is needed in order to carry out the proof in this case.

Namely, let $m \in (\lambda_k, \lambda_{k+1})$, and consider the equation

$$u_t - (\Delta u + mu) = f(t, x), \quad (t, x) \in R \times \overline{\Omega}, \tag{7.38}$$

under the Dirichlet boundary value condition $u|_{\partial\Omega} = 0$. If u and f are $L^2(\Omega)$-bounded (i.e., they satisfy a condition like formula (7.31)), then the following inequality holds:

$$\sup \int_\Omega u^2(t, x)dx \leq \max\{(m - \lambda_k)^{-2}, (\lambda_{k+1} - m)^{-2}\} \sup \int_\Omega f^2(t, x)dx. \tag{7.39}$$

In order to prove formula (7.39), we rely on the properties of $L^2(\Omega)$ as a Hilbert space. If one denotes by $u_j(x)$, $j \geq 1$, the eigenfunctions of the problem (7.27) corresponding to the eigenvalue λ_j, $j \geq 1$, we can assume, without loss of generality, that the system $\{u_j(x); j \geq 1\}$ is orthonormal in $L^2(\Omega)$. This means that

$$\int_\Omega u_j(x)u_k(x)dx = \delta_{jk}, \tag{7.40}$$

where δ_{jk} is Kronecker's symbol. Then, we can write for $(t, x) \in R \times \Omega$

$$u(t, x) = \sum_{j=1}^\infty a_j(t)u_j(x), \quad f(t, x) = \sum_{j=1}^\infty b_j(t)u_j(x), \tag{7.41}$$

the convergence being that of $L^2(\Omega)$. It is known that Parseval's formulas

$$\sum_{j=1}^\infty a_j^2(t) = \int_\Omega u^2(t, x)dx \leq M,$$

$$\sum_{j=1}^\infty b_j^2(t) = \int_\Omega f^2(t, x)dx \leq M, \tag{7.42}$$

hold, with $M > 0$ independent of $t \in R$. If we substitute equation (7.41) into equation (7.38), we obtain after proper identification the equations

$$\dot{a}_j(t) + (\lambda_j - m)a_j(t) = b_j(t), \quad j \geq 1. \tag{7.43}$$

Each equation (7.43) is of the type examined in Section 5.4, and we have found there the estimates

$$\sup |a_j(t)| \leq (m - \lambda_k)^{-1} \sup |b_j(t)|, \qquad t \in R, \ j \leq k. \tag{7.44}$$

It has to be noted that from equations (7.42) there results that all $a_j(t)$ and $b_j(t)$, $j \geq 1$, are bounded functions on R.

In a similar manner, we can obtain the inequalities

$$\sup |a_j(t)| \leq (\lambda_{k+1} - m)^{-1} \sup |b_j(t)|, \qquad t \in R, \ j > k. \tag{7.45}$$

Taking equations (7.42) and formulas (7.44) and (7.45) into account, we obtain formula (7.39).

We can now proceed further, and we denote $w = w(t, x) = u(t + \tau, x) - u(t, x)$, $(t, x) \in R \times \Omega$. It is easily seen that w is a solution of the equation

$$w_t - \Delta w = f_u(t + \tau, x, \tilde{u}(t, x))w + f(t + \tau, x, u(t, x)) - f(t, x, u(t, x)) \tag{7.46}$$

satisfying the boundary condition $w|_{\partial \Omega} = 0$. We have denoted by $\tilde{u}(t, x)$ an intermediary number between $u(t, x)$ and $u(t + \tau, x)$.

Let us now rewrite equation (7.46) in the equivalent form as follows:

$$w_t - (\Delta w + mw) = [f_u(t + \tau, x, \tilde{u}(t, x)) - m]w$$
$$+ f(t + \tau, x, u(t, x)) - f(t, x, u(t, x)).$$

The equation above for $w(t, x)$ is of the form (7.38), for which we have established formula (7.39). By applying formula (7.39) to $w(t, x)$, which makes sense if we also keep in mind the condition $w|_{\partial \Omega} = 0$, one obtains based on L_2-norm properties

$$\sup \left(\int_\Omega w^2(t, x) dx \right)^{1/2} \leq \max \left\{ (m - \lambda_k)^{-1}, (\lambda_{k+1} - m)^{-1} \right\}$$

$$\times \sup \left(\int_\Omega [(f_u^* - m)w + f(t + \tau, x, u(t, x)) - f(t, x, u(t, x))]^2 dx \right)^{1/2}.$$

One more step leads us to the estimate

$$\sup \left(\int_\Omega w^2(t, x) dx \right)^{1/2}$$

$$\leq \max \left\{ (m - \lambda_k)^{-1}, (\lambda_{k+1} - m)^{-1} \right\} \max \left\{ (m - p_k)^{-1}, (q_k - m)^{-1} \right\}$$

$$\times \left(\int_\Omega w^2(t, x) dx \right)^{1/2} + \max \left\{ (m - p_k)^{-1}, (q_k - m)^{-1} \right\}$$

$$\times \sup \left(\int_\Omega [f(t + \tau, x, u(t, x)) - f(t, x, u(t, x))]^2 dx \right)^{1/2}.$$

It is now obvious from hypothesis (d) that we can choose m, $p_k < m < q_k$, such that

$$\max\{(m - \lambda_k)^{-1}, (\lambda_{k+1} - m)^{-1}\}\max\{(m - p_k)^{-1}, (q_k - m)^{-1}\} < 1, \quad (7.47)$$

and then one obtains from the last inequality an inequality of the form

$$\sup\left(\int_\Omega w^2(t, x)dx\right)^{1/2}$$
$$\leq K \sup\left(\int_\Omega |f(t + \tau, x, u(t, x)) - f(t, x, u(t, x))|^2 dx\right)^{1/2}, \quad (7.48)$$

with $K > 0$, which implies the almost periodicity in $L^2(\Omega)$-norm of the solution $u(t, x)$. This is because $w(t, x) = u(t + \tau, x, u(t, x)) - u(t, x, u(t, x))$ and formula (7.48) expresses exactly this property. The periodic case is also covered by the inequality (7.48).

Remark 7.7. Assumption (d) in Theorem 7.5 does not appear in Y. Yang's paper [101], from which the second part of the proof is adapted. It is stated that one can always choose m, with $p_k < m < q_k$, such that hypothesis (d) is satisfied. In fact, this is not the case in general.

First, we shall note that assumption (d) is equivalent to the statement that at least one of the inequalities $(m - \lambda_k)(m - p_k) > 1$, $(m - \lambda_k)(q_k - m) > 1$, $(\lambda_{k+1} - m)(m - p_k) > 1$, or $(\lambda_{k+1} - m)(q_k - m) > 1$ possesses a solution in the interval (p_k, q_k).

If this fact were true, then from the inclusion $(p_k, q_k) \subset (\lambda_k, \lambda_{k+1})$ and each of the four inequalities above, we obtain $(\lambda_{k+1} - \lambda_k)^2 > 1$ or $\lambda_{k+1} - \lambda_k > 1$.

Therefore, a *necessary* condition for the existence of $m \subset (p_k, q_k)$ satisfying assumption (d) is $\lambda_{k+1} - \lambda_k > 1$. That this condition is also *sufficient* one can easily see from the first of the four inequalities above (any m in the interval is good). Let $(p_k, q_k) \subset (\lambda_k, \lambda_{k+1})$, $q_k - p_k > 1$, which is always possible. The first inequality means

$$m^2 - (\lambda_k + p_k)m + \lambda_k p_k - 1 > 0, \quad (7.49)$$

and the greatest root of the polynomial on the left-hand side of formula (7.49) is

$$r_k = \frac{1}{2}(\lambda_k + p_k) + \sqrt{\frac{1}{4}(p_k - \lambda_k)^2 + 1}. \quad (7.50)$$

One can easily see that this root is in the interval $(p_k, p_k + 1) \subset (p_k, q_k)$ when $q_k - p_k > 1$. Any $m \in (p_k, q_k)$ is a solution to our problem.

Therefore, a *necessary and sufficient* condition for the existence of $m = m(k)$, satisfying assumption (d), is

$$\lambda_{k+1} - \lambda_k > 1 \quad (7.51)$$

for the particular k such that $p_k < f_u < q_k$.

Condition (7.51) is not automatically verified for every $k \geq 1$. The distribution of eigenvalues of the problem (7.27) is dependent on the domain Ω.

In conclusion, hypothesis (d) cannot be eliminated from the statement of Theorem 7.5. On the other hand, it must be noted that formula (7.51) is not a severe restriction in many situations.

Remark 7.8. Another approach to the second part of the proof of Theorem 7.5 can be based on a paper by Amann. It is presented in a book by C. Corduneanu [25]. The space $L^2(\Omega)$ is the particular Hilbert space in which Amann's result must be applied.

We shall now elaborate on the case where the right-hand side of an equation similar to equation (7.25) is actually an operator acting on the space $L^2(\Omega)$ for each $t \in R$. For instance, such an operator could be of the form

$$f(t; u) = \int_\Omega k(t, x - \xi)u(t, \xi)d\xi, \tag{7.52}$$

which will make the equation

$$u_t = \Delta u + f(t; u), \qquad t \in R, \tag{7.53}$$

an integro–partial differential equation, of course under adequate conditions on k (we do not specify them here, but the reader can easily formulate such conditions).

Let us point out the fact that a better notation for the left-hand side of equation (7.53) would be $f(t; u)(x)$, $x \in \Omega$. In equation (7.53), x does not appear explicitly, but $u(t, x) \in L^2(\Omega)$ for each $t \in R$.

The following result can be stated for equation (7.53).

Theorem 7.6. *Consider equation* (7.53) *under the following hypotheses:*

(a) *Condition the same as in Theorem 7.5.*

(b) $u(t, x) \longrightarrow f(t; u)(x)$ *is a map from* $L^2(\Omega)$ *into itself, for each* $t \in R$, *verifying the condition (monotonicity)*

$$\int_\Omega (fu - fv)(u - v)dx \leq \mu \int_\Omega |u - v|^2 dx \tag{7.54}$$

for each $u, v \in L^2(\Omega)$, $\mu = const.$, $t \in R$.

(c) *For each* $u \in L^2(\Omega)$, *the map* $t \longrightarrow f(t; u)$ *is almost periodic from* R *into* $L^2(\Omega)$ *uniformly with respect to* u *in any bounded set of* $L^2(\Omega)$.

(d) $\mu < \lambda_1 =$ *the smallest eigenvalue from condition* (a).

Then any solution $u(t, x)$ *of equation* (7.53) *defined on* $R \times \overline{\Omega}$ *such that it satisfies formula* (7.31) *and vanishes on* $\partial\Omega$ *is in* $AP(R, L^2(\Omega))$.

Proof. Before we get into the details, a few comments are in order. Namely, we regard equation (7.53) as an equation in which the unknown $u(t, x)$ is a map from R into $L^2(\Omega)$. Hence, the derivative u_t must be understood as a limit in the convergence of $L^2(\Omega)$ of the ratio $h^{-1}[u(t + h, x) - u(t, x)]$ as $h \longrightarrow 0$. The equality of this limit with the derivative u_t when the latter exits in the usual sense (i.e., as the limit in R) and is uniformly continuous in $\overline{\Omega}$ necessarily takes place.

All we need to prove is the monotonicity of the operator $\Delta u + f(t; u)$ on the space $L^2(\Omega)$. This can be carried out easily if we rely on the well-known inequality

$$\int_\Omega (u - v)\Delta(u - v)dx \leq -\lambda_1 \int_\Omega |u - v|^2 dx, \tag{7.55}$$

where $u, v \in C^2(\overline{\Omega}) \subset L^2(\Omega)$. The inequality (7.55) is the consequence of the classical Green's formula and Poincaré's inequality (see the proof of Theorem 7.5). We assumed that both u and v vanish on $\partial\Omega$ (it is obviously sufficient to assume that they coincide on $\partial\Omega$).

Let us now combine formula (7.55) with condition (7.54) imposed on f. For the operator $A = \Delta u + f(t; u)$, we obtain the inequality

$$\int_\Omega (Au - Av)(u - v)dx \leq -(\lambda_1 - \mu) \int_\Omega |u - v|^2 dx. \tag{7.56}$$

The formula (7.56) shows that the operator $-A$ is monotone. This does not pose any problem because in qualitative inequalities on R one can change t into $-t$ and obtain the same result for the inequality. See also Remark 6.9 to Proposition 6.5.

According to Remark 6.25 to Proposition 6.13, Theorem 7.6 is proven.

Remark 7.9. The monotonicity of the right-hand side of equation (7.53) guarantees also the uniqueness of the almost periodic solution (if any) under the conditions of Theorem 7.6. We send the reader to Remark 6.22 of Proposition 6.13, which covers the monotonicity case.

We will consider in what follows an equation similar to equation (7.25) but in which the operator Δ is substituted by an operator of higher order; more precisely, an *elliptic* operator of order $2n$, $n \geq 1$. Using standard notation in the theory of partial differential equations, we shall consider the differential operator

$$A(D)u = \sum_{|j|=0}^{2n} a_j(x)D^j u, \qquad x \in \Omega, \tag{7.57}$$

where $a_j(x)$ are continuous and bounded in $\Omega \subset R^m$ together with their partial derivatives up to the order $2n$ $(n \geq 1)$.

The *ellipticity* condition is expressed by the inequality

$$(-1)^n \sum_{|j|=2n} a_j(x)\xi^j \geq c_0|\xi|^{2n}, \qquad c_0 > 0, \tag{7.58}$$

where $\xi = \text{col}(\xi_1, \xi_2, \ldots, \xi_m) \in R^m$ and $\xi^j = \xi_1^{k_1}\xi_2^{k_2}\ldots\xi_m^{k_m}$, $|j| = k_1 + k_2 + \cdots + k_m$ being the length of the multi-index j. We always have $k_p \geq 0$ and $k_p \leq 2n$ for those k_p, $p = 1, 2, \ldots, m$, involved in ξ^j. Let us point out that the concept just defined is also known as *strong ellipticity*.

For the notation used above and the definition of ellipticity, see, for instance, S. Mizohata [72].

We will now formulate an auxiliary result known as Gårding's inequality. Its proof can be found in the book by S. Mizohata quoted above as well as several books on partial differential equations.

Proposition 7.3. *Assume the differential operator A given by equation (7.57) is strongly elliptic in Ω. Then one can find real constants K and $C > 0$ such that the inequality (Gårding)*

$$\int_\Omega uAu dx + K\|u\|_0 \geq C\|u\|_m^2 \tag{7.59}$$

holds for any $u \in W_0^{2m,2}(\Omega)$.

The proof of Proposition 7.3 is carried out first for $u \in C_0^\infty(\Omega)$; i.e., for those $u : R^m \longrightarrow R$ infinitely differentiable and with compact support in Ω. Then, it is extended to the case $W_0^{2m,2}(\Omega)$, which is the Sobolev space obtained by the completion of the space $C_0^\infty(\Omega)$ with respect to the norm

$$\|u\|_m = \left(\int_\Omega \sum_{|j|=0}^{2m} |D^j u|^2 dx\right)^{1/2}. \tag{7.60}$$

It is understood that $D^0 u = u$, which implies the fact that $\|u\|_0$ is the $L^2(\Omega)$-norm on Ω.

Remark 7.10. The inequality (7.59) can be rewritten in the form

$$\int_\Omega u(A + KI)u \, dx \geq C\|u\|_m^2, \tag{7.61}$$

which is telling us that, when dealing with Gårding's inequality, we can always reduce the general case to the special one, corresponding to $K = 0$.

This remark will be used below. We shall now state and prove a result related to equation (7.53) with A given by equation (7.57). This result is similar to Theorem 7.6.

Theorem 7.7. *Consider equation (7.53) for $u = u(t,x)$, $(t,x) \in R \times \Omega$, under the following assumptions:*

(a) *The operator A given by equation (7.57) on the space $W_0^{2m,2}(\Omega)$ with $\Omega \subset R^m$ a bounded domain satisfies Gårding's inequality with $K = 0$ (see Remark 7.10 to Proposition 7.3).*

(b) *The map $f : L^2(\Omega) \longrightarrow L^2(\Omega)$ for each $t \in R$ satisfies the (monotonicity) condition*

$$\int_\Omega (fu - fv)(u - v)dx \geq \mu \int_\Omega |u - v|^2 dx \tag{7.62}$$

with $\mu \in R$ a constant.

(c) *The map $t \longrightarrow f(t, u)$ from R into $L^2(\Omega)$ is Stepanov almost periodic uniformly with respect to u in any bounded set of Ω.*

(d) *The constant C occurring in Gårding's inequality*

$$\int_\Omega uAu\, dx \geq C\|u\|_m^2 \tag{7.63}$$

and μ in formula (7.62) are such that

$$C + \mu > 0. \tag{7.64}$$

Then, any solution $u(t, x) \in W_0^{2m,0}(\Omega)$ of equation (7.53) such that inequality (7.31) takes place is almost periodic (Bohr).

Proof. Let $u = u(t, x)$, $(t, x) \in R \times \Omega$, be a solution of equation (7.53) satisfying condition (7.31), and denote by $\tau \in R$ an arbitrary number fixed in the following estimates. We will need to prove that the map $t \longrightarrow u(t, x)$ from R into $L^2(\Omega)$ is almost periodic; i.e., it belongs to the space $AP(R, L^2(\Omega))$. Some uniformity of the almost Stepanov periodicity in respect to u will also be proven.

It is clear that it will suffice to show the monotonicity of the right-hand side of equation (7.53) with respect to u uniformly with respect to $t \in R$. Indeed, if one denotes by Bu the right-hand side in equation (7.53), one obtains for $u, v \in L^2$

$$\int_\Omega [Bu - Bv](u - v)dx$$
$$= \int_\Omega [Au - Av](u - v)dx + \int_\Omega [fu - fv](u - v)dx, \tag{7.65}$$

which leads, from formulas (7.62), (7.63), and (7.64), to the inequality

$$\int_\Omega (Bu - Bv)(u - v)dx \geq C\|u\|_m^2 + \mu \int_\Omega (u - v)^2 dx. \tag{7.66}$$

But $\|u\|_m \geq \|u_0\|$ and

$$\|u\|_0^2 = \int_\Omega u^2 dx,$$

which allows us to write

$$\int_\Omega (Bu - Bv)(u - v)dx \geq (C + \mu) \int_\Omega (u - v)^2 dx. \qquad (7.67)$$

According to condition (d) of Theorem 7.7, $(C + \mu) > 0$, which means that the operator B is a monotone operator on $L^2(\Omega)$. Therefore, we can conclude that any solution of equation (7.53) satisfying formula (7.31) is almost periodic (Bohr–Bochner).

Remark 7.11. With more refined estimates, including a better form for Gårding's inequality (see, for instance, K. Yosida [103]), one can obtain also the almost periodicity of some derivatives of u, $D^j u$.

Remark 7.12. Condition (c) of Theorem 7.7 requires the Stepanov almost periodicity for the maps $t \longrightarrow (fu)(t)$ from R into $L^2(\Omega)$ for each $u \in L^2(\Omega)$. In Section 6.4, we have shown the validity of the statement in the case of Bohr almost periodicity for the term $f(t, u)$, $t \in R$, because Theorem 5.3 is obviously true for Hilbert spaces and not only for R^m. The key to this problem is the application of Proposition 6.1 with particular regard to the qualitative inequality (6.121) in Section 6.6. We leave the details to the reader.

7.4 Almost Periodic Waves; a Linear Case

In this section, we shall discuss the almost periodicity of solutions to the linear wave equation

$$u_{tt} = [p(x)u_x]_x - q(x)u, \qquad (7.68)$$

which is slightly more general than the classical wave equation $u_{tt} = u_{xx} - q(x)u$ and usually appears in nonhomogeneous structures (like a nonhomogeneous string, for instance).

The equations of the form (7.68) under various assumptions concerning the functions $p(x)$ and $q(x)$ have been investigated by many authors and appear in many books dedicated to partial differential equations. The solution is usually constructed in the form of a series obtained by applying the method of separation of variables.

Let us note that the special case $p(x) = 1$ and $q(x) = 0$ has been discussed in the introduction to this book. This case leads to the simplest form of the wave equation in the presence of a single spatial dimension.

We shall quote first a result of existence for equation (7.68) that can be found in the comprehensive treatise of V. I. Smirnov [91]. More precisely, we shall prove that the solution of equation (7.68) is given by a series of the form

$$u(t, x) = \sum_{k=1}^{\infty} (A_k \cos \sqrt{\lambda_k}\, t + B_k \sin \sqrt{\lambda_k}\, t)\varphi_k(x), \qquad (7.69)$$

in which λ_k and $\varphi_k(x)$, $k \geq 1$, are the eigenvalues and the eigenfunctions of the Sturm–Liouville problem

$$\frac{d}{dx}\left(p(x)\frac{d\varphi}{dx}\right) + [\lambda - q(x)]\varphi = 0, \qquad x \in [0,\ell], \qquad (7.70)$$

with the boundary conditions

$$\varphi(0) = 0, \quad \varphi(\ell) = 0. \tag{IC}$$

Without loss of generality, one can assume that $\{\varphi_k; \ k \geq 1\}$ is an orthonormal system in $L^2([0,\ell], R)$.

To obtain equation (7.69), one has to look for solutions of equation (7.68) in the form $u(t,x) = T(t)\phi(x)$, which when introduced in equation (7.68) leads to the following equations (after separating variables):

$$\frac{T''}{T} = \frac{(p\phi')' - q\phi}{\phi} = -\lambda = \text{const.} \tag{7.71}$$

From equation (7.71), one obtains

$$T'' + \lambda T = 0, \qquad t \in R, \tag{7.72}$$

and

$$(p\phi')' + (\lambda - q)\phi = 0, \qquad x \in [0,\ell]. \tag{7.73}$$

Since the eigenvalues λ_k, $k \geq 1$, are positive under assumptions to be specified later, equation (7.72) leads to the solutions $T_k(t) = A_k \cos\sqrt{\lambda_k}\,t + B_k \sin\sqrt{\lambda_k}\,t$. Taking into account that equation (7.73) is the same as equation (7.70), and denoting by $\varphi_k(x)$ the eigenfunction corresponding to $\lambda = \lambda_k$, there immediately follows that equation (7.69) is a series whose terms are solutions to equation (7.68).

Anytime this series is convergent (we shall specify later the type of convergence), we can expect that it represents a solution (classical or generalized) of equation (7.68) under boundary value conditions

$$u(t,0) = 0, \ u(t,\ell) = 0, \qquad t \in R. \tag{7.74}$$

It is well known that initial conditions of the form

$$u(0,x) = u_0(x), \ u_t(0,x) = u_1(x), \qquad x \in [0,\ell], \tag{7.75}$$

must be associated to equations (7.68) and (7.74) in order to obtain a unique solution.

The classical result, which appears, for instance, in V. I. Smirnov [91], can be stated as follows.

Theorem 7.8. *Consider the equation (7.68) under boundary condition (7.74) and initial conditions (7.75). Assume the following conditions on the data:*

(1) $p : [0,\ell] \longrightarrow R$ *is of class* $C^{(2)}$, *and* $p(x) > 0$ *for* $x \in [0,\ell;]$.
(2) $q : [0,\ell] \longrightarrow R$ *is of class* $C^{(1)}$, *and* $q(x) \geq 0$ *for* $x \in [0,\ell;]$.

(3) $u_0 : [0, \ell] \longrightarrow R$ is of class $C^{(3)}$ and such that

$$u_0(0) = 0, \qquad u_0(\ell) = 0, \tag{7.76}$$

and

$$\frac{d}{dx}\,[p(x)u_0'(x)] - q(x)u_0(x) = 0 \qquad at\ x = 0,\ x = \ell. \tag{7.77}$$

(4) $u_1 : [0, \ell] \to R$ is of class $C^{(2)}$ and vanishes at the ends $x = 0$ and $x = \ell$.

Then the function $u(t, x)$, $t \in R$, $x \in [0, \ell]$, given by the uniformly convergent series (7.69), is of class $C^{(2)}$ in both t and x and satisfies equation (7.68) inside the strip $R \times [0, \ell]$. The series in equation (7.69) is uniformly convergent in the strip $R \times [0, \ell]$ together with the series obtained by differentiation term by term up to the second order. Moreover, the solution given by equation (7.69), in which $A_k = \int_0^\ell u_0(x)\varphi_k(x)dx$ and $\sqrt{\lambda_k}\,B_k = \int_0^\ell u_1(x)\varphi_k(x)dx$, $k \geq 1$, is the unique function of class $C^{(2)}$ satisfying all the conditions stipulated above.

Remark 7.13. The uniform convergence of the series (7.69) is a strong property required for the existence (and representation) of the derivatives of $u(t, x)$ up to the second order. Such assumptions on the initial functions are rather restrictive, and the attempt to relax them will lead to a *generalized solution*. We shall illustrate this concept in the case of equation (7.68), constructing a solution that may not enjoy all the regularity properties mentioned in Theorem 7.8. Nevertheless, it will have some legitimacy, conferred by the properties it possesses.

Remark 7.14. What's the connection between the solution given by equation (7.69) and the concept of almost periodicity?

The question above has a very satisfactory answer. We express it in the following corollary to Theorem 7.8.

Corollary 7.1. The solution $u(t, x)$ given by equation (7.69) is almost periodic in t uniformly with respect to $x \in [0, \ell]$. This property holds also for the derivatives of $u(t, x)$ up to the second order.

Proof. As we know from the constructions of the series (7.69), each term

$$u_k(t, x) = (A_k \cos \sqrt{\lambda_k}\,t + B_k \sin \sqrt{\lambda_k}\,t)\varphi_k(x)$$

is a solution of equation (7.68) satisfying the boundary value conditions (7.74). Since the series (7.69) is uniformly convergent in the strip $R \times [0, \ell]$, there results that $u(t, x)$ is almost periodic in t uniformly with respect to $x \in [0, \ell]$ because each term is almost periodic in t uniformly with respect to $a \in [0, \ell]$.

Actually, each term of equation (7.69) is periodic in t. But the distribution of λ_k's on the positive semiaxis is, generally speaking, arbitrary. We invite the reader to reflect about the following statements:

(a) The solution $u(t, x)$ is *periodic* in t iff the set $\{\sqrt{\lambda_k};\ k \geq 1\}$ is such that $\sqrt{\lambda_k} = k\omega,\ k \geq 1,\ \omega > 0$ fixed.

(b) The solution $u(t, x)$ is *quasi-periodic* iff there exists an additive group of real numbers containing the set $\{\sqrt{\lambda_k};\ k \geq 1\}$ and having a finite number of linearly independent generators (with respect to the field Q of rationals).

(c) If neither (a) nor (b) is valid, then the solution $u(t, x)$ is almost periodic in the most general sense. More precisely, the set of periods of simple harmonics involved in equation (7.69) has an *infinite* basis.

The material presented in Section 4.6 will help in clarifying the statements above.

Remark 7.15. The boundary value conditions (7.70) could be replaced by other (linear) types of conditions. For instance, in the introduction to this book, we considered the boundary value conditions $u(t, 0) = 0$, $u_x(t, \ell) + hu(t, \ell) = 0$, where $h > 0$ is given. As seen there, the distribution of the eigenvalues is determined by the transcendental equation in λ, $h^{-1}\sqrt{\lambda} + \tan(\sqrt{\lambda}\,\ell) = 0$, and it appears highly unlikely (if not impossible!) to obtain periodic solutions or even quasi-periodic ones.

The result on the almost periodicity of solutions (all of them!) to equation (7.68) is an illustration of the fact that the wave equation is, by its own structure, a generator of almost periodic solutions.

We can surmise that this type of result prompted S. L. Sobolev [92] to entitle his series of papers quoted above "Sur la presque périodicité des solutions de l'équation des ondes."

We shall return now to the series (7.69) and introduce the concept of a *generalized solution* to equation (7.68). This is also related to Sobolev's approach to the subject of generalized solutions to partial differential equations, and we will pursue here a rather elementary aspect that just aims to illustrate these ideas.

Equation (7.68) will now be considered under boundary value conditions (7.74) and initial conditions (7.75). Instead of imposing the restrictions appearing in the statement of Theorem 7.8, we shall make the weaker assumptions on the data $p(x), q(x), u_0(x)$, and $u_1(x)$:

I. $p : [0, \ell] \longrightarrow R$ is positive and of class $C^{(1)}$.

II. $q : [0, \ell] \longrightarrow R$ is nonnegative and continuous.

III. $u_0 : [0, \ell] \longrightarrow R$ is such that $u_0' \in L^2([0, \ell], R)$ and $u_0(0) = u_0(\ell) = 0$.

IV. $u_1 : [0, \ell] \longrightarrow R$ is such that $u_1 \in L^2([0, \ell], R)$.

Under conditions I–IV, the series (7.69), in which A_k and B_k are chosen as in Theorem 7.8, does not converge uniformly in general. In the very special case where $\sqrt{\lambda_k} = k$, $k \geq 1$, for fixed $x \in (0, \ell)$, we obtain from equation (7.69) a classical Fourier series (the periodic case), or the convergence (pointwise or uniform) of Fourier series is a very intricate subject. There are continuous functions whose Fourier series are uniformly convergent. Also, there are continuous functions with Fourier series diverging at infinitely many points, even on sets of points that are uncountable. For details on this topic, the reader is sent to the book of R. E. Edwards [36], where more references are indicated.

If we note the fact that the series (7.69) is well defined under hypotheses I–IV stated above, it is natural to investigate what kind of convergence could be fit under such hypotheses. This is one possible path toward the introduction of generalized solutions for equations like equation (7.68). We shall now state the following result providing the existence of a generalized solution to equation (7.68) under boundary conditions (7.74) and initial conditions (7.75) as well as its almost periodicity.

Theorem 7.9. *The series* (7.69), *constructed as shown above, converges in the space* $L^2([0,\ell],R)$ *uniformly with respect to* $t \in R$. *The sum* $u(t,x)$ *is, by definition, a generalized solution of our problem. Each partial sum* $U_n(t,x) = \sum_{k=1}^{n} u_k(t,x)$ *of the series* (7.69) *is a classical solution of equation* (7.68), *verifying the boundary condition* (7.74). *The initial conditions* (7.75) *are satisfied in the following sense:*

$$\lim_{n\to\infty} \int_0^\ell [u_0(x) - U_n(0,x)]^2 dx = 0, \tag{7.78}$$

$$\lim_{n\to\infty} \int_0^\ell \left[u_1(x) - \frac{\partial}{\partial x} U_n(0,x)\right]^2 dx = 0. \tag{7.79}$$

Then the map $t \longrightarrow u(t,\cdot)$ *from* R *into* $L^2([0,\ell],R)$ *is almost periodic (Bohr).*

Proof. We shall prove first that the series (7.69) is convergent in the space $L^2([0,\ell],R)$ uniformly with respect to $t \in R$. This will enable us to have a series that defines a (real) function in the strip $R\times[0,\ell]$. Indeed, if one takes into account the fact that $\{\varphi_k(x); \ k \geq 1\}$ is an orthonormal system in $L^2([0,\ell],R)$, one obtains

$$\int_0^\ell \left\{\sum_{k=m}^{n} (A_k \cos \sqrt{\lambda_k}\, t + B_k \sin \sqrt{\lambda_k}\, t)\varphi_k(x)\right\}^2 dx$$

$$= \sum_{k=m}^{n} (A_k \cos \sqrt{\lambda_k}\, t + B_k \sin \sqrt{\lambda_k}\, t)^2 \tag{7.80}$$

$$\leq \sum_{k=m}^{n} (A_k^2 + B_k^2)$$

because $(a\cos x + b\sin x)^2 \leq a^2 + b^2$ for any a, b, and x real.

Now let us note that the series of real numbers

$$\sum_{k=1}^{\infty} (A_k^2 + B_k^2) < \infty. \tag{7.81}$$

From Parseval's identity, there follows

$$\sum_{k=1}^{\infty} A_k^2 = \int_0^{\ell} [u_0(x)]^2 dx, \qquad \sum_{k=1}^{\infty} \lambda_k B_k^2 = \int_0^{\ell} [u_1(x)]^2 dx. \tag{7.82}$$

But it is known that $\lambda_k \to \infty$ as $k \to \infty$, which means $\lambda_k > 1$ for sufficiently large $k > 0$. Therefore, from the second formula in equation (7.82), we derive that

$$\sum_{k=1}^{\infty} B_k^2 < +\infty. \tag{7.83}$$

Hence, taking into account equation (7.82) and formula (7.83), one obtains formula (7.81).

Now, returning to equation (7.80) and taking formula (7.81) into account, we can write

$$\int_0^{\ell} \left\{ \sum_{k=m}^{n} (A_k \cos \sqrt{\lambda_k}\, t + B_k \sin \sqrt{\lambda_k}\, t) \varphi_k(x) \right\}^2 dx$$

$$\leq \sum_{k=m}^{n} (A_k^2 + B_k^2) < \varepsilon,$$

provided $n, m \geq N(\varepsilon)$. This proves the convergence in $L^2([0, \ell], R)$ of the series (7.69) uniformly in $t \in R$.

We shall now prove that the series obtained from equation (7.69) by differentiating with respect to t is also convergent in $L^2([0, \ell], R)$ uniformly with respect to $t \in R$. This series is

$$\sum_{k=1}^{\infty} \sqrt{\lambda_k}\, (-A_k \sin \sqrt{\lambda_k}\, t + B_k \cos \sqrt{\lambda_k}\, t) \varphi_k(x),$$

from which we can see that it is sufficient to show that

$$\sum_{k=1}^{\infty} \lambda_k (A_k^2 + B_k^2) < \infty \tag{7.84}$$

in order to assure its convergence.

If we take equation (7.82) into account, we see that in order to obtain formula (7.84), it suffices to show

$$\sum_{k=1}^{\infty} \lambda_k A_k^2 < \infty. \tag{7.85}$$

This convergence can be obtained by starting from the inequality

$$\int_0^{\ell} [p(x)y'^2 + q(x)y^2] dx \geq 0, \tag{7.86}$$

valid for any $y \in C^{(1)}([0, \ell], R)$, and letting

$$y(x) = u_0(x) - \sum_{k=1}^{n} A_k \varphi_k(x). \tag{7.87}$$

Keeping in mind that $\{\varphi_k(x) : k \geq 1\}$ is an orthonormal system in $L^2([0, \ell], R)$, after elementary calculations, we obtain from equation (7.87)

$$\sum_{k=1}^{n} \lambda_k A_k^2 \leq \int_0^{\ell} [p(x)(u_0'(x))^2 + q(x)(u_0(x))^2] dx.$$

From condition I, the right-hand side in the inequality above is finite, which implies formula (7.85).

Therefore, we have proven that by differentiating equation (7.69) term by term with respect to t, we obtain a series that has the same type of convergence as equation (7.69).

Based on the discussion conducted above, we can infer the *existence* of the function $u(t, x)$ defined on $R \times [0, \ell]$ and taking real values. For each $t \in R$, $u(t, x)$ is in $L^2([0, \ell], R)$ only (generally), even though each $\varphi_k(x)$, $k \geq 1$, is of class $C^{(2)}$. But the type of convergence does not allow us to conclude that $u(t, x)$ possesses some regularity properties in x. Such properties may occur, as seen in Theorem 7.8, but under stronger assumptions. On the other hand, $u(t, x)$ is differentiable in t, regarded as a map from R into $L^2([0, \ell], R)$.

Concerning the initial conditions (7.78) and (7.79), we note that they are satisfied, this being a consequence of the fact that any function in $L^2([0, \ell], R)$ can be represented as the sum of its Fourier series with respect to the complete orthonormal system $\{\varphi_k(x); k \geq 1\} \subset L^2([0, \ell], R)$.

The last property of $u(t, x)$ we need to establish is its almost periodicity (as a function of t with values in $L^2([0, \ell], R)$). This property will follow from the fact that the limit of a uniformly convergent sequence of (Bohr) almost periodic functions with values in a Banach space is almost periodic (see Section 3.2, property III, for the case of complex-valued functions).

This ends the proof of Theorem 7.9.

Remark 7.16. From formula (7.84) there results that $u_t(t, x)$ does exist and

$$u_t(t, x) = \sum_{k=1}^{\infty} \sqrt{\lambda_k} (B_k \cos \sqrt{\lambda_k}\, t - A_k \sin \sqrt{\lambda_k}\, t) \varphi_k(x).$$

The convergence of the series representing $u_t(t, x)$ is in $L^2([0, \ell], R)$ uniformly with respect to $t \in R$.

We shall now conduct a brief discussion regarding the term *generalized solution* used above. As usually understood, this means a generalized solution in the sense of Sobolev or even a *distributional* solution.

It is interesting to know if the generalized solution defined above can also be interpreted as being of Sobolev type. The matter is rather intricate, and we will simply note that in the (very) special case where the series (7.69) is uniformly convergent in $R \times [0, \ell]$, in V. I. Smirnov [91] it is shown that the definition for a generalized solution adopted above is compatible with Sobolev's definition.

In concluding this section, we shall briefly illustrate the method used above in the case of waves that are linear but not necessarily one-dimensional. In other words, in equation (7.68) we will consider the case where $x = (x_1, x_2, \ldots, x_n) \in \Omega \subset R^n$, and to simplify the description we shall take $p(x) \equiv 1$, $q(x) \equiv 0$. This leads to the scalar equation

$$u_{tt}(t, x) = \Delta_x u(t, x), \tag{7.88}$$

which is the homogeneous wave equation in n spatial dimensions. Concerning the domain Ω, we shall consider that $\partial \Omega$ is a smooth enough boundary that the Dirichlet problem

$$\Delta v + \lambda v = 0 \quad \text{in } \Omega, \qquad v|_{\partial \Omega} = 0, \tag{7.89}$$

has a sequence $\{\lambda_k; \ k \geq 1\}$ of positive eigenvalues with $\lambda_k \to \infty$ as $k \to \infty$, while the corresponding eigenvalues $\{v_k(x); \ k \geq 1\}$ form an orthonormal complete sequence in $L^2(\Omega, R)$.

Each function of the form

$$u_k(t, x) = v_k(x) e^{i \sqrt{\lambda_k} t}, \qquad k \geq 1, \tag{7.90}$$

is a solution of equation (7.88) in Ω satisfying the boundary value condition

$$u|_{\partial \Omega} = 0. \tag{7.91}$$

Instead of the series (7.69), one has to deal with the series

$$\sum_{k=1}^{\infty} A_k e^{i \sqrt{\lambda_k} t} v_k(x), \tag{7.92}$$

which is a good candidate for representing an element of $AP(R, L^2(\Omega, R))$.

Discussions about the convergence of the series (7.92) and the connection with the concept of the classical or generalized solution to equation (7.88) can be found in the book by O. Ladyzhenskaya [59] as well as in many journal papers.

It is also possible to substitute for the classical Dirichlet condition (7.91) conditions of a different type (Neumann, Newton's radiation law).

7.5 Almost Periodic Waves; a Mildly Nonlinear System

We shall consider here the nonlinear system of the form

$$\begin{aligned} u_t + Au + Bv + g(t, v) &= 0, \\ v_t - Bu + cv &= 0, \end{aligned} \tag{7.93}$$

in which the variables u, v represent the unknowns $u, v : R \to H$ with H a real Hilbert space, while other quantities involved will be described below.

The nonlinear part in equation (7.93) is given by the map $g(t, v)$ from $R \times H$ into H, while A and B are assumed to be linear operators on H.

Let us point out that, by formal elimination, assuming $c = 0$ (scalar) and B invertible, from equations (7.93) one obtains for v the second-order equation

$$v_{tt} + BAB^{-1}v_t + B^2v + Bg(t, v) = 0, \tag{7.94}$$

which constitutes a typical equation for nonlinear oscillations (in our case, in infinite dimension). In order to obtain almost periodicity of the solution $U = \mathrm{col}(u(t, x), v(t, x))$ with respect to t, we shall rely on a result established in Section 6.4, where we dealt with differential equations whose right-hand sides are monotone.

Let us now formulate the condition under which we shall examine the almost periodicity of the solutions to equation (7.93) with the basic property that they are bounded in the real axis R.

1. $A : H \longrightarrow H$ is a linear operator such that $A - \alpha I$ is monotone for some number $\alpha > 0$.
2. $B : H \longrightarrow H$ is linear, self-adjoint, and invertible.
3. $g : R \times H \longrightarrow H$ is almost periodic (Bohr) with respect to t uniformly in the second argument on bounded sets of H; moreover, it satisfies a Lipschitz type condition

$$\|g(t, x) - g(t, y)\| \le m\|x - y\| \tag{7.95}$$

   for some $m > 0$.
4. The constants α, m, and c satisfy the inequality

$$m^2 < 4\alpha c. \tag{7.96}$$

Remark 7.17. It is obvious that c must be a positive number $(c > 0)$.

We can now formulate the result of almost periodicity for the solution $U = \mathrm{col}(u, v)$ of the system (7.93).

Theorem 7.10. *Let $U(t) = (u(t), v(t))$ be a solution of system (7.93) defined on R and bounded there. If conditions 1–4 formulated above are verified, then $U(t)$ is (Bohr) almost periodic from R into $H \times H$.*

Proof. As suggested by the statement of Theorem 7.10, the underlying space is the product $H \times H$. This is again a Hilbert space over the reals with a scalar product defined by the well-known relation

$$\langle (x, y), (x', y') \rangle = \langle x, x' \rangle + \langle y, y' \rangle.$$

We shall now find the first-order differential equation for $U(t) = \mathrm{col}(u(t), v(t))$, equivalent to equation (7.93). Let us denote

$$\mathcal{A}(t) = B_1 + B_2 + B_3 + G(t, \cdot), \tag{7.97}$$

where

$$B_1 = \begin{pmatrix} A & 0 \\ 0 & 0 \end{pmatrix}, \qquad B_2 = \begin{pmatrix} 0 & B \\ -B & 0 \end{pmatrix}, \qquad B_3 = \begin{pmatrix} 0 & 0 \\ 0 & \nu I \end{pmatrix},$$

and

$$G(t, U) = \mathrm{col}(g(t, v), 0). \tag{7.98}$$

Then equation (7.93) can be written in the form

$$U'(t) + \mathcal{A}(t)U(t) = 0, \qquad t \in R. \tag{7.99}$$

We agree to designate by 0 the null element in either space H or $H \times H$.

From equation (7.99), in the product space $H \times H$, we can easily derive the result on almost periodicity of the bounded solution $U(t)$, provided we can show the monotonicity of the operator $\mathcal{A}(t) : R \to H \times H$. Then, by applying Proposition 6.12 with $\phi(r) = Kr^2$, $K > 0$, and $\psi(r) \equiv 0$, one obtains the desired result.

Let us now prove that the operator $\mathcal{A}(t)$, given by equation (7.97), is a monotone operator on the space $H \times H$. More precisely, we have to show that an inequality of the form

$$\left\langle \mathcal{A}(t)U - \mathcal{A}(t)\tilde{U}, U - \tilde{U} \right\rangle \geq \mu \|U - \tilde{U}\|^2, \qquad \mu > 0, \tag{7.100}$$

is verified, where the scalar product is the one in $H \times H$.

The following inequalities are immediate consequences of our assumptions:

$$\langle B_1 U, U \rangle \geq \alpha \|u\|^2, \ \forall U \in H \times H, \tag{7.101}$$

$$\langle B_2 U, U \rangle = 0, \ \forall U \in H \times H, \tag{7.102}$$

$$\langle B_3 U, U \rangle \geq c \|v\|^2, \ \forall U \in H \times H, \tag{7.103}$$

$$\left\langle G(t, U) - G(t, \tilde{U}), U - \tilde{U} \right\rangle = \langle g(t, v) - g(t, \tilde{v}), u - \tilde{u} \rangle \geq -m \|u - \tilde{u}\| \|v - \tilde{v}\|. \tag{7.104}$$

If we rely on the elementary inequality $ab \leq \frac{1}{2}(a^2 \varepsilon + \frac{b^2}{\varepsilon})$, $a, b, \varepsilon > 0$, then equation (7.104) allows us to write

$$\left\langle G(t, U) - G(t, \tilde{U}), U - \tilde{U} \right\rangle \geq -\frac{m}{2}(\varepsilon \|v - \tilde{v}\|^2 + \varepsilon^{-1} \|u - \tilde{u}\|^2). \tag{7.105}$$

Summing up formulas (7.101)–(7.103) and (7.105), we obtain the following inequality:

$$\left\langle \mathcal{A}(t)U - \mathcal{A}(t)\tilde{U}, U - \tilde{U} \right\rangle \geq \min\left\{ \alpha - \frac{m}{2\varepsilon}, c - \frac{m\varepsilon}{2} \right\} \|U - \tilde{U}\|^2. \tag{7.106}$$

It is now obvious that we can satisfy formula (7.100), provided we can choose $\varepsilon > 0$ in formula (7.106) such that

$$\mu = \min\left\{\alpha - \frac{m}{2\varepsilon}, c - \frac{m\varepsilon}{2}\right\} > 0. \tag{7.107}$$

But equation (7.107) is verified, provided $\varepsilon > 0$ is chosen such that

$$\frac{m}{2\alpha} < \varepsilon < \frac{2c}{m}. \tag{7.108}$$

Taking into account condition (4) of Theorem 7.10, we note that this is always possible. This ends the proof of Theorem 7.10 because the validity of equation (7.107) implies inequality (7.100); i.e., the monotonicity of the operator $\mathcal{A}(t) : H \times H \to H \times H$.

The result of Theorem 7.10 is taken from the paper [31] by C. Corduneanu and J. Goldstein.

Remark 7.18. This result belongs to the category of Bohr–Neugebauer since it states the almost periodicity of any bounded solutions of equation (7.93). The first result of this nature is given in Proposition 5.12.

Remark 7.19. We have obtained the second-order differential equation for v, eliminating u from equation (7.93).

It is interesting to see what we obtain if one eliminates v. From the second equation (7.93), we obtain v in the form

$$v(t) = Ke^{-ct} + \int_0^t e^{-c(t-s)} Bu(s) ds,$$

where $K \in H$ is arbitrary. Substituting this in the first equation (7.93), one obtains

$$u_t + Au + B\left(Ke^{-ct} + \int_0^t e^{-c(t-s)} u(s) ds\right)$$
$$+ g\left(t, Ke^{-ct} + \int_0^t e^{-c(t-s)} u(s) ds\right) = 0. \tag{7.109}$$

Obviously, equation (7.109) is an integro-differential equation for $u(t)$. Since $K \in H$ is arbitrary, we have in fact a family of such equations involving the parameter $K \in H$.

If $g(t, v)$ is linear with respect to the second argument, then equation (7.109) becomes a linear integro-differential equation of the form

$$u_t + Au + \beta(t) \int_0^t e^{-c(t-s)} u(s) ds + \gamma(t) = 0, \tag{7.110}$$

where $\beta(t)$ and $\gamma(t)$ are known. More precisely, $\beta(t)$ is a family of operators on H, while $\gamma(t)$ is a function with values in H.

In the remaining part of this section, we shall consider the case of partial differential equations, which occurs, for instance, when A is a differential operator and H is chosen to be a function space such as $L^2(\Omega)$ with $\Omega \subset R^m$ a fixed domain. Then $u = u(t, x)$, $(t, x) \in R \times \Omega$, and $R \ni t \to L^2(\Omega)$ is the map defined by $t \to u(t, x)$, $x \in \Omega$. In other words, the map $t \to u(t, \cdot) \in L^2(\Omega)$ is the map from R into $L^2(\Omega)$. This is how we can consider a partial differential equation as an ordinary differential equation whose unknown variable takes a value in a Hilbert space (or another abstract space).

Let us briefly illustrate how the abstract equation (7.110) can be viewed as a *partial integro-differential equation* in the Hilbert space $L^2(\Omega)$. If we examine equation (7.110), we see that a choice can be made for the operator A in the class of partial differential operators. It turns out that one can take, for instance, $A = -\Delta$ = the Laplace operator. Indeed, using the same notation as in Section 6.3, under the same assumptions for $\Omega \subset R^m$, one easily finds

$$\int_\Omega (-\Delta u, u)dx = \int_\Omega |\text{grad } u|^2 dx \geq \lambda_1 \int_\Omega u^2 dx, \qquad (7.111)$$

the last inequality being known as Poincaré's inequality. Since $\lambda_1 > 0$, we see that $-\Delta$ is monotone. Then, $-\Delta - \alpha I$, for $0 < \alpha < \lambda_1$, is also monotone on $L^2(\Omega)$, as required in Theorem 7.10.

An equation related to equation (7.110) is the partial integro-differential equation

$$u_t = \Delta u + a(t, x)u + \int_0^t B(t, s, x)u(s, x)dx + f(t, x), \qquad (7.112)$$

in which $t \in R$ or $R_+ = [0, \infty)$, $\Omega \subset R^m$ being a bounded domain with smooth enough boundary $\partial \Omega$, and the boundary value condition (Neumann)

$$\frac{\partial u}{\partial n} = 0, \quad (t, x) \in R_+ \times \partial \Omega. \qquad (7.113)$$

The result of existence and almost periodicity we shall provide here, without proof, is due to S. Murakami and Y. Hamaya [74]. Some technicalities in the proof make it somewhat lengthy, and we will omit the details.

The following assumptions will be made on the functions appearing in equation (7.112).

(1) $a(t, x)$ and $f(t, x)$ are almost periodic in t uniformly with respect to $x \in \Omega$, while $B(t, t + s, x)$, assumed continuous for $-\infty < s \leq t < +\infty$, $x \in \Omega$, is almost periodic in t uniformly with respect to $(s, x) \in R_- \times \Omega$, $R_- = (-\infty, 0)$.
(2) $a(t, x)$, $f(t, x)$, and $B(t, s, x)$ satisfy a Lipschitz–Hölder type condition of the form

$$|a(t, x) - a(s, y)| \leq L(|t - s| + |x - y|^\alpha),$$
$$|f(t, x) - f(s, y)| \leq L(|t - s| + |x - y|^\alpha), \qquad (7.114)$$
$$\int_0^\tau |B(t, s, x) - B(\tau, s, y)|ds \leq L(|t - \tau| + |x - y|^\alpha),$$

in their domains of definition, with $L > 0$ a constant, and $0 < \alpha < 1$.

(3) For each $\varepsilon > 0$, one can find $S = S(\varepsilon) > 0$ such that

$$\sup_{t \in R} \int_{-\infty}^{t-S} |B(t, s, x)| ds = \sup_{t \in R} \int_{-\infty}^{t-S} |B(t, t + s, x)| ds < \varepsilon$$

for any $x \in \Omega$.

(4) There exists a classical solution of equation (7.112), $u = u(t, x)$, defined for $t \in R_+$ and $x \in \overline{\Omega}$ such that

$$\sup\{|u(t, x)|; \ t \in R_+, \ x \in \overline{\Omega}\} < +\infty.$$

Remark 7.20. Condition (4) requires the existence of a bounded solution defined only for $t \in R_+$. We note that this condition is similar to that encountered in the finite-dimensional case (see Section 6.2).

Before we state the basic result on the existence of an almost periodic solution to equation (7.112), we need one definition related to a certain kind of stability of the zero solution of the homogeneous equation corresponding to equation (7.112), namely

$$u_t = \Delta u + a(t, x)y + \int_0^t B(t, s, x)u(s, x)ds. \tag{7.115}$$

In what follows, $g : R_- \longrightarrow [1, \infty)$ is continuous and nonincreasing, and $g(t) \to \infty$ as $t \to -\infty$.

One says that the solution $u = 0$ of equation (7.115) is *g-uniformly stable* if the following property holds: For each $\varepsilon > 0$, there exists $\delta = \delta(\varepsilon) > 0$ such that any solution of equation (7.115) for which $\|u_{t_0}\|_g < \delta(\varepsilon)$, $t_0 \geq 0$, $x \in R$, satisfies $\|u(t, x)\| < \varepsilon$ for all $t \geq t_0$.

The norm $\|u_{t_0}\|_g$ above should be understood in the following sense: If $\varphi \in C(R_- \times \Omega, R^m)$ is a bounded map, then

$$\|\varphi\|_g = \sup_{\tau \in R_-} \{g^{-1}(\tau) \sup_{x \in \Omega} |\varphi(\tau, x)|\}. \tag{7.116}$$

u_{t_0} is the restriction of u to $(-\infty, t_0]$.

The result we have in mind can now be formulated as follows.

Theorem 7.11. *Assume conditions 1–4 above are verified for equation (7.112) under the boundary value condition (7.113). If the solution $u = 0$ of equation (7.115) is g-uniformly stable for a function g as described above, then there exists a solution $u = u(t, x)$, $t \in R$, $x \in \Omega$, of the problem such that $u(t, x)$ is almost periodic in t uniformly with respect to $x \in \Omega$.*

Remark 7.21. If one considers equation (7.115), in which $|B(t, s, x)| = k(t-s)|x|$ with $\int_0^\infty k(t)dt < 1$, then g can be chosen in such a way that

$$\int_0^\infty k(t)g(-t)dt < 1.$$

Obviously, the choice makes sense. For instance, when $k > 0$, one can take

$$g(-t) = [k(t)]^{-1}m(t)$$

with $m(t)$ in $L^1(R_+, R)$ and

$$\int_0^\infty m(t)dt < 1.$$

7.6 A Case with Data on Characteristics

As we know, the wave equation in two dimensions can be written in the form

$$u_{xy} = f(x, y, u, u_x, u_y) \tag{7.117}$$

because by the change of variables x and y according to the formulas $x = \eta + \xi$, $y = \eta - \xi$, the mixed derivative u_{xy} must be replaced by $-\frac{1}{4}(u_{\eta\eta} - u_{\xi\xi})$, which is the typical part for the wave equation.

A semilinear form associated with equation (7.117) is

$$u_{xy} + a(x, y)u_x + b(x, y)u_y = c(x, y, u), \tag{7.118}$$

which can be reduced by the substitution $u = v \exp\left\{-\int_0^x b(\xi, y)d\xi\right\}$ to the simpler form

$$v_{xt} + a(x, t)v_x = C(x, t, v). \tag{7.119}$$

We have changed y by t since in the interpretation given in gas dynamics (see Tikhonov and Samarskii [94]), t plays the role of time.

Let us point out the fact that by the reduction of equation (7.118) to the simpler form (7.119), we do not lose the generality as it appears in equation (7.118). Of course, some obvious assumptions have to be made on the coefficients a and b in equation (7.118) in order to operate the substitution.

What is really advantageous in working with equation (7.119) is the fact that by denoting $V(x, t) = v_x(x, t)$, equation (7.119) is reduced to a functional equation, namely

$$V_t + a(x, t)V = C\left(x, t, \varphi(t) + \int_0^x V(\xi, t)d\xi\right), \tag{7.120}$$

where

$$\varphi(t) = v(0, t). \tag{7.121}$$

This condition represents a datum on the characteristic $x = 0$ of equation (7.119).

We shall need some facts about the counterpart of equation (7.120), namely the linear equation

$$V_t + a(x, t)V = c(x, t), \tag{7.122}$$

under various assumptions to be specified below. Here the variable $x \in [0, \ell]$ plays the role of a parameter. The independent variable is $t \in R$.

We shall denote in subsequent text

$$D = \{(x,t);\ (x,t) \in [0,\ell] \times R\}, \tag{7.123}$$

which will be the domain of investigation of equations (7.119), (7.120), and (7.122). From our notation $V(x,t) = v_x(x,t)$, and taking into account equations (7.120) and (7.121), we will obtain in D

$$v(x,t) = \varphi(t) + \int_0^x V(\xi,t)d\xi, \tag{7.124}$$

which will represent the solution of equation (7.119) in D with condition (7.121). Of course, we need to make precise the conditions that allow us to perform the operations indicated above.

We shall look first at the auxiliary equation (7.122) in the domain D. The following assumptions will be made:

(1) $a(x,t)$ is continuous from D to R and such that

$$a(x,t) \geq m > 0, \qquad (x,t) \in D. \tag{7.125}$$

(2) $c(x,t)$ is continuous from D to R and bounded:

$$|c(x,t)| \leq A < \infty, \qquad (x,t) \in D. \tag{7.126}$$

We will prove the following result.

Proposition 7.4. *Under conditions* (1) *and* (2) *above, there exists a unique bounded solution in D of equation* (7.122), *$V = V(x,t)$, given by the formula*

$$V(x,t) = \int_{-\infty}^t \exp\left\{-\int_\tau^t a(x,\theta)d\theta\right\} c(x,\tau)d\tau. \tag{7.127}$$

Proof. Due to condition (7.125), the convergence of the improper integral on the right-hand side of equation (7.127) is absolute. One easily finds

$$|V(x,t)| \leq \frac{A}{m}, \qquad (x,t) \in D, \tag{7.128}$$

which proves the boundedness of $V(x,t)$ in D. By differentiating the right-hand side of equation (7.127), we obtain

$$V_t(x,t) = c(x,t) - a(x,t) \int_{-\infty}^t \exp\left\{-\int_\tau^t a(x,\theta)d\theta\right\} c(x,\tau)d\tau, \tag{7.129}$$

which shows that $V(x,t)$ satisfies equation (7.122). The differentiation under the integral sign is justified by the fact that one obtains again an integral that is absolutely and uniformly convergent (in x).

Remark 7.22. One can obtain regularity properties of $V(x, t)$ besides the existence of $V_t(x, t)$ if we make extra assumptions on the data $a(x, t)$ and $c(x, t)$.

Remark 7.23. The boundedness condition (7.126) can be improved considerably in order to obtain the boundedness of $V(x, t)$. Namely, one could assume $c(x, t) \in M(R, R)$ in t uniformly with respect to $x \in [0, \ell]$.

Corollary 7.2. *If we make the extra assumption that both $a(x, t)$ and $c(x, t)$ are almost periodic in t uniformly with respect to $x \in [0, \ell]$, then the unique bounded solution of equation (7.122) given by equation (7.127) is also almost periodic in t uniformly with respect to $x \in [0, \ell]$.*

Proof. Indeed, for $V(t, x)$ given by equation (7.122), one can write the equation involving the difference $V(x, t + T) - V(x, t)$, where $t \in R$ and T is a fixed real number:

$$[V(x, t+T) - V(x, t)]_t + a(x, t+T)[V(x, t+T) - V(x, t)]$$
$$= c(x, t+T) - c(x, t) - [a(x, t+T) - a(x, t)]V(x, t).$$

Now applying the estimate (7.128), one obtains the following inequality:

$$|V(x, t+T) - V(x, t)| \leq \frac{1}{m} \sup |c(x, t+T) - c(x, t)|$$
$$+ \frac{A}{m} \sup |a(x, t+T) - a(x, t)|. \tag{7.130}$$

Since both $a(x, t)$ and $c(x, t)$ are almost periodic in t uniformly with respect to $x \in [0, \ell]$, we will choose $T \in R$ to be a common $(m\varepsilon)/2$-almost period of $c(x, t)$ and $m\varepsilon/2A$-almost period of $a(x, t)$ corresponding to $\varepsilon = \min\{m\varepsilon/2, m\varepsilon/2A\}$. Then, for such T, we obtain from formula (7.130)

$$|V(x, t+T) - V(x, t)| < \frac{\varepsilon}{2} + \frac{\varepsilon}{2} = \varepsilon, \tag{7.131}$$

which holds true for $t \in R$ and $x \in [0, \ell]$.

We shall now consider equation (7.120) for $V(x, t)$, and based on Proposition 7.1, we shall try to prove the existence of a bounded solution. Conditions (1) and (2) above will be preserved, but we need to formulate adequate conditions on $\varphi(t)$ and $C(x, t, u)$.

Namely, it appears to be appropriate to assume the following:

(3) $\varphi(t) \in BC(R, R)$.

(4) $C(x, t, u)$ is a continuous function on $D \times R$ with values in R and satisfies a Lipschitz type condition in its domain,

$$|C(x, t, u) - C(x, t, \bar{u})| \leq L|u - \bar{u}|, \qquad (7.132)$$

for some constant $L > 0$. Moreover, $C(x, t, 0)$ is bounded in D.

In regard to equation (7.120), the following result holds true.

Theorem 7.12. *Consider equation (7.120) in the domain $D = [0, \ell] \times R$ under assumptions (1)–(4). Then there exists a unique solution $V = V(x, t)$ bounded on $[0, \ell] \times R = D$.*

Proof. Since we look for bounded solutions, it is natural to consider the integral equation

$$V(x, t) = \int_{-\infty}^{t} \exp\left\{ -\int_{\tau}^{t} a(x, \theta)d\theta \right\} C\left(x, \tau, \varphi(\tau) + \int_{0}^{x} V(\xi, \tau)d\xi \right) d\tau. \quad (7.133)$$

This equation has as support formula (7.127), which we have used to produce a bounded solution to the auxiliary equation (7.122). It can be seen easily that any bounded solution of equation (7.133) is also a solution to equation (7.122).

We can now proceed by successive approximations, keeping in mind that we need uniform convergence in D.

Let us choose $V_0(x, t) \equiv 0$ in D, which implies

$$V_1(x, t) = \int_{-\infty}^{t} \exp\left\{ -\int_{\tau}^{t} a(x, \theta) \right\} C\left(x, \tau, \varphi(\tau) \right) d\tau. \qquad (7.134)$$

We note that $C(x, \tau, \varphi(\tau))$ is bounded in D because of assumptions (3) and (4).

We now construct the sequence $\{V_k(x, t); \; k \geq 1\}$ of functions continuous on D according to the iterative formula ($k \geq 2$)

$$V_k(x, t) = \int_{-\infty}^{t} \exp\left\{ -\int_{\tau}^{t} a(x, \theta)d\theta \right\} C\left(x, \tau, \varphi(\tau) + \int_{0}^{x} V_{k-1}(\xi, \tau)d\xi \right) d\tau. \quad (7.135)$$

Formula (7.135) makes sense for every $k \geq 1$ and defines the sequence $\{V_k(x, t); \; k \geq 1\}$ consisting of bounded functions on D. This last property is a consequence of the fact that $C\left(x, \tau, \varphi(\tau) + \int_{0}^{x} V_{k-1}(\xi, \tau)d\xi \right)$ is bounded in D for $k = 1$, and if we assume it to be bounded for some $k > 1$, then we find from assumptions (3) and (4) that $C\left(x, \tau, \varphi(\tau) + \int_{0}^{x} V_k(\xi, \tau)d\xi \right)$ is also bounded on D.

According to the standard procedure for iteration, we can write the inequality

$$|V_{k+1}(x,t) - V_k(x,t)|$$
$$\leq L \int_{-\infty}^{t} \exp\left\{-\int_{\tau}^{t} a(x,\theta)d\theta\right\} \int_{0}^{x} |V_k(\xi,\tau) - V_{k-1}(\xi,\tau)|d\xi\, d\tau, \tag{7.136}$$

taking into account also the Lipschitz type condition for $C(x,t,u)$ with respect to u. Since the integral on the right-hand side of formula (7.136) is absolutely convergent, we can change the order of integration and obtain

$$|V_{k+1}(x,t) - V_k(x,t)|$$
$$\leq L \int_{0}^{x} \int_{-\infty}^{t} \exp\left\{-\int_{\tau}^{t} a(x,\theta)d\theta\right\} |V_k(\xi,\tau) - V_{k-1}(\xi,\tau)|d\tau\, d\xi. \tag{7.137}$$

Let us now denote $d_k(x) = \sup_{t \in T}\{|V_k(x,t) - V_{k-1}(x,t)|\}$. It is known that this definition makes sense due to the boundedness in D of each $V_k(x,t)$, $k \geq 1$. Moreover, each $d_k(x)$ is upper semicontinuous on $[0,\ell]$. Then formula (7.137) allows us to write

$$d_{k+1}(x) \leq L \int_{0}^{x} d_k(\xi)d\xi \int_{-\infty}^{t} \exp\left\{-\int_{\tau}^{t} a(x,\theta)d\theta\right\} d\tau$$
$$\leq L \int_{0}^{x} d_k(\xi)d\xi \int_{-\infty}^{t} e^{-m(t-\tau)}d\tau < \frac{L}{m} \int_{0}^{x} d_k(\xi)d\xi.$$

In other words, we have the following recurrent inequality, $k \geq 1$:

$$d_{k+1}(x) \leq \frac{L}{m} \int_{0}^{x} d_k(\xi)d\xi, \qquad x \in [0,\ell]. \tag{7.138}$$

By induction on k, one obtains in the usual manner from formula (7.138) the inequality

$$d_{k+1}(x) \leq \frac{M}{k!} \left(\frac{Lx}{m}\right)^k, \qquad k \geq 1, \tag{7.139}$$

with $M \geq d_1(x)$, $x \in [0,\ell]$.

Summing up the estimates above, we can write the inequalities $(k \geq 1)$

$$|V_{k+1}(x,t) - V_k(x,t)| \leq \frac{M}{k!} \left(\frac{L\ell}{m}\right)^k, \qquad (x,t) \in D. \tag{7.140}$$

From formula (7.140), we derive by applying the Weierstrass criterion on uniform convergence that

$$\lim_{k \to \infty} V_k(x,t) = V(x,t) \tag{7.141}$$

uniformly for $(x,t) \in D$. There results that $V(x,t)$ is a continuous function on D.

The fact that $V(x,t)$ given by equation (7.141) satisfies equation (7.120) can be easily checked if we note that

$$\lim_{k \to \infty} \int_{-\infty}^{t} \exp\left\{-\int_{\tau}^{t} a(x,\theta)d\theta\right\} \int_{0}^{x} |V_k(\xi,\tau) - V(\xi,\tau)|d\xi \, d\tau = 0$$

uniformly with respect to $(x,t) \in D$, just using simple estimates like those used above.

The uniqueness of the bounded solution can be obtained by means of the usual pattern with successive approximations.

Corollary 7.3. *If one assumes that the functions $\varphi(t), a(x,t)$, and $C(x,t,0)$ are almost periodic in t uniformly with respect to $x \in [0, \ell]$, then the unique bounded solution $V(x,t)$ of equation (7.120) with the condition $V(0,t) = \varphi(t)$, $t \in R$, is almost periodic in t uniformly with respect to $x \in [0, \ell]$.*

Proof. Indeed, the conditions imposed on the data $\varphi(t)$, $a(x,t)$, and $C(x,t,u)$ being satisfied, they imply the conditions of Theorem 7.12.

On the other hand, Corollary 7.2 to Proposition 7.4 assures the almost periodicity in t, uniformly with respect to $x \in [0, \ell]$, of each $V_k(x,t)$. Based on the properties of the functions in this category given in Section 3.6 and due to the uniform convergence in D of the sequence $\{V_k(x,t); \; k \geq 1\}$, one concludes that the limit of this sequence (i.e., the solution $V(x,t)$) is also almost periodic in t uniformly with respect to $x \in [0, \ell]$.

Since the usual interpretation of the wave equation is not applicable to the case considered in this section, it seems adequate to provide some explanations.

A physical model for which equation (7.118) is the key ingredient is provided by the passing of a gas through a tube or cylinder containing an absorbent substance. The gas has to be purified, the undesirable component being absorbed during the passage.

The unknown function $v(x,t)$ in equation (7.118) represents the concentration of the gas in the pores of the absorbent. Since t is the time, by x one denotes the abscissa of the section of the tube (the axis of the tube is taken as the x-axis).

What Corollary 7.3 above states can be rephrased in the following terms: If on the terminal cross section $x = 0$ of the tube we maintain a concentration of the gas that is varying almost periodically, then this property remains valid for any cross section of the tube ($x \in (0, \ell)$). In other words, the whole process is almost periodic (uniformly with respect to x).

We shall conclude this section with the remark that for $v(x,t)$ appearing in equation (7.118) we easily obtain the double integral equation

$$v(x,t) = \varphi(t) + \int_{0}^{x} \int_{-\infty}^{t} \exp\left\{-\int_{\tau}^{t} a(\xi,\theta)d\theta\right\} C(\xi,\tau,v(\xi,\tau))d\xi \, d\tau, \quad (7.142)$$

which can be treated also by the method of iteration to obtain the existence and uniqueness of a bounded solution in $D = [0, \ell] \times R$.

7.7 Miscellanea

In this section, we shall discuss some problems concerning the existence of almost periodic solutions or waves related to various classes of equations encountered in applied fields or just from an abstract point of view.

We shall start with an inhomogeneous wave equation of the form

$$u_{tt} - u_{xx} = f(t, x) \tag{7.143}$$

with $f \in C^{(1)}(R \times R, R)$.

It is known (from any classical book on partial differential equations) that the solution of equation (7.143) under zero initial conditions,

$$u(0, x) = u_t(0, x) = 0, \qquad x \in R,$$

is given by the simple formula

$$u(t, x) = \frac{1}{2} \int_0^t \int_{x-t+\tau}^{x+t-\tau} f(\tau, \sigma) d\sigma \, d\tau. \tag{7.144}$$

Formula (7.144) provides a $u(t, x)$ that is in $C^{(2)}(R^2, R)$ and suggests substituting into the nonlinear equation

$$u_{tt} - u_{xx} = f(t, x, u) \tag{7.145}$$

with zero initial conditions the following integral equation in two independent variables:

$$u(t, x) = \frac{1}{2} \int_0^t \int_{x-t+\tau}^{x+t-\tau} f(\tau, \sigma, u(\tau, \sigma)) d\sigma \, d\tau. \tag{7.146}$$

A solution of equation (7.145) that satisfies the zero initial conditions is also a solution of equation (7.146), and under some regularity conditions for $f(t, x, u)$, including being of class $C^{(1)}$ with respect to all its arguments, any solution of equation (7.146) is a $C^{(2)}$-solution of equation (7.145) under zero initial conditions.

In Section 7.6, we actually dealt with an equation similar to equation (7.146) corresponding to a partial differential equation of hyperbolic type, namely

$$u_{tx} + a(t, x)u_x = f(t, x, u),$$

with characteristic data and considered in the strip $t \in R$, $x \in [0, \ell]$. We have proven the existence of almost periodic solutions with respect to t, which may constitute an indication that equation (7.146) is worth investigating in regard to the existence of solutions (almost periodic or otherwise).

We also consider that equation (7.146) could be a source defining some kind of "generalized" solution for equation (7.145). A difficulty that may occur in investigating nonlinear integral equations such as equation (7.146) is certainly related to the fact that the domain of integration changes with (t, x).

We suggest that the reader try the direct investigation of equation (7.146) in order to obtain valid results for the wave equation (7.145).

Obviously, more general wave equations than equation (7.145) can be considered, and it is right to mention the fact that they occur in applications (for instance, damping forces involving the velocity u_t). Such equations have the form

$$u_{tt} - u_{xx} = f(t, x, y, u, u_t, u_x), \tag{7.147}$$

or in higher dimension with respect to x,

$$u_{tt} - \Delta u = f(t, x, u, u_t, \nabla u), \tag{7.145'}$$

where f is a known function, generally nonlinear. It has to be almost periodic in t if we want to get almost periodic solutions in t.

If we preserve the zero Cauchy data for u, then equation (7.147) can be substituted by an integro–partial differential equation similar to equation (7.146) but more difficult to handle. In the case of equation (7.146), we will need a function space for the solution that will admit a norm expressed only in terms of u, while if $f = f(t, x, u, u_t, u_x)$, we will need a norm of Sobolev type.

We must say that the literature does not offer a large number of contributions with nonlinear f in the almost periodic case. Nevertheless, some results are available, and we shall dwell here on the paper [104] by Rong Yuan.

Equation (7.147) will be considered in the strip $R \times [0, 1]$ under dissipative boundary value conditions

$$u(t, 0) = 0, \ u_x(t, 1) + hu(t, 1) = 0, \tag{7.148}$$

where $h > 0$ is a real parameter.

We shall consider *classical* solutions to equation (7.147) under boundary value conditions (7.148). The term classical means that

$$f : R \times [0, 1] \times R^3 \longrightarrow R \tag{7.149}$$

is of class $C^{(2)}$ with respect to all five of its arguments (t, x, u, v, w), while the derivatives involved exist and are continuous.

The almost periodicity of f with respect to $t \in R$ uniformly in regard to $(x, u, v, w) \in [0, 1] \times R^3$ is understood in the sense made precise in Section 3.6.

The following existence result for almost periodic solutions to equation (7.147), is given in the paper [104] by Rong Yuan.

Theorem 7.13. *Consider equation (7.147) in the domain $R \times [0, 1] \times I^3$ with f of class $C^{(2)}$, and assume the following hypotheses are verified:*

(1) *f is almost periodic in t uniformly with respect to $(x, u, v, w) \in [0, 1] \times I^3$.*
(2) *$\|f\|_{C^{(2)}}$ is small enough. More precisely,*

$$\|f\|_{C^{(2)}} = \max\{\sup |D^\alpha f|; \ 0 \le |\alpha| \le 2\} < \gamma, \tag{7.150}$$

each supremum being taken for $(t, x, u, v, w) \in R \times [0,1] \times I^3$, *where* $I = [-1,1]$,

$$\gamma = \bar{h}/600 e\bar{t}, \tag{7.151}$$

with

$$\bar{h} = \begin{cases} |h-1||h+1|^{-1}, & h \neq -1, \\ 1/2, & h = 1, \end{cases}$$

and

$$\bar{t} = 2(1 - \ln 20/\ln \bar{h}).$$

Then, there exists a classical solution $u = u(t,x)$, $(t,x) \in R \times [0,1]$ *of equation (7.147) such that:*

(a) $u(t,x)$ *is almost periodic (Bohr) in* $t \in R$ *uniformly with respect to* $x \in [0,1]$;
(b) $\mathrm{mod}(u) \subset \mathrm{mod}(f)$; *and*
(c) u *is uniformly asymptotically stable on* R_+.

The proof of Theorem 7.13 is entirely based on methods of classical and functional analysis and the theory of almost periodic functions (as presented in the preceding sections). It does require lengthy consideration and we shall omit it, sending the reader to the paper of Rong Yuan quoted above.

We shall only mention that the proof relies on the estimate of the solution of a linearized problem. Namely, the auxiliary equation is $u_{tt} - u_{xx} = g(t,x)$, $(t,x) \in [0,T] \times [0,1]$, and the boundary value conditions are from equation (7.148). The estimate has been obtained by Y. Shi [90], who dealt with the case of a periodic f (covered by Theorem 7.13).

Remark 7.24. Condition (7.150) is a very restrictive one, and it shows that the theorem is valid only for f satisfying a "smallness" condition. The nonlinear case leads in some situations to results of existence of the so-called small solutions.

We shall consider now a problem of almost periodicity related to equations of *elliptic* type. In cases like the wave equation (hyperbolic) or heat transfer diffusion, equations of parabolic type can possess almost periodic solutions. The almost periodicity is meant with respect to t, the variable representing the time. The elliptic equations usually describe stationary phenomena, and an interpretation of one of the variables as time is not accommodating. Nevertheless, one can provide reasonable conditions that secure the existence of almost periodic solutions to such equations.

The type of elliptic (nonlinear) equation we will consider is of the form

$$u_{tt} + \Delta u = f(t,x,u) \tag{7.152}$$

in a domain $R \times \Omega \subset R^{m+1}$, where Δu is the Laplacian in m variables,

$$\Delta u = u_{x_1 x_1} + u_{x_2 x_2} + \cdots + u_{x_n x_n},$$

and $f(t, x, u) : R \times R^m \times R \longrightarrow R$ is a given function whose properties will be specified below. The boundary value condition associated to equation (7.152) is

$$u(t, x)|_{x \in \partial \Omega} = 0, \quad t \in R, \tag{7.153}$$

the more general case where the right-hand side in equation (7.153) is a function of $t \in R$ and $x \in \partial \Omega$ being reducible by a substitution like $u = v + \tilde{u}$ to the special one in equation (7.153). $\tilde{u}$ must be a regular enough function, such that $u|_{\partial \Omega} = \tilde{u}|_{\partial \Omega}$.

The following result, of Bohr–Neugebauer type, can be proven by using the same approach as in the proof of the first part of Theorem 7.5, which dealt with the heat equation.

Theorem 7.14. *Consider equation (7.152) in $R \times \Omega$, with $\Omega \subset R^m$ a bounded domain with a smooth boundary. Assume further that condition (b) of Theorem 7.5 is satisfied. Moreover, assume that $f(t, x, u)$ is continuous in $R \times \overline{\Omega} \times R$, of class $C^{(1)}$ with respect to u, and f_u satisfies the inequality $f_u \geq \mu$, $\lambda_1 + \mu > 0$, where λ_1 is the first eigenvalue of the Laplacian in Ω. Then any $C^{(2)}$-solution of equation (7.152) bounded in the mean on R, i.e., such that*

$$\int_{\Omega} u^2(t, x) dx \leq M < \infty \tag{7.154}$$

is almost periodic as a map from R into $L^2(\Omega) : t \longrightarrow u(t, \cdot)$ (i.e., it is Bohr almost periodic and $L^2(\Omega)$-valued).

Proof. We shall closely follow the proof of the first part of Theorem 7.5. The only difference is that we shall rely on the second-order qualitative inequality from Section 6.4.

Let us denote $v(t, x) = u(t + \tau, x) - u(t, x)$, where $\tau \in R$ is a fixed number and $u(t, x)$ is the solution of equation (7.152) whose existence is postulated in Theorem 7.14. The function $u(t + \tau, x)$ satisfies the equation

$$u_{tt}(t + \tau, x) + \Delta u(t + \tau, x) = f(t + \tau, x, u(t + \tau))$$

for $t \in R$, $x \in \Omega$. From the equation above and equation (7.152), we obtain (for v) by subtraction

$$v_{tt} + \Delta v = f(t + \tau, x, u(t + \tau, x)) - f(t, x, u(t, x)),$$

which we rewrite as

$$\begin{aligned} v_{tt} + \Delta v = {} & f(t + \tau, x, u(t + \tau, x)) - f(t + \tau, x, u(t, x)) \\ & + f(t + \tau, x, u(t, x)) - f(t, x, u(t, x)). \end{aligned}$$

We now multiply both sides of the last equation by $v(t, x)$ and obtain

$$\begin{aligned} v v_{tt} + v \Delta v = {} & v[f(t + \tau, x, u(t + \tau, x)) - f(t + \tau, x, u(t, x))] \\ & + v[f(t + \tau, x, u(t, x)) - f(t, x, u(t, x))]. \end{aligned}$$

Integrating both sides of the last equation on Ω, we obtain after elementary estimates

$$\frac{1}{2}\frac{d^2}{dt^2}\int_\Omega v^2(t,x)dx - \int_\Omega v_t^2(t,x)dx + \int_\Omega v(t,x)\Delta v(t,x)dx$$
$$= \int_\Omega v[f(t+\tau,x,u(t+\tau,x)) - f(t+\tau,x,u(t,x))]dx$$
$$+ \int_\Omega v[f(t+\tau,x,u(t,x)) - f(t,x,u(t,x))]dx.$$

According to Green's formula, we can write

$$\int_\Omega v(t,x)\Delta v(t,x)dx = -\int_\Omega |\text{grad } v|^2 dx,$$

and Poincaré's inequality says

$$\lambda_1 \int_\Omega v^2(t,x)dx \leq \int_\Omega |\text{grad } v(t,x)|^2 dx,$$

with $\lambda_1 > 0$ the smallest eigenvalue of the Laplacian in Ω.

Combining the last three relations and using the mean value theorem as well as Cauchy's inequality, one obtains the differential inequality

$$\frac{1}{2}\frac{d^2}{dt^2}\int_\Omega v^2(t,x)dx \geq \lambda_1\int_\Omega v^2(t,x)dx + \int_\Omega f_u(t+\tau,x,u^*)v^2(t,x)dx$$
$$-\left(\int_\Omega v^2(t,x)dx\right)^{1/2}\left(\sup\int_\Omega |f(t+\tau,x,u) - f(t,x,u)|^2 dx\right)^{1/2},$$

where the supremum is taken with respect to t, $u \in R$.

Let us now denote

$$V(t) = \int_\Omega v^2(t,x)dx \tag{7.155}$$

and keep in mind that $f_u \geq \mu$. Then the last inequality leads to

$$\frac{1}{2}\frac{d^2}{dt^2}V(t) \geq (\lambda_1 + \mu)V(t)$$
$$-\sqrt{V(t)}\left(\sup\int_\Omega |f(t+\tau,x,u) - f(t,x,u)|^2 dx\right)^{1/2}, \tag{7.156}$$

which represents a qualitative inequality of the form treated in Section 6.4. Since $V(t)$ is bounded in R, according to formulas (7.154) and (7.155), from formula (7.156) we derive for $t \in R$

$$V(t) \leq (\lambda_1 + \mu)^{-2}\sup_{t,u}\int_\Omega |f(t+\tau,x,u) - f(t,x,u)|^2 dx.$$

The last inequality above and equation (7.155) lead to the concluding inequality

$$\int_\Omega |u(t+\tau,x) - u(t,x)|^2 dx$$

$$\leq (\lambda_1 + \mu)^{-2} \sup_{t,w} \int_\Omega |f(t+\tau,x,w) - f(t,x,w)|^2 dx,$$

with the supremum taken for $t, w \in R$.

Remark 7.25. The supremum taken for $w \in R$ is a strong requirement. Unfortunately, it is not possible to substitute into it an assumption of the form $w \in I \subset R$ with I a bounded interval. Indeed, a function $u(t,x)$ satisfying formula (7.154) may not be bounded on $R \times \Omega$.

Results concerning the almost periodicity of solutions of elliptic equations are available in the literature (see, for instance, Zaidman [107]).

We shall conclude this section with some considerations regarding almost periodicity in parabolic (heat) equations of the form encountered in Section 7.3 with only one spatial dimension,

$$u_{xx} = u_t + f(x,t,u), \tag{7.157}$$

in the strip $(x,t) \in R \times [0,T]$, $T > 0$.

To equation (7.157) one attaches the initial condition

$$u(x,0) = u_0(x), \qquad x \in R, \tag{7.158}$$

and since we look for solutions that can be candidates for almost periodicity (with respect to $x \in R$), we do not impose boundary value conditions. Or, as sometimes appears in the literature, the boundary conditions are at $\pm\infty$, and they require boundedness of $u(x,t)$.

The treatment of the problem is inserted in our book [23], and it relies on the so-called method of lines. This method is a reduction procedure of problems in partial differential equations to similar problems for systems of ordinary differential equations.

In the case of equation (7.157), this method can be presented in the following manner. First, we discretize in equation (7.157) with respect to t, substituting into u_t the quotient $h^{-1}[u(x,t+h) - u(x,t)]$ with $nh = T$. We fix h as shown in $nh = T$, which means $h \to 0$ as $n \to \infty$. We denote $t_k = kh$, $k = 0, 1, \ldots, n$. The "approximate" value for $u(x, t_k)$ will be denoted by $u_k(x)$, $k = 1, 2, \ldots, n$. The ordinary differential system for $u_k(x)$, $k = 1, 2, \ldots, n$, will be

$$\frac{d^2 u_k}{dx^2} = h^{-1}[u_k - u_{k-1}] + f(x, t_k, u_{k-1}), \tag{7.159}$$

where $u_0(x)$, $x \in R$, is assigned according to equation (7.158). The system (7.159) is a recurrent one, and since $u_0(x)$ is assigned, we can determine each

$u_k(x)$, $k = 1, 2, \ldots, n$. We need to consider only bounded solutions on R for equation (7.159) due to the fact that an almost periodic solution of equation (7.157) is bounded in the strip (almost periodicity in x uniformly with respect to $t \in [0, T]$).

In order to prove that $u_k(x)$ is indeed approximating $u(x, t_k)$, $k = 1, 2, \ldots, n$, we will examine the "errors" occurring in the process. Namely, let us denote for $k = 1, 2, \ldots, n$,

$$\varepsilon_k(x) = u(x, t_k) - u_k(x). \tag{7.160}$$

One can find by means of elementary calculations the system for $\varepsilon_k(x)$,

$$\frac{d^2\varepsilon_k(x)}{dx^2} - h^{-1}\varepsilon_k(x) = h^{-1}\varepsilon_{k-1}(x) + r_k(x), \tag{7.161}$$

taking into account equation (7.157) for $t = t_k$ and system (7.159). The term $r_k(x)$ is given by the formula

$$\begin{aligned} r_k(x) =\ & u_t(x, t_k) - h^{-1}[u(x, t_k) - u(x, t_{k-1})] + f(x, t_k, u(x, t_k)) \\ & - f(x, t_k, u_{k-1}(x)), \quad k = 1, 2, \ldots, n. \end{aligned} \tag{7.162}$$

In order to estimate $\varepsilon_k(x)$, we need to state the hypotheses on the data. These hypotheses are:

1. $f(x, t, u)$ is almost periodic in x uniformly with respect to $(t, u) \in [0, T] \times [-A, A]$, $A > 0$, and is Lipschitz continuous in u.
2. $u_0(x)$ is (Bohr) almost periodic.
3. $u(x, t)$ is a solution of equation (7.157) in the strip $R \times [0, T]$ satisfying equation (7.158), and such that both u and u_t are uniformly continuous in the strip with modulus $\omega(h)$.

Theorem 7.15. *If conditions* (1), (2), *and* (3) *above are verified, then the solution* $u(x, t)$ *of equation* (7.157), $|u(x, t)| \leq A$, $(x, t) \in R \times [0, T]$, *is almost periodic in* x *uniformly with respect to* $t \in [0, T]$. *The approximation procedure is valid, and for some* $M > 0$ *one has*

$$\sup |\varepsilon_k(x)| \leq M\omega(h), \qquad k = 1, 2, \ldots, n, \tag{7.163}$$

M being independent of n.

Proof. We shall first prove the convergence of the approximation method (of lines). Each equation (7.161) has the form

$$\frac{d^2y}{dx^2} - \alpha^2 y = f(x), \qquad x \in R, \tag{7.164}$$

with $\alpha > 0$ a positive number. When $f(x)$ is bounded on R, there exists only one solution of equation (7.164) bounded on R. It is given by the formula

$$\widetilde{y}(x) = -\frac{1}{2\alpha}\left\{e^{\alpha x}\int_x^\infty e^{-\alpha t}f(t)dt + e^{-\alpha x}\int_{-\infty}^x e^{\alpha t}f(t)dt\right\},$$

which can be obtained by variation of parameters, and it satisfies the estimate

$$\|\widetilde{y}\| \le \alpha^{-2}\|f\|, \tag{7.165}$$

where $\|\cdot\|$ stands for the supremum norm (on the whole R).

If we apply these simple facts to equation (7.161), there results

$$\|\varepsilon_k\| \le \|\varepsilon_{k-1}\| + h\|r_k\|, \qquad k = 1, 2, \ldots, n. \tag{7.166}$$

Based on our assumptions, including the Lipschitz type condition for $f(x, t, u)$, namely

$$|f(x, t, u) - f(x, t, v)| \le L|u - v|, \tag{7.167}$$

$(x, t, u), (x, t, v) \in R\times[0, T]\times[-A, A]$, and using in estimates the modulus of (uniform) continuity $\omega(h)$, we obtain for $\|\varepsilon_k\|$ the inequality

$$\|\varepsilon_k\| \le (1 + hL)\|\varepsilon_{k-1}\| + (1 + L)h\omega(h), \tag{7.168}$$

which holds true for $k = 1, 2, \ldots, n$. Of course, $\|\varepsilon_0\| = 0$. By recursion, from formula (7.168) we obtain the estimate

$$\|\varepsilon_k\| \le \omega(h)(1 + L)L^{-1}[(1 + hL)^n - 1] \tag{7.169}$$

for $k = 1, 2, \ldots, n$. From formula (7.169), we get the final estimate (7.163), taking into account that $nh = T$ and choosing $M = (1+L)L^{-1}(e^{LT} - 1)$. The procedure based on the method of lines is thus justified.

We shall now use the result above in order to prove the almost periodicity of the solution $u(x, t)$, whose existence is postulated in the statement of Theorem 7.15.

First, let us note the simple fact, directly following from the estimate (7.165) for each difference $u_k(x+\xi) - u_k(x)$, $x \in R$ and $\xi \in R$ fixed, that each $u_k(x)$, $k = 0, 1, \ldots, n$, is almost periodic.

Then, we rely on the following elementary inequality and the validity of the method of lines (see above). We can write the inequalities

$$|u(x + \xi, t) - u(x, t)| \le |u(x + \xi, t) - u(x + \xi, t_k)|$$
$$+ |u(x + \xi, t_k) - u_k(x + \xi)| + |u_k(x + \xi) - u_k(x)|$$
$$+ |u_k(x) - u(x, t_k)| + |u(x, t_k) - u(x, t)|$$
$$\le 2(M + 1)\omega(h) + |u_k(x + \xi) - u_k(x)|.$$

In the inequality above, for each $t \in [0, T]$ one must choose a k such that $|t - t_k| \le h$. Then h must be "small enough" or n large enough (because $nh = T$). Finally, ξ must be chosen to be a common almost period to all $u_k(x)$, $k = 0, 1, \ldots, n$, for the large n fixed as indicated above. In this way, the left hand side in the inequalities above can be done arbitrarily small, which proves the almost periodicity of $u(x, t)$. The argument above in fact shows the almost periodicity with respect to x uniformly with respect to $t \in [0, T]$.

This ends the proof of Theorem 7.15.

Remark 7.26. The method of lines can be used to prove the existence of solutions to partial differential equations. An illustration can be found in V. I. Smirnov [91].

7.8 Quasi-periodic Waves

We have defined the quasi-periodic functions in Section 4.6 and noted that they are almost periodic functions whose set of Fourier exponents contains a finite basis. The elements of any basis are linearly independent, and in this case we mean linear independency over the field Q of rationals. If the basis consists of a single element, we obtain the periodic functions. It appears natural to deal first with quasi-periodic solutions and waves on the scale of complexity from periodic to almost periodic.

It appears, when examining the recent literature, that this is really the case. The papers dealing with quasi-periodic waves outnumber those related to the general concept of almost periodicity. In the general case of almost periodicity, one has to deal with modules of Fourier exponents having infinite (but countable!) basis, a feature that complicates the discussion of the properties.

The quasi-periodic functions were discovered and their theory was developed before Bohr presented his theory of almost periodicity during the years 1923–1925. It is admitted that the creators of quasi-periodicity were P. Bohl and E. Esclangon (see our book [23] for more details). In regard to applications, these authors were concerned only with ordinary differential equations.

We shall review in this section some results obtained during the last 20–30 years, leaving aside the proofs. Usually, they are rather lengthy, and may rely on facts or results not included in our exposition.

Besides the classical wave and heat equations, known for about two centuries and closely investigated in relation to their physical interpretation, we shall also consider the Schroedinger equation, which constitutes the basic tool in quantum mechanics. Some authors claim that the Schroedinger equation plays the same role in quantum mechanics as Newton's equation $ma = F$ in classical mechanics.

The Schroedinger equation is not deducible in the same way as wave or heat equations but should be regarded as a postulate in developing quantum mechanics. It is motivated by its consequences, such as Schroedinger's famous paper of 1926 "Quantisierung als Eigenwertproblem," published in volume 79 of *Annalen der Physik*, in which Planck's theory of quanta is theoretically founded.

The Schroedinger wave equation in the three-dimensional case can be written in the form

$$\frac{h^2}{8m\pi^2}\,\Delta\psi - P\psi = \frac{h}{2\pi i}\,\frac{\partial\psi}{\partial t} \tag{7.170}$$

with

$$\Delta\psi = \frac{\partial^2\psi}{\partial x^2} + \frac{\partial^2\psi}{\partial y^2} + \frac{\partial^2\psi}{\partial z^2},$$

h standing for Planck's constant (in many sources, one finds $\hbar$), and P denotes the potential energy of the particle associated with the wave, $P = P(x, y, z)$. The function ψ has a probabilistic interpretation.

In one spatial dimension, the Schroedinger wave equation becomes

$$\frac{h^2}{8m\pi^2}\,\frac{\partial^2\psi}{\partial x^2} - P\psi = \frac{h}{2\pi i}\,\frac{\partial\psi}{\partial t}, \tag{7.171}$$

and this form is often used in studying problems in quantum mechanics. The higher-dimensional case presents difficulties, and we do not attempt to discuss it in this section.

In the mathematical literature, the Schroedinger equation is "reduced" to the standard form

$$iu_t = u_{xx} - V(x)u - N(u), \qquad 0 \le x \le \pi, \tag{7.172}$$

under boundary value conditions

$$u(t,0) = 0, \quad u(t,\pi) = 0, \qquad t \in R. \tag{7.173}$$

The term $N(u)$, generally nonlinear, can be regarded as a perturbation of the linear Schroedinger equation $iu_t = u_{xx} - V(x)u$.

The problem of constructing almost periodic solutions to equations of the form (7.172) is treated by many authors. We shall mention here a paper by J. Pöschel [81] and another by J. Bourgain [16], both authors being interested in almost periodic solutions and quasi-periodic ones. The latter are viewed as "approximate" solutions, a fact that agrees with the property stating the possibility of approximating any almost periodic solution by means of trigonometric polynomials (and these are the simplest quasi-periodic functions).

In J. Pöschel's paper [81], the potential $V(x)$ is chosen in $L^2([0, \pi], R)$ with strictly positive Dirichlet eigenvalues. This set of potentials is open in $L^2([0, \pi], R)$. The nonlinearity $N(u)$, which should normally be of the form $f(|u|^2)u$, will be replaced by

$$N(u) = \Psi(f(|\psi u|^2)\psi u), \tag{7.174}$$

where ψ stands for a smoothing convolution operator $(u \longrightarrow \psi * u)$ with ψ an even function of a special type, while u is extended to R to become an odd 2π-periodic function. The function f is real analytic in a neighborhood of zero with $f(0) = 0$. The nonlinearity (7.174) is chosen such that equation (7.172) becomes Hamiltonian with

$$H = \frac{1}{2} \langle Lu, u \rangle + \frac{1}{2} \int_0^\pi F(|\psi|^2) dx,$$

where $L = -d^2/dx^2 + V$ and $F = \int_0^x f(\xi) d\xi$.

The solution to equation (7.174) under the conditions mentioned above is sought in the form of a series, namely

$$u(t, x) = \sum_k U_k(x) e^{i(k \cdot \omega)t}, \tag{7.175}$$

where $\omega = (\omega_1, \omega_2, \ldots, \omega_m, \ldots)$ is an infinite sequence of rationally independent frequencies, while $k = (k_1, k_2, \ldots, k_m, 0, 0, \ldots)$ stands for any sequence of integers with only finitely many nonzero elements. $k \cdot \omega$ stands for a finite sum: $k_1 \omega_1 + \cdots + k_p \omega_p$ with $k_m = 0$ for $m > p$.

We can now state the following result of existence for almost periodic solutions to equation (7.172).

Theorem 7.16. *Consider equation (7.172) with boundary conditions (7.173) and nonlinearity (7.174), the potential $V(x)$ being as described above. Then, there exist uncountably many almost periodic (in t) solutions of the problem of the form (7.175) for "almost all" potential $V(x)$. Such solutions belong to the class of "small" solutions.*

Remark 7.27. It is necessary to provide some explanations for the terms in quotation marks. First, by "almost all" potentials $V(x)$, of course in the class described above we mean that the set of these potentials for which the conclusion of Theorem 7.16 may not be true has (probabilistic) measure zero in $L_+^2([0, \pi], R)$. Such measures are defined in the paper of J. Pöschel [81].

Regarding the term "small," we can make it more precise by saying that any neighborhood of $u \equiv 0$ in the Sobolev space $H_0^1([0, \pi], \mathcal{C})$ contains such solutions. This implies, of course, that not only is u small but its first derivatives also.

Remark 7.28. In Bourgain's paper [16], the nonlinearity has a slightly different form, namely

$$N(u) = \varepsilon f(u), \tag{7.176}$$

where $f(u)$ stands for a polynomial with adequate properties. The proof of existence of an almost periodic solution is based on the construction of some quasi-periodic approximate solutions. The case of a one-dimensional spatial dimension is dealt with, and different methods are displayed. The potential $V(x)$ is assumed even and periodic.

A recent contribution to the existence of a quasi-periodic solution to the Schroedinger nonlinear equation is due to Jiansheng Geng and Yingfei Yi [44].

They also deal with the polynomial potential. The paper contains a good deal of references to this type of equation, covering various cases for the potential function and the nonlinearity.

Let us now consider the wave equation

$$u_{tt} - u_{xx} + V(x)u + \varepsilon u^3 = 0 \qquad (7.177)$$

under conditions (7.173). The solution is sought in the semistrip $0 \le x \le 1$, $t \ge 0$, with the potential function $V \in L^2([0, \pi], R)$. It is assumed that ε is a small parameter.

By means of a celebrated method known as the KAM (Kolmogorov, Arnold, Moser) method, C. E. Wayne [99] constructed periodic and quasi-periodic solutions to equation (7.177). The equation is reduced to an equivalent Hamiltonian system (see above in the case of the Schroedinger equation). This system is infinite-dimensional, and the solutions are represented by series similar to those encountered in classical texts (separation of variables) but made much more complicated by the nature of the problem.

Equation (7.177) for $\varepsilon = 1$ is studied in a recent paper by Xiaopin Yuan [105]. The boundary value conditions are again of the form (7.173), and the author proves the existence of a family of quasi-periodic solutions for equations (7.177) and (7.173). One uses the reduction to Hamiltonian systems and the KAM method to construct such solutions.

The existence result for L^2-potentials V is also valid for more complex equations than equation (7.177), namely

$$u_{tt} - u_{xx} + V(x)u = \pm u^3 + \sum_{m \ge k \ge 2} a_k u^{2k+1}, \qquad (7.178)$$

in which a_k's are some real numbers and $m > 2$ is a positive integer.

Another recent significant contribution in constructing quasi-periodic solutions to certain wave equations is due to J. Pöschel [82]. More precisely, he considers the wave equations of the form

$$u_{tt} = u_{xx} - mu - f(u) \qquad (7.179)$$

under boundary value conditions (7.173). The author constructs families of quasi-periodic solutions of the form (7.175) but somewhat simpler. Indeed, in this case there are only finitely many rationally independent frequencies, say $\omega_1, \omega_2, \ldots, \omega_n$, and the solution (if any) must be representable as

$$u(t, x) = \sum_{lk \in Z^m} e^{ik \cdot \omega t} U_k(x), \qquad (7.180)$$

where $k \cdot \omega = k_1 \omega_1 + k_2 \omega_2 + \cdots + k_n \omega_n$ with $(k_1, \ldots, k_n) \in Z^n$.

The quasi-periodic solutions constructed are of "small" amplitude, as mentioned in Remark 7.28 to Theorem 7.16 above.

The nonlinear function $f(u)$ in equation (7.179) is analytic and somewhat more general than the right-hand side of equation (7.178):

$$f(u) = au^3 + \sum_{k \geq 5} f_k u^k, \quad a \neq 0. \tag{7.181}$$

The main result of J. Pöschel [82] states that, for all $m > 0$, the quasi-periodic solutions to equation (7.179) can be regarded as forming a Cantor manifold that has full density at the origin. Moreover, this set of solutions is contained in the space of real analytic functions on $[0, \pi]$ with respect to x. Further properties of the manifold of solutions are established.

Remark 7.29. Equation (7.179), for $m > 0$, has also been investigated recently, by Zhenguo Lian and Jiangong You [64]. This paper, besides furthering the investigation of equation (7.179), contains results on the existence of quasi-periodic solutions for the fourth-order beam equation (1-dim)

$$u_{tt} + u_{xxxx} + mu + f(u) = 0, \quad m > 0,$$

the nonlinearity f having the form

$$f(u) = au^{2p+1} + \sum_{k \geq p+1} f_{2k+1} u^{2k+1}, \quad a \neq 0, \ p \geq 1.$$

This paper contains a list of references in which most major contributors to the problem of existence of quasi-periodic solutions to the wave or Schroedinger equations are included (J. Bourgain, L. Chierchia, W. Craig, C. E. Wayne, S. B. Kuksin, J. Pöschel, Q. Qiu, J. Xu, and others). An inspection of the Web will help the reader get plenty of items related to this type of problem.

An interesting contribution to the problem we discuss in this section is due to P. A. Vuillermot [97], who dealt with quasi-periodic soliton solutions to the nonlinear Klein–Gordon equation

$$u_{tt} = u_{xx} - g(u) \tag{7.182}$$

in the case of two spatial dimensions: $x = (x_1, x_2)$, $x_1, x_2 \in R$, $u = u(t, x_1, x_2)$. One associates a higher-order dynamical system to equation (7.182), and by a delicate and deep approach without using KAM methods, one obtains the existence of quasi-periodic solutions to equation (7.182). First, from the associated dynamical systems, one obtains multiple periodic solutions and then one takes their values on a line (see Section 4.6, where quasi-periodic functions are defined).

To conclude the discussion concerning the quasi-periodic or almost periodic solutions to wave type equations, we will emphasize the fact that this kind of solution can be obtained by writing the partial differential equation as an equation in a Banach or Hilbert space. We have briefly illustrated this aspect in Section 7.5.

The mathematical literature, in book form, offers the following items: L. Amerio and G. Prouse [3], A. Haraux [48], and Y. Hino et al. [50].

An early paper of C. M. Dafermos [33] treats the problem of almost periodicity for the wave equation in this manner by studying it for the abstract equation in a real Hilbert space H,

$$(A\dot{u})^{\cdot} + B\dot{u} + Cu = \theta, \quad u = u(t), \quad t \in R,$$

where A, B, C are linear operators. A and B are acting on H, while C is acting from $V \subset H$ to V', V standing for a Hilbert space $V \subset H$, the inclusion being topologic and algebraic.

In the remaining part of this section, we shall consider only parabolic equations. As illustrated in Section 7.3, for such equations (heat, diffusion), the problem of almost periodicity makes sense, a fact that is in agreement with physical reality.

We point out that the literature is not as rich for these evolution equations as in the case of wave equations.

First, we shall consider the classical equation of heat (or diffusion) of the form

$$u_t(x,t) = Du_{xx}(x,t) + f(u(x,t),t), \tag{7.183}$$

where $D > 0$ is constant, while $f(u,t)$ is almost periodic in t, $t \in R$, uniformly with respect to u in any bounded set of R (see Section 3.6). A less common hypothesis is that equation (7.183) is *bistable*. This means that the auxiliary equation $du/dt = f(u,t)$ has exactly three almost periodic solutions, say $u_-^f(t), u_+^f$, and $u_0^f(t)$, such that

$$u_-^f(t) < u_0^f(t) < u_+^f(t), \quad t \in R. \tag{7.184}$$

Moreover, it is assumed that $u_-^f(t)$ and $u_+^f(t)$ are stable, while $u_0^f(t)$ is unstable. It is obvious that all three solutions of formula (7.184) are also solutions of equation (7.183) independent of x.

A classical solution that has the form

$$u(x,t) = U(x + c(t), t), \tag{7.185}$$

where $U(x,t)$ and $c'(t)$ are almost periodic with respect to t, seems to be the right candidate for a generalization of a traveling wave. $U(x,t)$ is the wave profile, and $c'(t)$ is the velocity of propagation.

A solution of the form (7.185) with the properties stated above is called an *almost periodic traveling wave* solution of equation (7.183).

The aim of the author is to prove the existence of such solutions, connecting the two stable solutions of equation (7.183). More precisely, the term connecting means

$$\lim_{x \to \pm\infty} U(x,t) = u_\pm^f(t), \quad t \in R,$$

uniformly in $t \in R$.

Wengsian Shen [89; I, II] studies in depth the basic problems related to equation (7.183) and its almost periodic solutions that constitute periodic traveling waves (in particular, those called wave-like solutions). The basic aspects dealt with are existence, stability, and uniqueness. Some similar equations are also investigated.

Theorem 7.17. *Under the above-mentioned conditions on $f(u,t)$ and the existence of three almost periodic solutions satisfying formula* (7.184), *there exists an almost periodic traveling wave solution of equation* (7.183), $u(x,t) = U(x + c(t),t)$, *connecting the two stable almost periodic solutions. Also, the module containment relations*

$$m(c'(t)), m(U(x,t)) \subset m(f)$$

take place for each $x \in R$.

In a paper by Fengxin Chen [21], a more complex equation than equation (7.183) is investigated that preserves the almost periodicity assumption as well as the existence of bistable almost periodic solutions to $u_t = \widetilde{f}(u,t)$. Equation (7.183) is substituted by

$$u_{tt} = Du_{xx} + f(u, J_1 * s^1(u), \ldots, J_p * s^p(u), t),$$

with $J_k * s^k(u)(x,t) = \int_R J_k(x - y)s^k(u(y,t))dy$, $k = 1,2,\ldots,p$, $J_k \in C^{(1)}$, and nonnegative, while $s^k(u)$ are given smooth functions. $\widetilde{f}(u,t)$ is defined by $\widetilde{f}(u,t) = f(u, s^1(u), \ldots, s^p(u), t)$.

The definition of an almost periodic traveling wave solution being the same as above, the aim of the author is to construct such solutions connecting the bistable almost periodic solutions $u = u_-^{\widetilde{f}}(t)$ and $u = u_+^{\widetilde{f}}(t)$.

The case where $f(u, v_1, \ldots, v_p, t)$ is almost periodic in t uniformly with respect to other variables in bounded sets is conducive to a positive answer for the problem of existence. This paper follows step by step the papers of Wengsian Shen [89; I, II], but the dependencies of the right-hand side have a global character.

Early results concerning parabolic evolution equations were obtained starting in the 1960s by the Milano school of almost periodic functions and their applications (L. Amerio, G. Prouse, M. Biroli, and others).

References

[1] A. R. Aftabizadeh, Bounded solutions for some gradient type systems, *Libertas Mathematica*, **2** (1982), 121–130.

[2] L. Amerio, Soluzioni quasi-periodiche, o limitate, di sistemi differenziali non lineari quasi-periodici, o limitati, *Annali di Matematica Pura ed Applicata*, **39** (1955), 97–119.

[3] L. Amerio and G. Prouse, *Almost Periodic Functions and Functional Equations*, Van Nostrand, New York, 1971.

[4] J. Andres, A. M. Bersani, and R. F. Grande, Hierarchy of almost periodic function spaces, *Rendiconti di Matematica, Ser. VII*, **26** (2006), 121–188.

[5] V. Barbu, Solutions presque-périodiques pour un système d'équations linéaires aux dérivées partielles, *Ricerche Matematice*, **15** (1966), 207–222.

[6] N. K. Bary, *A Treatise on Trigonometric Series*, vols. I, II, Pergamon Press, New York, 1965.

[7] M. S. Berger and Y. Y. Chen, Forced quasiperiodic and almost periodic solutions for nonlinear systems, *Nonlinear Analysis*, **21** (1993), 949–965.

[8] Bolis Basit, Some problems concerning different types of vector valued almost periodic functions, *Disertationes Mathematicae (Warszawa)*, **338** (1995), 1–26.

[9] Bolis Basit and H. Günzler, Generalized Esclangon-Landau results and applications to linear difference-differential systems in Banach spaces, *Journal of Difference Equations and Applications*, **10** (2004), 1005–1023.

[10] A. S. Besicovitch, *Almost Periodic Functions*, Cambridge University Press, Cambridge, 1932.

[11] G. Birchoff and G. C. Rota, *Ordinary Differential Equations* (second ed.), John Wiley and Sons, New York, 1969.

[12] J. Blot, P. Cieutat, and J. Mawhin, Almost periodic oscillations of monotone second order systems, *Advances in Differential Equations*, **2** (1997), 693–714.

[13] P. Bohl, *Ueber die Darstellung von Funktionen einer Variablen durch trigonometrische Reihen mit mehrer einer Variablen proportionalen Argumenten*, Dorpat, 1893.

[14] P. Bohl, Ueber die Differentialgleichung der Störungstheorie, *Journal Reine Angewandte Mathematik*, **131** (1906), 268–321.

[15] M. Bostan, Almost periodic solutions for first-order differential equations, *Differential and Integral Equations*, **19** (2006), 91–120.

[16] J. Bourgain, Construction of approximate and almost periodic solutions of perturbed linear Schroedinger and wave equations, *Geometric and Functional Analysis*, **6** (1996), 201–230.

[17] F. Brauer and J. A. Nohel, *The Qualitative Theory of Ordinary Differential Equations*, W. A. Benjamin, New York, 1968.

[18] S. G. Braun, D. J. Ewins, and S. S. Rao, *Encyclopedia of Vibration*, vols. I, II, III, Academic Press, New York, 2002.

[19] H. Butazzo, M. Giaquinta, and S. Hildebrandt, *One-Dimensional Variational Problems*, Oxford Scientific Publicator, Oxford, 1985.

[20] C. Carminati, Forced systems with almost periodic and quasiperiodic forcing term, *Nonlinear Analysis, 32* (1998), 727–739.

[21] Fengxin Chen, Almost periodic travelling waves of nonlocal evolution equations, *Nonlinear Analysis, 50* (2002), 807–818.

[22] E. A. Coddington and N. Levinson, *Theory of Ordinary Differential Equations*, McGraw-Hill, New York, 1955.

[23] C. Corduneanu, *Almost Periodic Functions* (second English ed.), Chelsea, New York, 1989 (Distributed by AMS and Oxford University Press).

[24] C. Corduneanu, *Integral Equations and Stability of Feedback Systems*, Academic Press, New York, 1973.

[25] C. Corduneanu, *Integral Equations and Applications*, Cambridge University Press, Cambridge, 1991.

[26] C. Corduneanu, Almost periodic solutions of infinite delay systems. In *Spectral Theory of Differential Operators* (I. W. Knowles and R. T. Lewis, Eds.), North-Holland, Amsterdam, 1981, 99–106.

[27] C. Corduneanu, Two qualitative inequalities, *Journal of Differential Equations, 64* (1986), 16–25.

[28] C. Corduneanu, *Functional Equations with Causal Operators*, Taylor and Francis, London and New York, 2002.

[29] C. Corduneanu, Almost periodic solutions to nonlinear elliptic and parabolic equations, *Nonlinear Analysis, 7* (1983), 357–363.

[30] C. Corduneanu, *Principles of Differential and Integral Equations*, Chelsea, New York, 1977 (now distributed by American Mathematical Society).

[31] C. Corduneanu and J. Goldstein, Almost periodicity of bounded solutions to nonlinear abstract equations. In *Differential Equations* (I. W. Knowles and R. T. Lewis, Eds.), North-Holland, Amsterdam, 1984, 115–122.

[32] C. Corduneanu and H. Poorkarimi, Qualitative problems for some hyperbolic equations. In *Trends in Theory and Practice of Nonlinear Analysis*, (V. Lakshmikantham, Ed.), North-Holland, Amsterdam, 1985, 107–113.

[33] C. M. Dafermos, Wave equation with weak damping, *SIAM Journal of Applied Mathematics*, **18** (1970), 759–767.

[34] A. M. Davis, Almost periodic extension of band-limited functions and its applications to nonuniform sampling, *IEEE Transactions on Circuits and Systems*, **33** (1986), 933–940.

[35] N. Dunford and J. Schwartz, *Linear Operators*. Vol. I, John Wiley and Sons (Interscience), New York, 1961.

[36] R. E. Edwards, *Fourier Series*, Springer-Verlag, Berlin, 1979.

[37] R. Engelking, *General Topology*, PWN, Warszawa, 1977.

[38] E. Esclangon, Nouvelles recherches sur les fonctions quasi-périodiques, *Annales Observatoire de Bordeaux*, **16** (1919), 1–174.

[39] J. Favard, *Leçons sur les fonctions presque-périodiques*, Gauthier-Villars, Paris, 1933.

[40] A. M. Fink, *Almost Periodic Differential Equations*, Lecture Notes in Mathematics, vol. 377 Springer-Verlag, Berlin 1974.

[41] M. Fréchet, Les fonctions asymptotiquement presque-périodiques continues, *Comples Rendus Academic des Sciences Paris*, **213** (1941), 607–609.

[42] M. Fréchet, Les fonctions asymptotiquement presque-périodiques, *Revue Scientifique*, **79** (1941).

[43] A. Friedman, *Foundations of Modern Analysis*, Dover, New York, 1982.

[44] Jiansheng Geng and Yingfei Yi, Quasi-periodic solutions in a nonlinear Schroedinger equation, *Journal of Differential Equations*, **233** (2007), 512–542.

[45] E. A. Grebenikov, Yu. A. Mitropolsky, and Yu. A. Ryabov, *Asymptotic Methods in Resonance Analytical Dynamics*, Stability and Control vol 21, Chapman & Hall (CRC), London, 2004.

[46] A. Halanay, *Differential Equations: Stability, Oscillations, Time Lag*, Academic Press, New York, 1966.

[47] J. K. Hale, *Oscillations in Nonlinear Systems*, McGraw-Hill, New York, 1963.

[48] A. Haraux, *Nonlinear Evolution Equations*, Lecture Notes in Mathematics, vol. 841, Springer-Verlag, Berlin, 1981.

[49] A. Haraux, A simple almost periodicity criterion and applications, *Journal of Differential Equations*, **66** (1987), 51–61.

[50] Y. Hino, T. Naito, Nguyen Van Minh, and Jong Son Shin, *Almost Periodic Solutions of Differential Equations in Banach Spaces*, Taylor and Francis, London, 2002.

[51] A. O. Ignatyev and O. A. Ignatyev, On the stability in periodic and almost periodic systems, *Journal of Mathematical Analysis and Applications*, **313** (2006), 678–688.

[52] K. U. Ingard, *Fundamentals of Waves and Oscillations*, Cambridge University Press, Cambridge, 1988.

[53] J. P. Kahane, *Séries de Fourier absolument convergentes*, Springer-Verlag, Berlin, 1970.

[54] Y. Katznelson, *An Introduction to Harmonic Analysis*, Dover, New York, 1976.

[55] M. A. Krasnoselskii, V. S. Burd, and Yu. S. Kolesov, *Nonlinear Almost Periodic Oscillations*, John Wiley and Sons, New York, 1973.

[56] N. N. Krasovskii, *Stability of Motion*, Stanford University Press, Stanford, CA, 1963.

[57] S. Kuksin and J. Poschel, Invariant Cantor manifolds of quasi-periodic oscillations for a nonlinear Schroedinger equation, *Annals of Mathematics*, **142** (1995), 149–179.

[58] G. Ladas and I. Gyori, *Oscillation Theory of Delay Differential Equations with Applications*, Clarendon Press, Oxford, 1991.

[59] O. A. Ladyzhenskaya, *The Mixed Problem for Hyperbolic Equations* (Russian), Gostehizdat, Moscow 1953.

[60] C. E. Langenhop and G. Seifert, Almost periodic solutions of second order nonlinear differential equations with almost periodic forcing, *Proceedings of the American Mathematical Society*, **10** (1959), 425–432.

[61] G. Lellouche, A frequency criterion for oscillatory solutions, *SIAM Journal of Control*, **8** (1970), 202–206.

[62] B. M. Levitan, *Almost Periodic Functions* (Russian). Moscow, 1953.

[63] B. M. Levitan and V. V. Zhikov, *Almost Periodic Functions and Differential Equations*, Cambridge University Press, Cambridge, 1982.

[64] Zhenguo Liang and Jiangong You, Quasi-periodic solutions for $1D$ nonlinear wave equation with a general nonlinearity (27 pp).

[65] L. H. Loomis, *Abstract Harmonic Analysis*, New York, 1953.

[66] I. G. Malkin, *Theorie der Stabilität einer Bewegung*, Oldenbourg, München, 1959.

[67] V. Mañosa, Periodic travelling waves in nonlinear reaction-diffusion equations via multiple Hopf bifurcation, *Chaos, Solitons and Fractals*, **18** (2003), 241–257.

[68] J. L. Massera and J. J. Schäffer, *Linear Differential Equations and Function Spaces*, Academic Press, New York, 1966.

[69] J. Mawhin, Bounded and almost periodic solutions of nonlinear differential equations: Variational vs nonvariational approach, Research Notes in Mathematics, vol. 410, Chapman and Hall/CRC, London 2000, 167–184.

[70] D. Mentrup and M. Luban, Almost periodic wave packets and wave packets of invariant shape, *American Journal of Physics*, **71** (2003), 580–584.

[71] A. R. Mickelson and D. L. Jaggard, Electromagnetic wave propagation in almost periodic media, *IEEE Transactions on Antennas and Propagation*, **27** (1979), 34–40.

[72] S. Mizohata, *The Theory of Partial Differential Equations*, L. Shoten, 1965.

[73] C. F. Muckenhoupt, Almost periodic functions and vibrating systems, *Journal of Mathematical Physics*, **8** (1928–1929), 163–199.

[74] S. Murakami and Y. Hamaya, Periodic solutions and almost periodic solutions to an integro-partial differential equation, In *Integral Methods in Science and Engineering* (C. Constanda, Ed.), Longman, 1994, 23–30.

[75] O. Onicescu and V. Istrăţescu, Approximation theorems for random functions, *Rendiconti di Mathematica*, (VI), **8** (1975), 65–81.

[76] R. Ortega, Degree theory and almost periodic problems (to appear).

[77] R. Ortega and M. Tarallo, Almost periodic linear differential equations with non-separated solutions (manuscript).

[78] H. J. Pain, *The Physics of Vibrations and Waves*, John Wiley and Sons, London, 1976.

[79] D. Pennequin, Existence of almost periodic solutions of discrete time equations, *Discrete and Continuous Dynamical Systems*, **7** (2001), 51–60.

[80] H. Poorkarimi and Joseph Wiener, Almost periodic solutions of nonlinear hyperbolic equations with delay, *Electronic Journal of Differential Equations*, Conference 07, 2001, 99–102.

[81] J. Pöschel, Quasi-periodic solutions for nonlinear wave equation, *Commentarii Mathematici Helvetici*, **71** (1996), 269–296.

[82] J. Pöschel, On the construction of almost periodic solutions for a nonlinear Schroedinger equation, *Ergodic Theory and Dynamical Systems*, **22** (2002), 1537–1549.

[83] A. Precupanu, On the almost periodic functions in probability, *Rendiconti di Matematica ed Applicazioni, series VII*, **2** (1982), 613–626.

[84] Cora Sadosky, *Interpolation of Operators and Singular Integrals*, Marcel Dekker, New York, 1979.

[85] I. W. Sandberg and G. J. J. van Zyl, Harmonic balance and almost periodic inputs, *IEEE Transactions on Circuits and Systems*, **49** (2002), 459–464.

[86] G. Sansone and R. Conti, *Non-linear Differential Equations*, Pergamon Press, New York, 1964.

[87] S. Schwabik, M. Tvrdy, and O. Vejvoda, *Differential and Integral Equations*, Kluwer, Dordrecht, 1979.

[88] L. Schwartz, *Théorie des distributions* (second ed.), Herman, Paris, 1975.

[89] Wengsian Shen, Travelling waves in time almost periodic structures governed by bistable nonlinearities. I, II, *Journal of Differential Equations*, **159** (1999), 1–54, 55–101.

[90] Y. Shi, Exponential stability of semilinear wave equation with boundary dissipation, *Nonlinear Analysis*, **36** (1999), 35–43.

[91] V. I. Smirnov, *A Course of Higher Mathematics*, vol. 4, Addison-Wesley, New York, 1955.

[92] S. L. Sobolev, *Partial Differential Equations of Mathematical Physics*, Pergamon Press, New York, 1964.

[93] O. Staffans, On almost periodicity of solutions of an integrodifferential equation, *Journal of Integral Equations*, **8** (1985), 249–260.

[94] A. N. Tikhonov and A. A. Samarskii, *Equations of Mathematical Physics* (German), Deutsche Verlag fuer Wissenschaften, Berlin, 1959.

[95] V. Trénoguine, *Analyse Fonctionelle*, Mir, Moscow, 1985.

[96] Yu. V. Trubnikov and A. I. Perov, *Differential Equations with Monotone Nonlinearities* (Russian). Nauka Technika, Minsk, 1986.

[97] P. A. Vuillermot, Quasiperiodic soliton solutions to nonlinear Klein-Gordon equation on R^2, *Mathematischa Zeitschrift*, **203** (1990), 235–253.

[98] A. Walther, Fastperiosche Folgen und ihre Fouriersche Analyse, *Atti del Congresso Internazionale dei Matematici, Bologna*, **2** (1928), 289–298.

[99] C.-Eugene Wayne, Periodic and quasi-periodic solutions of nonlinear wave equations via KAM theory, *Communications in Mathematical Physics*, **127** (1990), 479–528.

[100] H. Weyl, Beweis des Fundamentalsatzes in der Theorie der fastperiodischen Funktionen, *Sitzungsberocjterer Preussschen Akademic Wissenschaften*, (1926), 211–214.

[101] Yisong Yang, Almost periodic solutions of nonlinear parabolic equations, *Bulletin of the Australian Mathematical Society*, **38** (1988).

[102] T. Yoshizawa, *Stability Theory and the Existence of Periodic Solutions and Almost Periodic Solutions*, Springer-Verlag, New York, 1975.

[103] K. Yosida, *Functional Analysis* (fifth ed.), Springer-Verlag, Berlin, 1978.

[104] Rong Yuan, Almost periodic solution of a class of semilinear wave equations with boundary dissipation, *Nonlinear Analysis*, **50** (2002), 749–761.

[105] Xiaopin Yuan, Quasi-periodic solutions of nonlinear wave equations with a prescribed potential (manuscript).

[106] S. Zaidman, *Almost Periodic Functions in Abstract Spaces*, Pitman, London, 1985.

[107] S. Zaidman, Quasi-periodicità per equazione di Poisson, *Accademia Nazionale dei Lincei (Rendiconti)*, **24** (1963), 241–254.

[108] Chuanyi Zhang, Pseudo almost periodic solutions of some differential equations, *Journal of Mathematical Analysis and Applications*, **181** (1994), 62–76.

[109] Chuanyi Zhang, Vector valued almost periodic functions, *Czechoslovak Mathematical Journal*, **47** (1997), 385–394.

[110] E. Zeidler, *Applied Functional Analysis*, Applied Mathematical Sciences vols. 108, 109, Springer-Verlag, New York, 1995.

Index